Bioinorganic Chemistry

Bioinorganic Chemistry: A Survey

Eiichiro Ochiai

AMSTERDAM • BOSTON • HEIDELBERG • LONDON • OXFORD • NEW YORK
PARIS • SAN DIEGO • SAN FRANCISCO • SINGAPORE • SYDNEY • TOKYO

Academic Press is an imprint of Elsevier

Cover Design: Joanne Blank
Cover Image © Ei-Ichiro Ochiai

Academic Press is an imprint of Elsevier
30 Corporate Drive, Suite 400, Burlington, MA 01803, USA
525 B Street, Suite 1900, San Diego, California 92101-4495, USA
84 Theobald's Road, London WC1X 8RR, UK

Library of Congress Cataloging-in-Publication Data
Application Submitted

British Library Cataloguing-in-Publication Data
A catalogue record for this book is available from the British Library.

ISBN: 978-0-12-088756-9

For information on all Academic Press publications
visit our Web site at www.books.elsevier.com

Typeset by Charon Tec Ltd., A Macmillan Company
(www.macmillansolutions.com)

Printed and bound by CPI Group (UK) Ltd, Croydon, CR0 4YY

Transferred to digital print 2013

Working together to grow
libraries in developing countries

www.elsevier.com | www.bookaid.org | www.sabre.org

ELSEVIER   BOOK AID
            International   Sabre Foundation

Contents

Preface

Bioinorganic chemistry as a discipline is now two and a half decades old, if we consider its birth to be the first International Conference on Bioinorganic Chemistry (ICBIC) held in Florence, Italy in 1983. I wrote one of the earliest books on the subject, *Bioinorganic Chemistry, an Introduction* (Allyn and Bacon) in 1977. The emphasis there was on physicochemical data and their interpretations. Since then the scope of research on bioinorganic chemistry has greatly expanded, diversified, and deepened, and the quantity of the literature has exploded.

This book does not attempt to summarize the current status of research in all the various areas of the field. It also omits discussion of the methods of research, and hardly deals with biophysical data. For experimental methods, readers are referred to works such as L. Que's *Physical Methods in Bioinorganic Chemistry* (2000) and an ACS publication, "Spectroscopic Methods in Bioinorganic Chemistry" (2003); and some textbooks including the ones by J. Cowan, *Inorganic Biochemistry*, 3rd ed., Wiley-VCH (2007) and R. M. Roat-Malone, *Bioinorganic Chemistry*, Wiley-Interscience (2002). Instead of research details, this book aims to provide general readers as well as specialists with an understanding of the basic chemistry of interactions of inorganic substances with biological systems at the molecular level (as much as possible), and also a perspective on the subject, as the subtitle "A Survey" implies.

Since living organisms and the biosphere they constitute are both open systems, they are in contact with and constantly exchanging energy and material with their environments. The environments experienced by a single organism include all the other organisms and the physical environment, which is composed of inorganic substances. This fundamental situation inevitably entails the inclusion of almost all inorganic elements in living organisms. Hence bioinorganic chemistry, or rather what it implies, is not a novelty but a necessity, and should be universal. Indeed we are increasingly being made aware of the fact that inorganic compounds are intimately associated with all biological phenomena.

My second book (*The General Principles of Biochemistry of the Elements*, Plenum Press, 1987) attempted to lay down several basic principles in order to answer some fundamental questions such as why iron is used for this purpose, and not copper. The present work is to try to extend these ideas while incorporating some of the more recent discoveries.

This book looks at the entire picture of the existence of organisms on Earth in terms of chemistry—molecules/compounds, their interactions and reactions, and the roles that inorganic elements play and why. The Introduction and the first two chapters give fundamentals of interactions between the biosphere and inorganic compounds from the molecular level to the geochemical level. The Introduction is a brief review of basic concepts of biochemistry, and can be skipped or can be used as a Glossary. Chapter 1 is an overview of biogeochemical aspects of bioinorganic chemistry, and Chapter 2 gives a brief introduction to relevant inorganic chemistry.

Chapter 3 is a discussion of the basic issue of why enzymes are so efficient. The information discussed here is not indispensable for understanding the rest of the book, so it can be skipped if circumstances require. Chapters 4 through 9 are concerned mainly with the chemical bases of functions of inorganic elements associated with enzymes and proteins. Chapter 10 deals with workings of inorganic compounds at physiological levels, and Chapter 11 treats the environmental issues including toxicity associated with inorganic elements. The last chapter deals with the medicinal applications of inorganic elements.

Two different kinds of exercises are provided at the end of each chapter—review questions and problems to explore. Both types of exercises are in line with the spirit of this book; they concern chemical principles of the interactions of inorganic chemicals with the biological systems at a molecular level, without discussing experimental data. Therefore, no question is provided regarding experimental data such as kinetic and spectroscopic data. Such questions, if necessary and appropriate, need to be provided by the instructors.

As the volume of relevant literature has become enormous and more details have become available, a single author surely has the difficulty of being sufficiently versed in each and every area of current research. No single book would be able to cover every aspect of the bioinorganic chemistry of even a single element such as calcium in full detail, let alone that of all the significant elements. However, that is in essence what I have tried to do, but not in full detail. In writing this book, I have relied heavily on some excellent summaries in the form of review articles written by experts in each field. I am much indebted to them, and would like to express my gratitude to all those researchers who have contributed to a better understanding of these fascinating aspects of chemical biology.

One of the results of technical advances made in recent decades is an increase in the structural data on proteins (and nucleic acids), and their accessibility through the PDB (Protein Data Bank). Individual researchers and students alike can now explore protein structures in detail. The availability of the PDB has been of enormous help in writing this book. In addition, several other Internet sites have proven very useful; in particular, the general literature search engines called Scopus and ScienceDirect have been indispensable in writing a book of this nature. I would like to express my gratitude to those individuals and organizations that are responsible for providing such readily accessible databases.

This book concludes my efforts over the last three decades to survey the fascinating subject of bioinorganic chemistry. I gratefully acknowledge the comments and suggestions by reviewers: R. J. P. Williams of Oxford University, C. Frank Shaw, III. of Illinois State University, and Murray S. Davies of James Cook Univeristy, Australia. Their comments have improved the manuscript. I also acknowledge gratefully the assistance of J. Woodling of Juniata College Library in finding literature.

Introduction: Basics of Bio/Ecosystems and Biochemistry, and Other Basic Concepts

A brief description of the chemical and biological basis of bio/eco-systems on the planet Earth is presented here in order to lay foundations for understanding of the material discussed in this treatment. Living organisms constitute the biosphere, which is embedded in the geological system of the planet. The overall picture of the interactions of the biosphere and its other components of the planet is briefly discussed in Chapter 1. Here will be discussed the basic features of the ecosystem and the living organisms. No thorough treatment is possible or intended. This chapter can be used as a glossary of most basic concepts.

BIOSPHERE (ECOSYSTEM)

Several million (at least) species of living organisms exist on the Earth. The entire system that includes all the living organisms is termed the *biosphere*. The biosphere in which different species interact with each other in complicated manners is itself interacting with its environments (in the broadest sense); the entirety of these interacting systems is the *ecosystem*. This book looks at the ecosystem in terms of *inorganic elements*.

Components of the Biosphere—Living Organisms

Several million species can be grouped into several categories. Different categorizations are possible depending on which features of an ecosystem are considered. We will look at it in terms of how an organism obtains its food—the so-called "food chain." The first distinction is based on how an organism obtains its own nutrition. Plants and some other organisms produce their own food from very simple chemical compounds such as H_2O and CO_2 and other simpler inorganic compounds; these are called *autotrophs* (self-nutrition acquiring). Animals, fungi, and many bacteria cannot produce their own food; they have to eat plants or other organisms (or their decayed substance). These are called *heterotrophs*.

All plants, including those in hydrosphere, use sunlight as the energy source to produce carbohydrates and other organic compounds from CO_2 and H_2O through photosynthesis; hence they are called *photoautotrops*. Some bacteria use H_2S instead of H_2O for the process; these are also photoautotrophs, but are called sulfur photosynthetic bacteria.

There is a variety of special bacteria that utilize chemical energy to produce carbohydrates, called chemoautotrophs. They may obtain energy from a number of mostly oxidative processes, as the oxidation reactions are typically exothermic. Some of them oxidize such compounds as NH_3, CH_4 (or other simpler organic compounds), Fe(II) or reduced sulfur compounds using O_2 as the oxidant, and use the energy released from the oxidative processes to effect synthesis of carbohydrates (from CO_2 and H_2O).

A generalized food chain is as follows:

(1) Producers (autotrophs) are eaten by heterotrophs, animals, fungi, and bacteria; these organisms are herbivores.

(2) Other types of heterotrophs are carnivores (animal eaters) and omnivores (eaters of both plants and animals).

(3) Dead body or its parts of organisms are utilized by fungi and bacteria; these are called *saprophytes*. These organisms tend to decompose large organic compounds into smaller compounds; they are decomposers in this sense.

Another fundamental category is the distinction between prokaryotes and eukaryotes. Organisms in the former group are rather primitive, consisting of *eubacteria* and *archeae* (bacteria); these organisms do not have a distinct separate nucleus (i.e., chromosome is lying in a cell). Eukaryotes are organisms with eukaryotic (truly nucleated) cells that have a nucleus containing chromosomes in a separate confinement in a cell. There is another type of life or rather half-life—the virus. A virus cannot live by itself, and has to live in another cell. It is made of a genetic material, either DNA or RNA enveloped by proteins, and it usurps the host cell's machinery to multiply.

Yet another distinction is that between anaerobic and aerobic organisms. The former cannot sustain life in the presence of air (oxygen), whereas the latter needs air to live. The organisms of the former type do not breathe oxygen, and they are sensitive to it. There are organisms that can adapt to either condition; they are facultative. For example, a typical bacterium *Escherichia coli* (bacteria in gut) can grow under either condition.

Bodily Structures of Living Organisms

Organisms can be made of a single cell (unicellular) or of many cells (multicellular). The handling of inorganic substances is different between a unicellular and a multicellular organism. A unicellular organism is quite small, and transport of substance in and out of the cell is relatively simple, mostly based on diffusion, and active transport across membrane. Multicellular organisms such as humans are structured, consisting of a number of organs and tissues. Hence even absorption of an element can be a multistep process.

For example, a substance is carried into the stomach through the mouth and esophagus. It may be absorbed through the stomach wall, or it may be carried further down the intestine, and then absorbed by epithelial cells of the intestine. It will go through several layers of cells to reach the circulating system (blood vessel). It will then be carried by the circulating system, probably bound with a carrier protein. When it is reached at the target cell, the carrier protein unloads the inorganic element, which is taken up by the cell.

The entire process of a multicellular organism (i.e., physiology) is complicated and its description is beyond the scope of this short introduction. However, some physiological processes involving inorganic elements will be discussed in Chapter 10.

CELLS, THE BASIC FUNCTIONAL UNITS OF LIVING ORGANISMS

Most biochemical reactions are conducted in an enclosed system, in the cell and on its surface, though some reactions are effected outside of it, for example, in circulating systems. A generalized structure of a cell (of animal) is schematically shown in Figure I.1. A plant cell has different organelles, such as chloroplast and vacuole. A prokaryotic cell (of bacteria) does not have specialized organelles as seen in this figure. Bacterial cells and plant cells have a cell wall just outside the plasma membrane.

■ **Figure I.1.** A generalized cell structure of an animal.

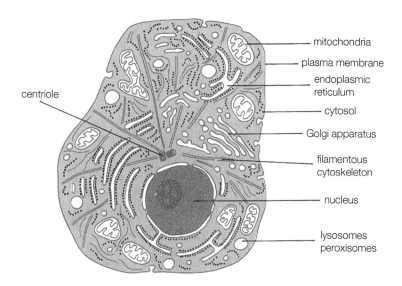

A few prominent organelles/vesicles are nucleus, mitochondria, Golgi apparatus, endoplasmic reticulum, and lysosome. Mitochondrion is where the later part of energy metabolism, TCA cycle (Krebs cycle) and electron-transfer (respiratory) processes, take place. It is believed to have originated from aerobic bacteria, and it has a DNA different from that in the parent cell nucleus. Endoplasmic reticulum is connected with the nucleus, and studded with a number of small globular matter, ribosome. Ribosome proved to be a catalytic entity where proteins are synthesized from amino acids. Golgi apparatus packages and delivers molecules synthesized. Lysosome is a vesicle in which proteins and others are decomposed.

A substance is transported into and out of a cell by several mechanisms: pores in membrane gated or ungated, passive diffusion (along the gradient) through membranes, and active transport across membranes. The energy for an active transport can be provided by ATP or a counteracting concentration gradient (of proton, Na(I), etc.). If a cell is covered by a cell wall, as in plant cells and bacterial cells, a substance needs to go through it as well.

BIOCHEMICAL COMPOUNDS ESSENTIAL TO LIFE

There are essentially three types of chemical compounds that are required by living organisms: (1) bodily constituents, (2) chemical compounds associated with energy metabolism, and (3) accessory compounds.

Bodily constituents include proteins, lipids, polysaccharides, and DNA/RNA. Different types of energy-producing/consuming compounds are involved. Photosynthesis produces carbohydrates and some lipids with energy from sunlight, which is the ultimate energy source of all the living organisms on Earth.

Carbohydrates and lipids have carbon atoms in lower oxidation states, the oxidation of which produces negative free energy used in animals and other heterotrophs. A large variety of other chemical compounds also are involved in producing chemical compounds and in conducting chemical processes of producing biological energy, as well as in other physiological processes. An overall life system in terms of these biochemical chemicals and processes are summarized in Figure I.5. Some typical examples of basic biochemical compounds follow. This is not intended to be a comprehensive discourse.

Carbohydrates

Monosaccharides

Carbohydrates are produced by autotrophs, the most important one being the photosynthesis in plants. Chloroplasts in green leaves produce carbohydrates like $C_6H_{12}O_6$ (as a representative) from CO_2 and H_2O. The reaction is nominally:

$$6CO_2 + 6H_2O \longrightarrow C_6H_{12}O_6 + 6O_2 \text{ (endothermic, energized}$$
$$\text{by sunlight)}$$

In addition, the photosynthetic process produces ATP (see later). Photosynthesis is discussed somewhat in Chapter 5, in connection with water oxidase.

Heterotrophs eat carbohydrates and essentially burn them; the reaction is nominally the reverse of the preceding reaction:

$$C_6H_{12}O_6 + 6O_2 \longrightarrow 6CO_2 + 6H_2O$$

This is not a simple one-step process; it consists of several metabolic processes starting with glycolysis, then TCA cycle, and then respiratory (electron transfer) process. ATPs are produced using the free energy released from these gradual oxidation processes.

■ **Figure I.2.** Carbohydrates.

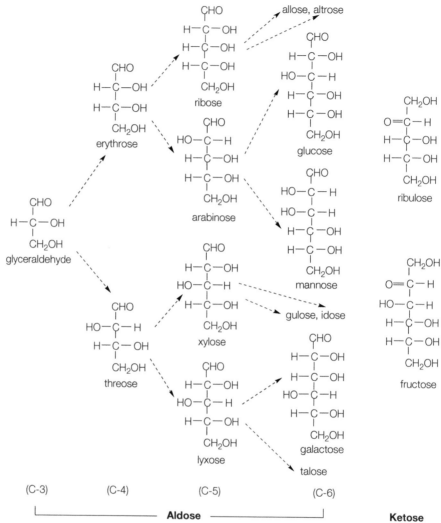

(C-3) (C-4) (C-5) (C-6)

Aldose **Ketose**

Carbohydrates are a collective term for chemical compounds with $C_m(H_2O)_n$. The smallest (m=n=1), HCHO (formaldehyde) is not considered as a carbohydrate, but carbohydrates can be regarded as polymers of formaldehyde. One with m=n=3 is glyceraldehyde, which can be regarded as the smallest carbohydrate, and an oxidized form of glycerol ($C_3H_8O_3$).

C-4 carbohydrates ($C_4H_8O_4$) are D-erythrose and D-threose (see Fig. I.2); these are aldose form; that is, with an aldehyde at the end

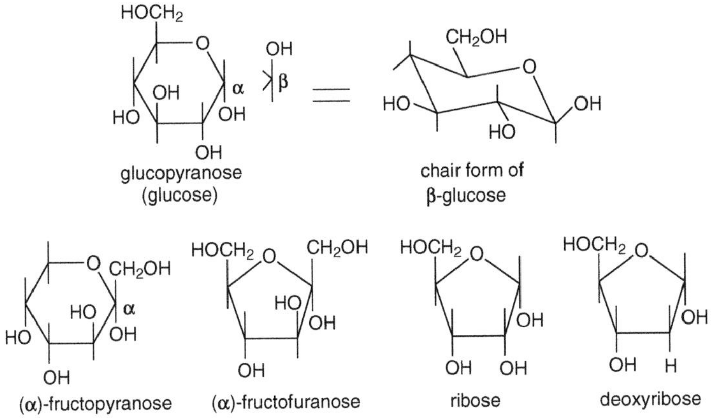

■ **Figure I.3.** Ring forms of pentose and hexose.

(the alternative ketone can exist). The issue of enantiomers (D- or L-, optical isomerism) is omitted here. Several geometrical isomers are possible for C-5 carbohydrates (collectively called pentoses): ribose, arabinose, xylose, and lyxose (see Figure I.2). The most common carbohydrates in the aldose form are C-6 carbohydrates (hexoses) that include glucose, mannose, and galactose (see Fig. I.2). One ketose with C-6 is fructose, which has a ketone at 2-C atom (see Fig. I.2).

The structures shown in Figure I.2 are open chain, but in reality the predominant form of pentoses and hexoses are in self-condensed ring form as shown in Figure I.3. Glucose forms a six-membered ring glucopyranose (α and β depending on the way OH attaches to 1-C). Fructose forms fructofuranose or fructopyranose (see Fig. I.3). Ribose in ring form is one of the components of ribonucleotides that constitute RNA, whereas the sugar entity in deoxyribonucleotides is deoxyribose (see Fig. I.3).

Polysaccharides and Derivatives

Carbohydrates mentioned so far are monosaccharides, consisting of a single ring. Two or more monosaccharides can condense to form a disaccharide or a higher polysaccharide. A few examples are given in Figure I.4.

The OH group of carbohydrates are subject to reactions, typically esterification, with phosphate, sulfate, acetate, borate, among others. Another kind of derivative is the oxidized product of 6-C's CH_2OH;

■ **Figure I.4.** Disaccharide and polysaccharides.

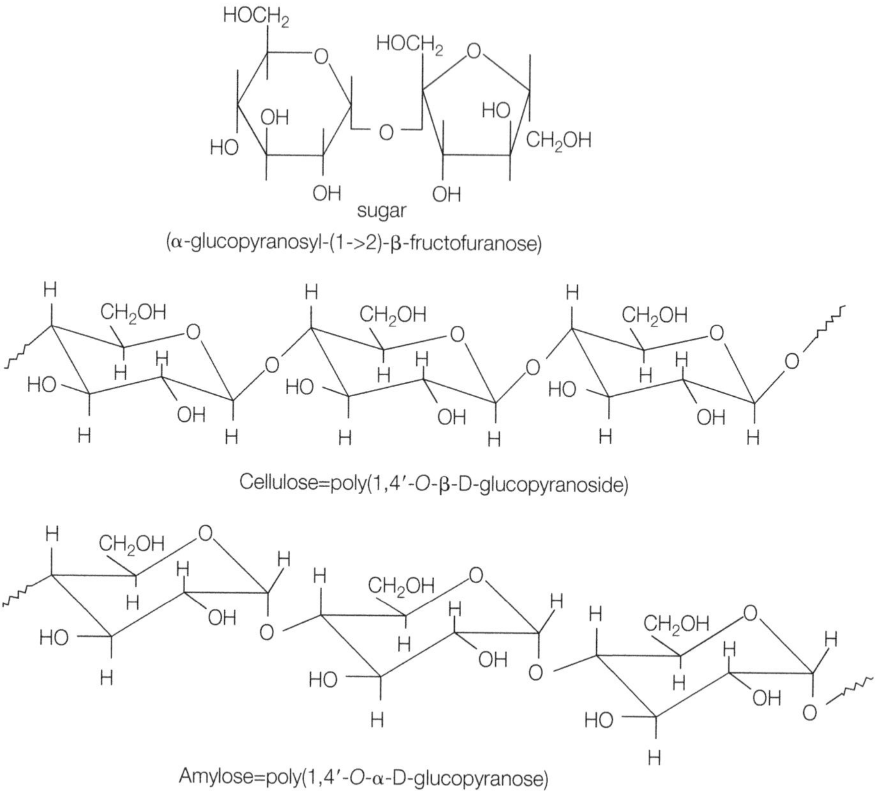

HOCH$_2$

(α-glucopyranosyl-(1->2)-β-fructofuranose)

Cellulose=poly(1,4'-O-β-D-glucopyranoside)

Amylose=poly(1,4'-O-α-D-glucopyranose)

that is, COOH. The OH on 2-C of hexose (or pentose) can be replaced by NH$_2$ (the resulting compound is called glucosamine), and this NH$_2$ is acetylated in many special polysaccharides such as chondroitin and hyaluronate. Some examples of such derivatives are shown in Figure I.5. The OH groups in pentoses and hexoses, being a Lewis base, can coordinate to a metallic ion, especially O-loving cation such as Mg(II), Ca(II), Fe(III), and so on. COOH (COO$^-$), phosphate, sulfate, and the amine or imine of acetylglucosamine in various derivates can coordinate or bind (through ionic interaction) a variety of metallic ions.

Lipids

Different types of lipid are found in living systems: fat, steroid, prostaglandin, and terpene. Terpenes are dimers and trimers of isoprene, and are responsible for many fragrant oils such as those of citrus and

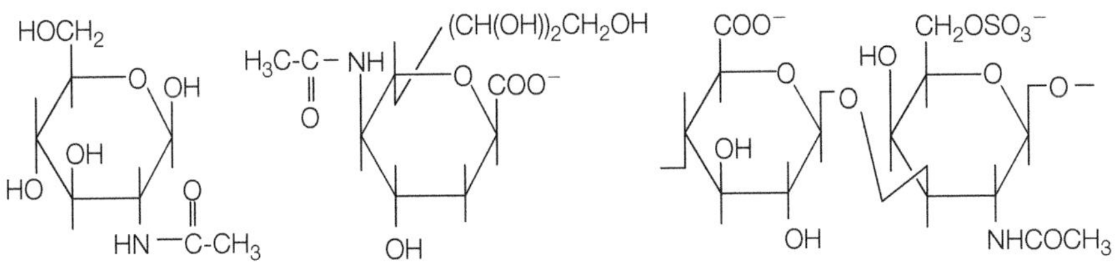

■ **Figure I.5.** Derivative of sugar.

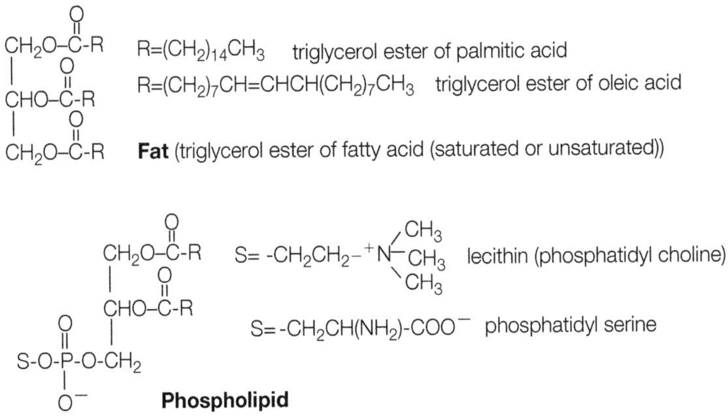

α-acetylglucosamine   N-acetyl neuramic acid (sialic acid)   chondroitin sulfate

■ **Figure I.6.** Fats and phospholipids.

R=(CH$_2$)$_{14}$CH$_3$ triglycerol ester of palmitic acid

R=(CH$_2$)$_7$CH=CHCH(CH$_2$)$_7$CH$_3$ triglycerol ester of oleic acid

Fat (triglycerol ester of fatty acid (saturated or unsaturated))

S= -CH$_2$CH$_2$–$^+$N(CH$_3$)(CH$_3$)(CH$_3$)   lecithin (phosphatidyl choline)

S= -CH$_2$CH(NH$_2$)-COO$^-$ phosphatidyl serine

Phospholipid

rose. Prostaglandins will be mentioned in Chapter 6. Only typical lipids will be mentioned here: fats and steroids.

Fats and Phospholipids

A basic type of fats is glycerol ester of fatty acids. A fatty acid of biological importance has a relatively long hydrocarbon chain (most of —CH$_2$—), which is lipophilic and hydrophobic, and the one end is a carboxylic group, which is hydrophilic. The hydrocarbon chain is saturated or unsaturated (see Fig. I.6).

Phospholipids has two fatty acid chains bound to glycerol through ester bonds and another OH on glycerol bound to one of a variety of ionic phosphate derivatives (see Fig. I.6). Phospholipids constitute plasma membrane and other membranous structures. They form a double layer (lipid bilayer) in an aqueous medium in the following manner. The hydrophobic (hydrocarbon) double chains of phospholipid align themselves in a parallel manner with other double chains; this interaction forms a layer consisting of a large number of hydrophobic double chains associated through hydrophobic interactions. Another such layer associates with it through the oily (hydrocarbon) surface, hence forming a double layer. The hydrophilic heads (phosphate derivative) are arranged along the surface of the bilayer in contact with aqueous medium. Often this bilayer takes a spheric structure, called micelle, with the surfaces inside and outside lined with ionic hydrophilic groups. This is the structural basis of the cell membrane. The hydrophilic, ionic heads often are associated with opposing ionic entities. Ca(II) often binds such anionic groups (as phosphate), and hence stabilizes the membrane structure.

Steroids

Another type of lipid is steroid, and a variety of steroids are used in biological systems such as cell membrane components and hormones. Typical examples of steroids are shown in Fig I.7.

Proteins and Amino Acids

A very large number of proteins are involved in living systems. Their roles are to provide physical structures (such as hair, muscle), catalytic (enzymatic) functions, signal transduction, gene regulation, and others.

Structures

A protein consists of about 20 amino acids $RCH(NH_2)COOH$ with various Rs. A polypeptide is a linear polymer of these 20 or so amino acids through a condensed bond called a peptide bond —CO—NH—. The number of amino acid residues in polypeptide varies widely, from small (20 or so) to several hundred. Small polypeptides (perhaps up to 20 amino acid residues) usually are called *polypeptides*. Larger polypeptides are proteins. There is no strict

■ **Figure I.7.** Steroids.

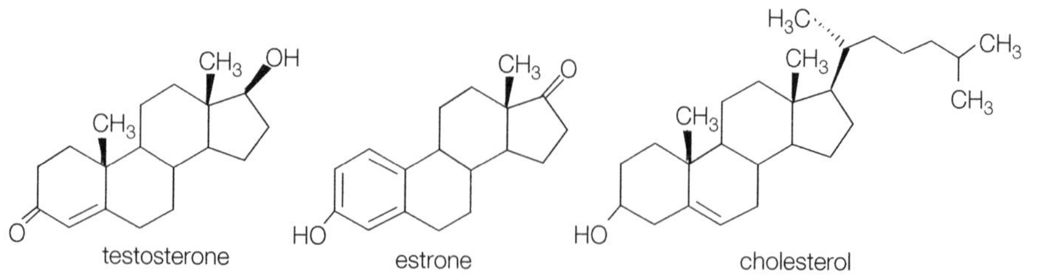

testosterone   estrone   cholesterol

distinction, though. A protein may consist of a single polypeptide chain or of more than two polypeptides. Each polypeptide chain in the latter type of protein is called a subunit. For example, myoglobin, the oxygen-binding protein in muscle cells, is made of a single chain; hemoglobin, the oxygen-carrying protein, consists of two sets each of two subunits α and β, and usually is expressed as $\alpha_2\beta_2$.

The amino acids commonly found in natural proteins are listed in Table I.1. Different types of R are found: hydrophobic (aliphatic and aromatic), hydroxylic (—OH), carboxylic (—COOH), amine, amide, imidazole, sulfhydryl, thioether, and proline. Nature employs these amino acids of different characteristics to provide a catalytic ability and an appropriate ambiance for a specific protein's function. For example, if the active site needs to be hydrophobic, the active site may be surrounded by those hydrophobic amino acid residues. Some of these residues, especially imidazole (of His), carboxyl group (Asp and Glu), sulfhydryl (Cys), hydroxyl group (Ser and Thr), and amine group (Lys and Arg), can coordinate metallic ions, and hence play important roles in the metallic active site of metalloenzymes. All the proteins have evolved to be suitable for their functions.

The structure of a protein is critical for its function; it has to take a specific structure. The protein structure can be defined at several levels. The basic level, *primary* structure, is the amino acid sequence of the polypeptide. For example, spinach ferredoxin is a relatively small protein consisting of 97 amino acids. A portion of this ferredoxin is written as:

$$H_2NCH(CH_3)CO—NHCH(CH_3)CO—NHCH(C_5H_4OH)CO—$$
$$—NHCH(C_2H_4OH)CO—NH—CH(CH_3)COOH$$

The first two are alanine, the third tyrosine; the second from last is threonine, and the last is alanine. The amino acid sequence is shown to start with the amine group and ends with carboxylic group, as this example shows. The entire amino acid sequence of this ferredoxin (i.e., the primary structure of spinach ferredoxin) is:

Ala-Ala-Tyr-Lys-Val-Thr-Leu-Val-Thr-Pro-Thr-Gly-Asn-Val-Glu-Phe-Gln-Cys-Pro-Asp-
Asp-Val-Tyr-Ile-Leu-Asp-Ala-Ala-Glu-Glu-Glu-Gly-Ile-Asp-Leu-Pro-Tyr-Ser-Cys-Arg-
Ala-Gly-Ser-Cys-Ser-Ser-Cys-Ala-Gly-Lys-Leu-Lys-Thr-Gly-Ser-Leu-Asn-Gln-Asp-Asp-
Gln-Ser-Phe-Leu-Asp-Asp-Asp-Gln-Ile-Asp-Glu-Gly-Trp-Val-Leu-Thr-Cys-Ala-Ala-Tyr-
Pro-Val-Ser-Asp-Val-Thr-Ile-Glu-Thr-His-Lys-Glu-Glu-Leu-Thr-Ala

Table I.1. Natural Amino Acids ($RCH(NH_2)COOH$)

Amino acid	Abbreviation	R
Hydrophobic R		
Glycine	Gly (G)	$H-$
Alanine	Ala (A)	CH_3-
Valine	Val (V)	$(CH_3)_2CH-$
Leucine	Leu (L)	$(CH_3)_2CHCH_2-$
Isoleucine	Ile (I)	$(C_2H_5)(CH_3)CH-$
Phenylalanine	Phe (F)	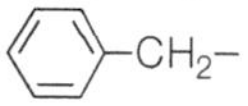
Inert heteroatom containing R		
Tryptophan	Trp (W)	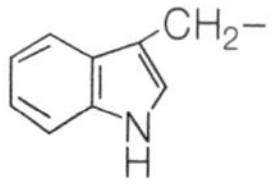
Hydroxylic R		
Serine	Ser (S)	$HOCH_2-$
Threonine	Thr (T)	$HO(CH_3)CH-$
Tyrosine	Tyr (Y)	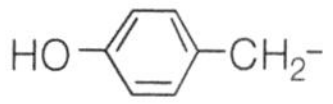
Carboxylic R		
Aspartic acid	Asp (D)	$HOOCCH_2-$
Glutamic acid	Glu (E)	$HOOCCH_2CH_2-$
Amine R		
Lysine	Lys (K)	$H_2NCH_2CH_2CH_2CH_2-$
Arginine	Arg (R)	$H_2NC(=NH)-NHCH_2CH_2CH_2-$
Histidine (imidazole base)	His (H)	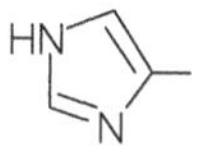
Amide R		
Asparagines	Asn (N)	$H_2NC(=O)-CH_2-$
Glutamine	Gln (Q)	$H_2NC(=O)-CH_2CH_2-$
Sulfur containing R		
Cysteine	Cys (C)	$HS-CH_2-$
Methionine	Met (M)	$CH_3SCH_2CH_2-$
Other		
Proline	Pro (P)	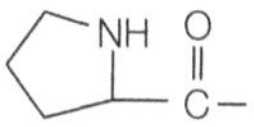

A polypeptide chain can take several characteristic arrangements: α-helix, β-strand, and random coil. This segmental structure in a protein is defined as *secondary* structure. α-Helix and β-strand (and β-pleated sheet) are illustrated in Figure I.8. α-Helix is a coiled structure (conformation) within a single peptide chain stabilized by hydrogen bonds between a C=O (of one peptide bond) and HN (of another peptide bond several units apart), as illustrated. β-Strand is a planar arrangement of amino acid residue chain, and this structure is stabilized by hydrogen bonds to another segment of the polypeptide with β-strand conformation. Hence β-strands arrange themselves side by side; the arrangement can be parallel or anti-parallel. The segments to connect these distinct segments (helix or β-strand) do not have a specific structure, and are called *random coil*.

A polypeptide (of protein) will have a specific three-dimensional structure; this is called the *tertiary* structure. Figure I.9 illustrates

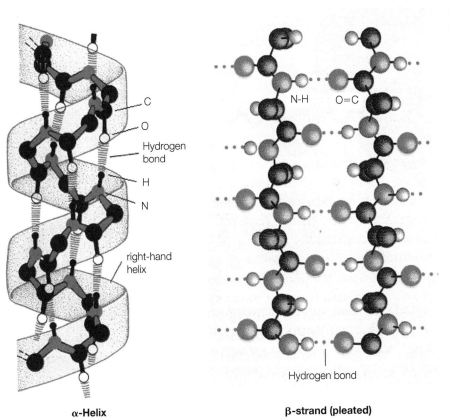

■ **Figure I.8.** Secondary structures of protein.

C

O

Hydrogen bond

H

N

right-hand helix

N-H   O=C

Hydrogen bond

α-**Helix**   β-**strand (pleated)**

■ **Figure I.9.** (a) & (b) The (x-ray crystallographic) structure of spinach ferredoxin.

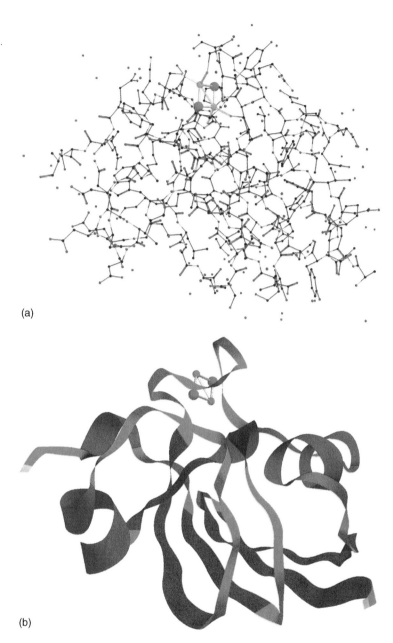

(a)

(b)

(c)

these structures for spinach ferredoxin as an example as determined by x-ray crystallography. It shows:

■ Primary structure: amino acid sequence showing all atoms connected by bars except for hydrogen (a)

■ Secondary structure: α-helices as coiled ribbons and β-strands as planar tapes (b)

■ Tertiary structure: the overall 3D structure (c)

The overall structure is believed to represent a minimum of energy associated with the interactions within the polypeptide, its interaction with cofactors (small additional entities), and the environment (water molecules and other subunits). Interactions are mostly electrostatic in nature—hydrogen bond and electrostatic attraction and repulsion between charged entities—however, it can include covalent bonding as in S—S bond (formation) between two cysteine residues. The formation of this bond requires an oxidation reaction. As illustrated in Figure I.9, this ferredoxin molecule contains a cofactor: a binuclear Fe_2S_2 active site in which two Fe ions are bound to four cysteines, as seen in Figure I.9(a). This binding also contributes to the overall 3D structure; this is the tertiary structure. A number of water molecules are seen (as red dots) in Figure I.9(a), which are also contributing to the structure.

As it was determined by x-ray crystallography, the structure represents that in the solid crystalline state. The molecule in a biological system (e.g., in cytosol medium) would not necessarily be the same as that in the solid crystalline state. NMR technique can provide some structural information of a protein in solution. The positions of amino acid residues in a protein change rapidly and constantly. This dynamic must be taken into account when considering the protein's performance, but the details are yet to be explored.

The final level of protein structure is *quaternary* structure. It is a multisubunit structure. As discussed earlier, myoglobin consists of a single polypeptide chain and hence its quaternary structure is mute. Hemoglobin, on the other hand, consists of four subunits: α, α, β, and β. How these subunits assemble themselves is the quaternary structure. The structure of hemoglobin is found in any standard textbook.

The secondary structure seems to be related to the flexibility of the protein; β-strands create β-sheet, which is rather rigid, whereas random coil and α-helix are rather flexible. Hence the change of relative position of α-helices seems to be involved in such a process of switching based on conformational change.

As far as the catalytic function is concerned, the arrangement and temporal change of amino acid residues surrounding the active site play important roles. That is, they provide auxiliary but essential roles (such as donating and accepting protons from substrate and the catalytic center).

Reactions—Formation and Hydrolysis of Protein

The formation of a protein by condensing amino acids is a rather complicated process, and has been unraveled only recently (since 2000). On the other hand, hydrolysis of polypeptides is straightforward and catalyzed by a number of peptide bond-breaking enzymes such as metal free peptidases and proteinases (pepsin, chymotrypsin, etc.) and metalloenzymes (thermolysin, carboxypeptidase, etc.). The metalloenzymes are discussed in Chapter 4.

Here is a brief description of protein synthesis. The code for the primary structure of a protein is written in the exon portion of a gene (a part of a chromosome-DNA). First the entire gene (exon/intron) for a protein is transcribed from that portion of DNA onto a messenger RNA (mRNA). A RNA polymerase (a protein) is employed in this reaction. mRNA contains both exon and intron, of which only exon contains the sequence information for the protein. A RNA enzyme ribozyme cleaves out the exon from intron (see Chapter 4). This RNA (exon) is now used as a template to synthesize the protein. Individual amino acids first need to bind to specific (transfer) tRNAs. The condensation reaction is then catalyzed by a specific RNA (rRNA). A model (see Fig. I.10) is provided to show a mechanism of this condensation on rRNA (Nissen *et al.*, 2000).

Vitamins (Coenzymes), Nucleotides, and Others

Coenzymes

Enzymes often require a small extraneous entity for activity; a compound of this kind is called a *coenzyme*. Most vitamin Bs are precursors

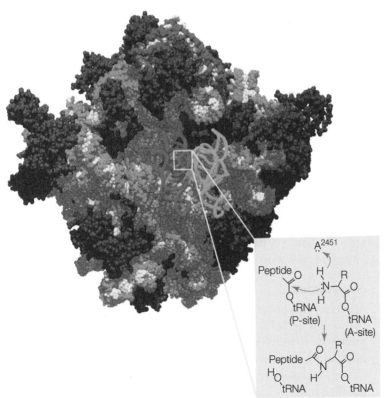

■ **Figure I.10.** A model of protein synthesis on rRNA (Nissen et al., 2000).

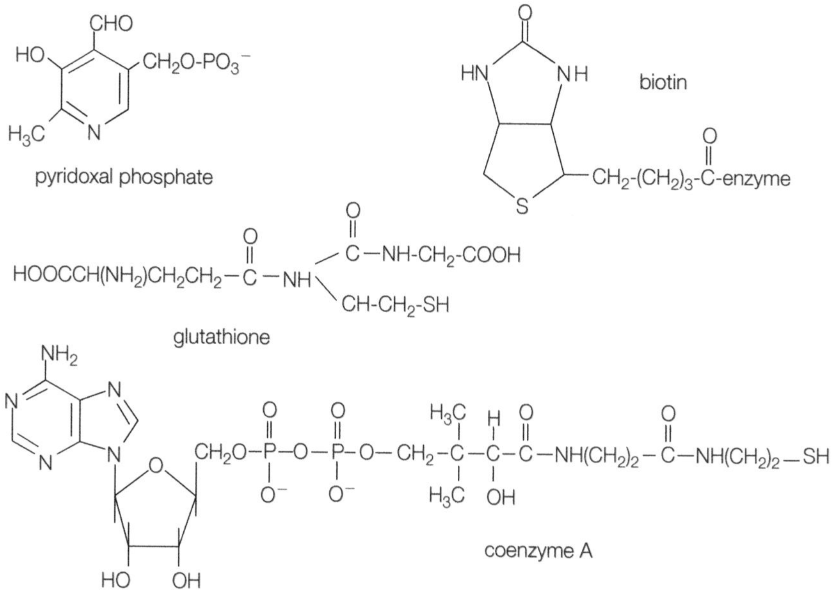

■ **Figure I.11.** Coenzymes.

of coenzymes; B_1 is chemically thiamine, B_2 is riboflavin, B_3 niacin or nicotinic acid, B_6 pyridoxal phosphate, and B_{12} cobalamin. Cobalamin and enzymes dependent on it will be discussed in Chapter 7. Riboflavin constitutes coenzymes such as FAD (flavin adenine dinucleotide), and nicotinic acid is a component of NADH and NADPH (NAD=nicotine adenine dinucleotide). These nucleotides are discussed in the next section. Pyridoxal phosphate is involved in a number of transferases. Coenzymes biotin and tetrahydrofolate come from, respectively, biotin and folic acid that are classified in the vitamin B category.

Aside from vitamin B-related ones, a variety of compounds play the roles of coenzyme. A few more examples are coenzyme A, coenzyme Q_{10}, S-adenosyl methionine, and glutathione (see Fig. I.11). Coenzyme A is a derivative of a nucleotide.

Nucleotides

A nucleotide consists of a base such as purine (adenine, guanine), pyrimidine (cytosine and uracil or thymine), and flavin, a monosaccharide (typically ribose or deoxyribose) and phosphate. The nucleotides that contain two bases are called dinucleotides. A few typical examples of nucleotides are shown in Figure I.12. Nucleotides involved in DNA and RNA will be mentioned in the next section.

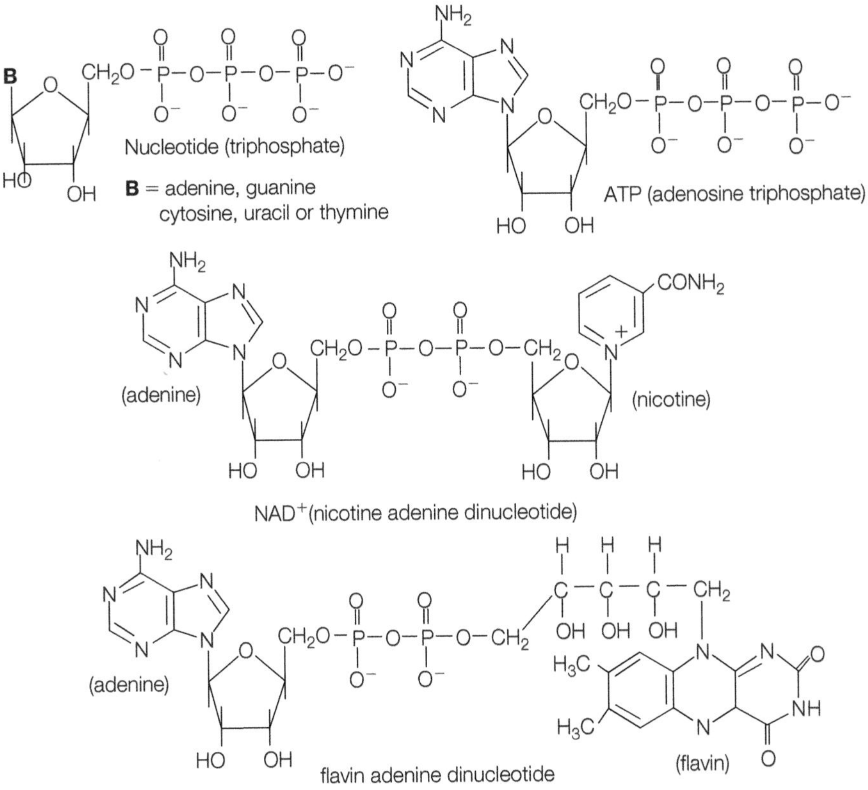

■ **Figure I.12.** Nucleotides.

Other Vitamins

Vitamin A is a precursor of eye pigment retinal, vitamin C is a reducing agent in a wide range of biocompounds (resulting often in preventing oxidative damages), and vitamin D is related to a steroid hormone governing bone growth. Vitamin E is a component of signal transmitting neuronal membrane, and vitamin K is involved in blood clotting.

DNA/RNA (Polynucleotide)

Structures

A nucleic acid is a linear polymer of nucleotides (hence polynucleotide). A nucleotide consists of a base bound to ribose or deoxyriboses with C-5 of triphosphate ester. The two types of base are purine (adenine and guanine) and pyrimidine (cytidine, uracil, and thymine). DNA consists of four deoxyribonucleotides: adenosine

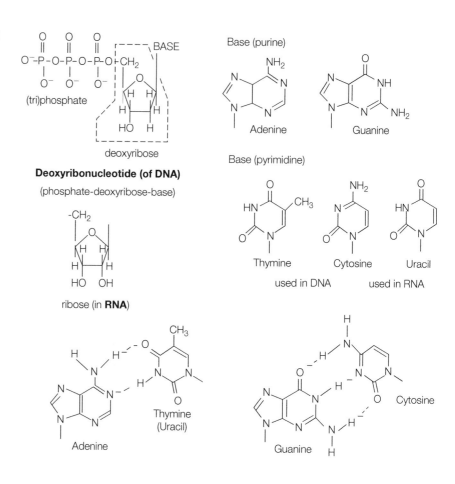

■ **Figure I.13.** Nucleotides and Watson-Crick type hydrogen bonding between purine and pyrimidine bases.

(A), guanosine (G), cytosine (C), and thymine (T). RNA is made of ribonucleotides: adenosine (A), guanosine (G), cytosine (C), and uracil (U) (see Fig. I.13). Nucleic acid is a linear copolymer of the four nucleotides bound through the phosphate group (the binding is essentially a phosphate ester to OH of ribose or deoxyribose), as seen in Figure I.14.

DNA usually exists as a double helix (see Fig. I.15); that is, except when DNA messages are read, either for replication or transcription. The most common double helix structure is the so-called B-DNA, in which the base pairs are of Watson-Crick type. There is a "major" groove and a "minor" groove along the double helix. The bases are directed inward, as they are hydrogen-bonded to the counterparts on the other strand, and the backbone is lined with riboses connected

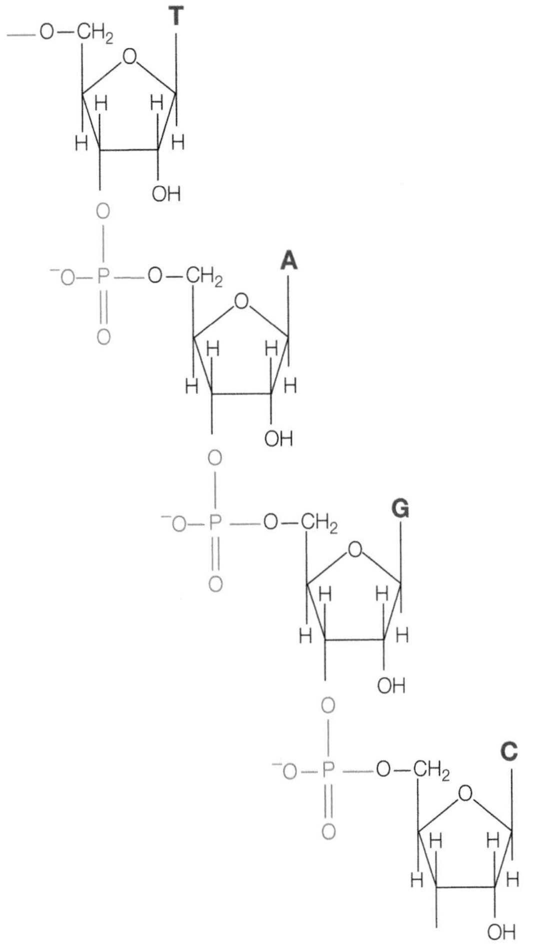

■ **Figure I.14.** Polynucleotide chain.

by phosphate groups. A-DNA is also right-handed and is a little fatter than A-DNA. The major and the minor grooves are of about the same size in opening, but the major groove is much deeper. Another double helix form is Z-DNA, which is left-handed and much more stretched than B-DNA.

A variety of RNA is found in cells: transfer tRNA, messenger mRNA, ribosomal rRNA, of which there are several different sizes (16S, etc.), and small nuclear (snRNAs) and nucleoproteins in the nucleus. The structures of RNA are quite diverse. tRNA takes the so-called clover-leaf shape in its secondary structure, which in turn twists and

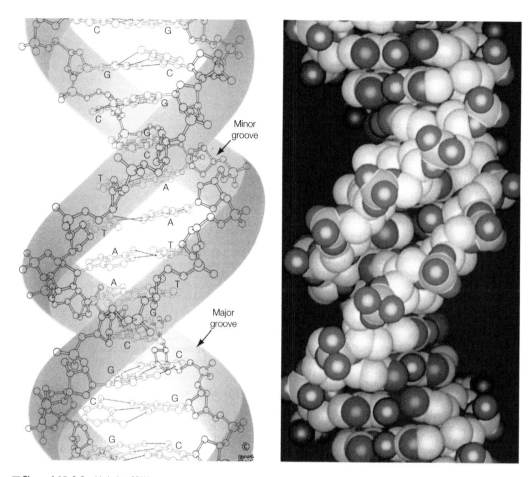

■ **Figure I.15.** B-Double helix of DNA.

folds to form its tertiary (3D) structure (see Fig. I.16). The double helix portion of tRNA seems to be of the type of A-DNA structure, and there are a number of non-Watson-Crick type base pairs.

A pre-mRNA, when first transcribed from the DNA, contains not only the functional segment (i.e., code for a protein, called *exon*), but also the noncoding segments (introns). Introns need to be excised before the mRNA is translated (into a protein) (see Chapter 4 for details). A primary rRNA transcript also requires removal of introns before it becomes functional rRNA, such as 5S, 16S, and 23S rRNA (in *E. coli*).

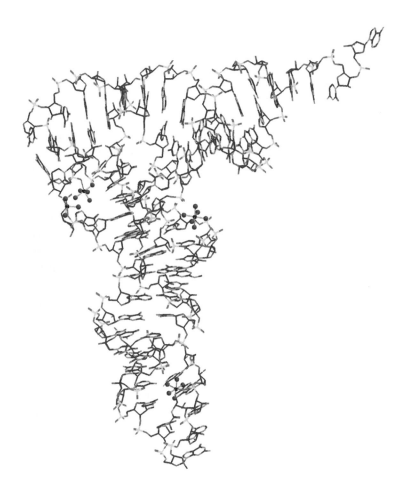

■ **Figure I.16.** Clover leaf structure of t-RNA. (PDB ID:4TRA: Westhof, E., Dumas, P., Moras, D. 1988. Restrained refinement of two crystalline forms of yeast aspartic acid and phenylalanine transfer RNA crystals. *Acta Crystallogr.*, Sect. A **44**, 112–123.) Note that four Mg(II)s (purple dots) are bound to it either directly or through hydrogen bonds.

Reactions

Deoxyribonucleotides, the components of DNA, are produced from the corresponding ribonucleotides through reduction (removal of an O-atom) by an enzyme, ribonucleotide reductase (see Chapter 7). A DNA is synthesized by a DNA polymerase using these nucleotides.

A DNA then is replicated and/or transcribed, repaired when an error is made in these processes, and cleaved as by restriction enzymes. A double helix first needs to be unwound and changed to single-stranded in order to be read. DNA polymerases are used for replication, and RNA polymerases transcribe the DNA sequence into its corresponding RNA sequence. In both of these processes, a strand of DNA is used as a template. Reverse transcriptase as found (e.g., in HIV-I), however, polymerizes deoxyribonucleoside triphosphates to form a DNA using a RNA as the template. The chemical reaction involved in all these processes is essentially esterification of the phosphate group of the incoming nucleotide (nucleoside triphosphate) by the 3′-OH group of the growing end.

The reverse reaction is hydrolysis, which is carried out by a variety of nucleases, exo-DNases, endo-DNases, and RNases. DNase and RNase are deoxyribonuclease and ribonuclease, respectively. Some of endo-deoxyribonucleases are called restriction enzymes, and are used for gene recombination procedures. The hydrolysis reaction is employed in the metabolisms of polynucleotides, cleaving DNAs, and repairing DNAs and RNAs. Endonucleases, particularly restriction enzymes, specifically cleave the middle portions of a polynucleotide. Repairing often is conducted as polymerization progresses, and the effecting enzyme, exonuclease, is often a part of a polymerase complex. Removal of the noncoding segments (introns) of pre-RNAs is carried out by certain types of RNA themselves (ribozymes) and/or nucleoproteins (RNA-protein composite) (see Chapter 4).

TYPES OF BIOCHEMICAL REACTIONS

Three different types of chemical reactions are recognized: acid-base type, oxidation-reduction type, and free radical type. All biochemical reactions can also be grouped into these three categories.

Reactions of Acid-Base Type

A reaction of this type involves simultaneous heterolytic bond cleavage and formation, and all the atoms in the molecules involved do not change their formal oxidation states. For example, the simplest acid-base reaction $(H_3O^+ + OH^- \rightarrow 2H_2O)$ involves the splitting of an H—O bond (of H_3O^+) and the formation of an H—O bond (to form H_2O). This reaction is accompanied by a large negative enthalpy change, because the bond energy of H—O in H_3O^+ is relatively small, whereas the energy of the H—O bond to be formed (H—O in the resulting H_2O) is large. Another acid-base reaction: $H_2O + NH_3 \rightarrow OH^- + NH_4^+$ involves H—O bond splitting and H—N bond formation (to result in NH_4^+), and its enthalpy of reaction (ΔH) is positive (and ΔG positive) because of a large bond energy of O—H (in H_2O) and a relatively small bond energy of N—H (in NH_4^+). One of the biochemical reactions, that catalyzed by carbonic anhydrase, is an example of acid-base reaction: $CO_2 + H_2O \rightarrow H^+ + HCO_3^{2-}$.

Many biochemical reactions involve simultaneous heterolytic bond cleavage and formation; they are thus of acid-base type, though not all the reactants can be recognized as acid or base. In organic chemistry, such reactions are called either *nucleophilic* or *electrophilic* substitution reactions. A nucleophilic substitution reaction is illustrated by:

$$H_3CH_2C—Br + {}^-OH \rightarrow H_3CH_2C—OH + Br^-$$

Here the C—Br bond splits heterolytically, formally resulting in a carbonium entity ($H_3CH_2C^+$), and then a C—O(H) bond forms heterolytically. This description does not necessarily imply the reaction mechanism.

A few examples from biochemical reactions are:

hydrolysis of peptide bond:

$$—(C{=}O)— NH— +OH^- \rightarrow —(C{=}O)—OH + {}^-NH—$$

transphosphorylation:

$$ADP—O—PO_3^2 + H—O—R \rightarrow$$
$$ADP(—O^-) + {}^{2-}O_3P—O—R + H^+$$

In the formal sense, the N—C (of C=O) bond splits heterolytically and the resulting $C^+(C{=}O)$ binds with OH^- heterolytically in the first example. Likewise, in the second reaction a heterolytical splitting occurs in the bond $O—P(O_3^{2-})$ (in ATP) and the lone pair on the O (of HOR) binds the resulting (formal) $P^+(O_3^{2-})$. This description does not imply the reaction mechanisms.

Reactions of Oxidation-Reduction Type

Removal of electron(s) from a molecular entity or others is oxidation, and the entity that removes electron(s) from another will receive electron(s). The latter is reduction, and hence such a reaction is oxidation-reduction. In such a reaction, electron(s) are transferred from one (reducing agent=reductant) to another (oxidizing agent=oxidant); hence this can also be called electron transfer reaction.

The Idea of Oxidation State

The idea of (formal) oxidation state has been devised in order to keep track of the movement of electron(s). In other words, an oxidation state is defined for each atom (or in certain cases, a group of atoms) in a molecule. Each atom in a homonuclear neutral diatomic molecule (such as H_2, N_2, O_2, and F_2) is defined to be in an oxidation state of zero. In heteroatomic molecules, oxidation states are defined based on the electronegativity of atoms. For example, "F" (the most electronegative element) is in the oxidation of -1 ($-I$) in all molecules. In most molecules "O" is assigned an oxidation state of -2 ($-II$). It is $-I$ each in the case of HOOH or ROOH. "H", except for H bound directly with a metallic atom, is in the oxidation state of $+1$ ($+I$). "H" in a metallic hydride such as NaH is in the state of $-I$. "N" can be in a variety of oxidation states from $-III$ to $+V$. N in NH_3 is in the oxidation state of $-III$, whereas N in NO_3^- is in the state of $+V$.

Nontransition metallic elements take their usual highest oxidation states—K($+I$), Ca($+II$), Al($+III$)—although some of them can take different oxidation states (e.g., Sn($+II$) and Sn($+IV$), Pb($+II$) and Pb($+IV$), and Tl($+I$) and Tl($+III$)). The "$+$" sign often is omitted; thus, Sn(IV) instead of Sn($+IV$). The oxidation states of transition elements vary (e.g., Fe(II), Fe(III), and others, and Mo(II)/Mo(III)/Mo(IV)/Mo(V)/Mo(VI)). The oxidation states of metallic elements are not "conceptual," as in the case of the oxidation state of "C" atoms in organic compounds (to be discussed in the next section), but substantial. The oxidation state of an atom does not necessarily represent its actual electric charge, though it is close to the real electric charge in simple ionic compounds such as NaCl and $CaBr_2$.

Note: In this book, Roman numerals such as III and IV are used to express the oxidation states (or oxidation number) of atoms in order to distinguish between the oxidation state and the electric charge of an entity, for which Arabic numerals are used; for example, $[Mn^{II}Cl^{-I}_4]^{2-}$.

The Oxidation State of C in Organic Compounds and Recognition of Oxidation-Reduction Reactions

It is not essential to recognize the oxidation state of "C" in an organic compound, in order to understand the oxidation-reduction reaction involving it, but the idea of oxidation state would make it clear how the oxidation-reduction takes place.

The overall reaction of formation of glucose from CO_2 and H_2O in photosynthesis can be written as:

$$6C^{+IV}O^{-II}{}_2 + 6H^{+I}{}_2O^{-II} \rightarrow C^0{}_6H^{+I}{}_{12}O^{-II}{}_6 + 6O^0{}_2$$

The oxidation states of atoms involved can be defined as written, based on the rules mentioned earlier. The "C" atoms change their oxidation state from $+IV$ to 0; hence each "C" atom gains four electrons; hence it is reduced. Some "O" atoms change their oxidation state from $-II$ (on the right-hand side) to 0 (in O_2); hence those "O's" are oxidized. This indicates that the reaction is an oxidation-reduction reaction overall, but does not necessarily imply that CO_2 is reduced directly by a water molecule. All oxidation-reduction reaction involves overall simultaneous oxidation and reduction reaction as in this case. However, a reaction to the major substrate often is emphasized, and is assigned a name of oxidation or reduction accordingly. In the preceding example, CO_2 is considered to be the major reactant and hence the reaction often is characterized as a reduction reaction. (From the point of view of another reactant, H_2O, it is an oxidation.)

A reaction by alcohol dehydrogenase is expressed as:

$$C^{-II}H_3C^{-II}H_2OH + NAD^+ \rightarrow C^{-I}H_3(C^{-I}{=}O)H + NADH + H^+$$

Nominally, the oxidation state of each "C" in ethanol can be assigned as $-II$, and that of each "C" in the product acetaldehyde is $-I$, and hence the carbon atoms in the reactant is oxidized (from $-II$ to $-I$). Strictly speaking, an ambiguity exists in assigning an oxidation state to each of the different C-atoms in a molecule; for example the following local assignment would be appropriate: $C^{-III}H_3C^{-I}H_2OH$ and $C^{-III}H_3(C^{+I}{=}O)H$ in this case. However only the average value of the oxidation states of all Cs in a molecule is sufficient to recognize oxidation-reduction. The two electrons removed are accepted by NAD^+ in the form of H^- (hydride) in this case. In other words, NAD^+ removes H^- from ethanol; hence the reaction is called *dehydrogenation*, which is equivalent to removal of two electrons and hence is a kind of oxidation. It is not practical to count the oxidation states of each atom in a complex molecule such as NAD^+ (see Fig. I.12). However, the fact that NAD^+ is indeed reduced in this process is obvious from the comparison of NAD^+ and NADH. In this particular reaction, the entity that is to be transferred is a hydride, which has a lone pair and can be regarded as a base. Hence the enzyme that catalyzes this type of dehydrogenation reaction uses an acid catalyst Zn(II) in contrast to other oxidation-reduction enzymes.

Many biological oxidation reactions are in fact dehydrogenation, though not necessarily always a hydride removal. It can involve two-step hydrogen (H) removal or two-step simultaneous removal of an electron and a proton (H^+). On the contrary, addition of "H" to a substrate is called *hydrogenation*, which is a kind of reduction.

A major type of biochemical oxidation reaction involves the removal of electron(s) (often together with removal of H^+) by an "O"-containing entity such as O_2, HOOH, and ROOH, though these entities may not be the immediate electron remover in reaction mechanism. An example is the reaction of a copper enzyme, laccase:

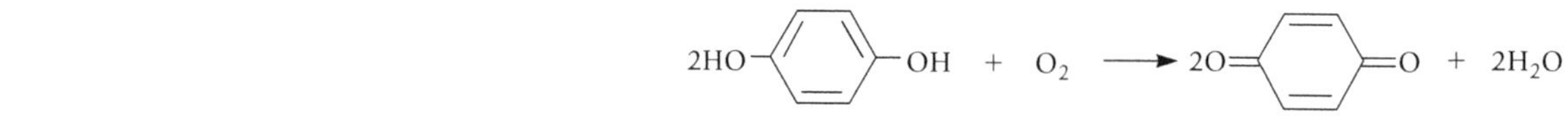

$$2HO-\text{〈benzene〉}-OH \; + \; O_2 \;\longrightarrow\; 2O=\text{〈benzene〉}=O \; + \; 2H_2O$$

The idea of oxidation state is not particularly useful in this case. But if applied, the average oxidation state of "C" in p-hydroxyben-zene (hydroquinone) is $1/3$ ($+II$ among C_6) and that in the product (p-quinone) is $2/3$ ($+IV$ among C_6), and thus, the reactant is oxidized. Another example is peroxidase reaction, for example,

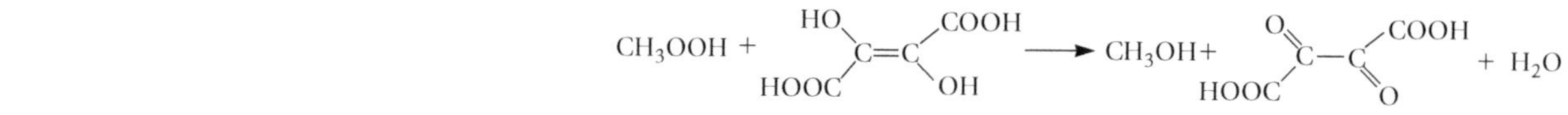

$$CH_3OOH + \;\; \underset{HOOC}{\overset{HO}{>}}C=C\underset{OH}{\overset{COOH}{<}} \;\longrightarrow\; CH_3OH+ \;\; \underset{HOOC}{\overset{O}{>}}C-C\underset{O}{\overset{COOH}{<}} \;\; + \; H_2O$$

The O—O in the peroxide is $R-O^{-I}-O^{-I}-H$ and ends up in $R-O^{-II}-H$ and H_2O^{-II}. That is, the O—O acts as the oxidizing agent. The oxidation number of "C" in the dihydroxyfumarate is $+VIII$ per C_4 and that in the product is $+X$ per C_4; hence the reactant is oxidized. In the enzymatic reaction, peroxide is not the direct oxidizing entity. A third example is the reaction of aldehyde oxidase (a Mo-enzyme):

$$CH_3(C=O)H + O_2 + H_2O^* \rightarrow CH_3(C=O)-O^*H + H_2O_2$$

In this reaction O_2 is reduced to H_2O_2 and acetaldehyde is oxidized to acetic acid. The "O" incorporated comes from H_2O rather than O_2, as indicated by O*.

In some reactions "O" atom(s) from O_2 or ROOH may be incorpo-rated directly into the product. This reaction is a kind of oxidation as shown here, but the fact that "O" is incorporated is emphasized, and such a reaction is called *oxygenation* reaction. The following reaction is called *monooxygenation,* as only one of the two Os of O_2 is incorporated into the product:

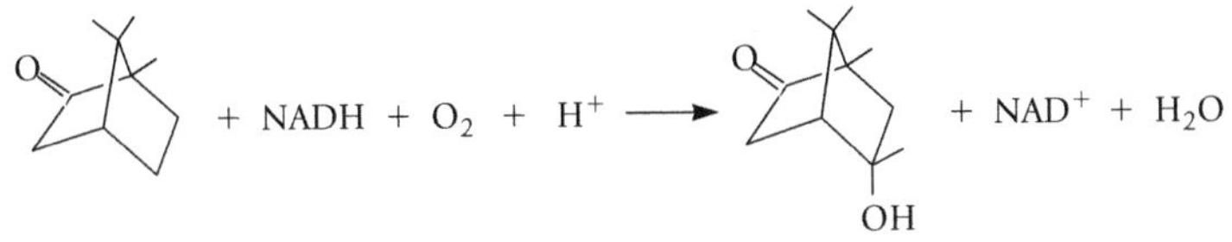

The oxidation state of the "C" to which "O" is attached changes from $-II$ to 0 (zero) (formally) upon reaction, hence being oxidized. NADH is also oxidized simultaneously. The other "O" ends up in H_2O, being reduced. A typical dioxygenation is illustrated by:

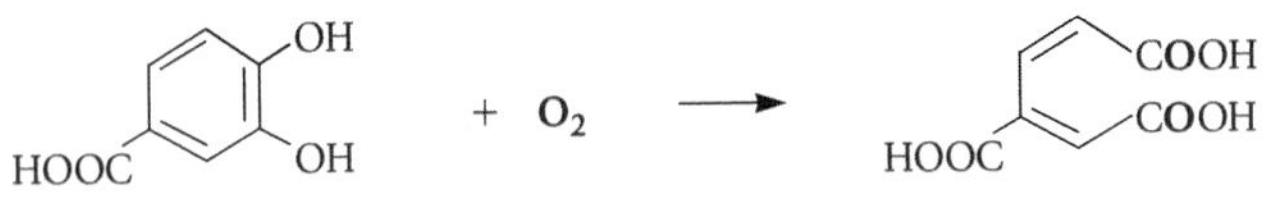

Formally the oxidation state of C_7 in the reactant can be calculated to be $+II(/C_7)$, whereas that in the product is $+VI(/C_7)$. An assignment of oxidation state for each of the carbon atoms can be as follows.

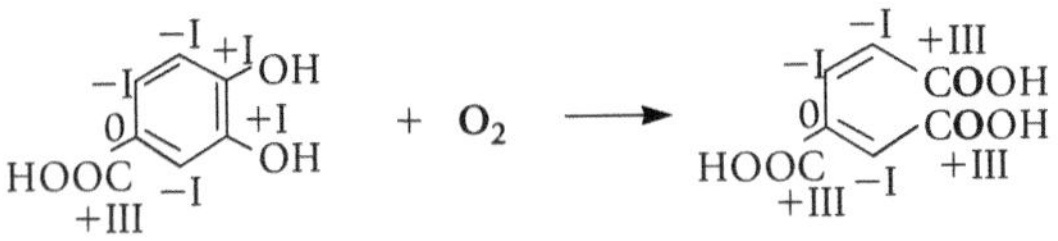

It is obvious that an oxidation reaction takes place here. The oxidation state of Cs in a C-containing compound is regarded to go up when it combines with "O" atom(s).

Other Kinds of Oxidation-Reduction Reactions

A whole variety of other kinds of oxidation-reduction reactions take place in biological systems. Simple nitrogen compounds undergo biochemical oxidation and reduction reactions: involved are anything from between the most reduced $N^{-III}H_3$ to the highest oxidized state $N^{VI}O_3^-$ including N_2 and NO_2^-; this is reviewed in Chapter 8. Sulfate SO_4^{2-} is reduced to sulfite, and then to sulfide S^{2-}. A brief biochemistry of sulfur is given in Chapter 9. Other types of oxidation-reduction reactions are dealt with in Chapter 5.

Free Radical Reactions

Free radical reactions involve entities that have unpaired electron(s). Many of biologically relevant free radicals are associated with oxygen.

O_2 is a biradical (with two unpaired electrons) in the ground state, and its direct reaction can take place with an entity with an unpaired electron (free electron, free radical, metallic elements, etc.). The product of one-electron reduction is superoxide O_2^- with a single unpaired electron; another free radical. O_2 might react with a free radical $R^{\cdot}$; the result is the formation of $RO_2^{\cdot}$; another free radical. One of the characteristic reactions of many free radicals is abstraction of a hydrogen atom from a hydrocarbon or the like. For example:

$$R'OO^{\cdot} + H—R \rightarrow R'OOH + {\cdot}R; \ Cl^{\cdot} + H—R \rightarrow Cl—H + {\cdot}R$$

Hydrogen abstraction thus produces another free radical entity, which then abstracts a hydrogen atom; hence such a reaction becomes a chain reaction. The hydroperoxide R'OOH readily reacts with a transition metal ion such as Fe(II) in the manner of:

$$R'OOH + Fe(II) \rightarrow R'O^{\cdot} + OH^- + Fe(III)$$

or

$$R'OOH + Fe(III) \rightarrow R'OO^{\cdot} + Fe(II) + H^+$$

It is very interesting that biological systems do use free radicals created on an amino acid residue in an enzyme protein as a catalytic entity. Such amino acids include tyrosine and glycine. This fact implies that certain biochemical reactions cannot be effected without invoking a free radical. Chapter 7 is devoted to reactions involving free radicals.

TRANSITION STATE THEORY OF REACTION, AND ENZYME KINETICS

An enzyme is a catalyst in living organisms. A great variety of enzymes have been discovered and cataloged. Each enzyme has been assigned a specific code number and systematic name (by enzyme commission (EC)). However, most of the enzymes are still designated by conventional names. The basics of reaction energetics and enzyme kinetics are discussed here.

Energy Profile and Transition State Theory of Reaction

From the point of view of the energy of reaction system, a reaction, catalyzed or uncatalyzed, proceeds over an energy surface, the height of which varies as the reaction proceeds, because of changes in interaction energies among entities involved in the reaction. This energy surface is multidimensional, but often it is simplified to a two-dimensional curve, in a way to go along the main reaction path.

It typically is shown as in Figure I.17. The vertical axis represents the total energy of the system. The initial state represents the total energy of reactants, and the final state, that of products. The horizontal axis represents the degree of reaction progress. The energy of this reaction system is thought to go along the curve such as shown in the figure. This often is called the energy profile of reaction. A reaction has to go over an energy barrier, enthalpy (or free energy) of activation, often designated as ΔH^* or ΔG^*. The highest point is called transition state. Eyring (and his coworkers) proposed a theory called *activated complex theory*. He assumed the following reaction:

$$A \text{ (reactant)} + B \text{ (reactant)} \Leftrightarrow (AB)^* \rightarrow \text{products}$$

He hypothesized that this activated complex $(AB)^*$ represents the transition state, and that it is in equilibrium with the reactants. If κ (called transmission factor) represents the probability that the activated complex (transition state) goes over to the product state, and thermodynamic principles are applied to the equilibrium between

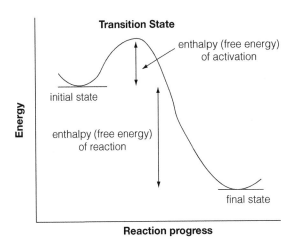

■ **Figure I.17.** Reaction energy profile.

the initial state and the activated complex, the following expression (Eyring equation) for the reaction rate constant k is obtained:

$$k = \kappa \exp(\Delta S^*/R)\exp(-\Delta H^*/RT)$$

An empirical expression for the temperature dependence of rate constant k was obtained by a Swedish chemist Arrhenius. The following is one version of the Arrhenius equation:

$$k = A\exp(-\Delta H^*/RT)$$

This is to be compared with the theoretically derived Eyring equation; they are identical if A (an empirical factor) is equated to $\kappa \exp(\Delta S^*/R)$. A further discussion on this issue as applied to enzyme reactions will be found in Chapter 3.

Enzyme Kinetics

Enzyme reactions are complicated, consisting of multiple steps, but they often can be described by a simple two-step reaction expressed by:

$$\text{S (substrate)} + \text{E (enzyme)} \underset{k_{-1}}{\overset{k_1}{\rightleftharpoons}} \text{ES (complex)}$$
$$\xrightarrow{k_2} \text{E} + \text{P (products)}$$

The overall reaction rate can be expressed as rate = $k_2[\text{ES}]$, and it can further be rendered to the so-called Michaelis-Menten equation, assuming the steady state of ES complex; that is:

$$\text{rate} = k_2[\text{E}]_\text{T}[\text{S}]/(((k_{-1} + k_2)/_{k1}) + [\text{S}] = k_2[\text{E}]_\text{T}[\text{S}]/K_\text{m} + [\text{S}])$$
$$= V_{\max}[\text{S}]/K_\text{m} + [\text{S}])$$

$[\text{E}]_\text{T}$ is the total concentration of enzyme, and K_m is called the Michaelis constant, and $V_{\max} = k_2[\text{E}]_\text{T}$. K_m can be approximated to $K_\text{m} = 1/K_1 = k_{-1}/k_1$, that is the inverse of the binding constant of S to E. This is justified if $k_2 \ll k_{-1}$. If we take the inverse of both sides of the Michaelis-Menten equation, we get:

$$1/r \text{ (rate)} = (1/V_{\max}) + (K_\text{m}/V_{\max})(1/[\text{S}])$$

Hence if we plot $(1/r)$ against $(1/[\text{S}])$, we would obtain a line with a slope of $K_\text{m}/V_{\max}$; this is called a Lineweaver-Burk plot, from which we can obtain kinetic parameters, K_m and k_2.

When an inhibitor I is added to the reaction system, rate expression likewise can be obtained. If I is a competitive inhibitor, that is, if I competes with S for the catalytic site, rate is:

$$\text{rate} = k_2[E]_T[S]/((1 + [I]/K_1)K_m + [S])$$

where K_I is the inverse of binding constant of I to E. If the inhibitor does not compete with S, rate is expressed as follows.

$$\text{rate} = k_2[E]_T[S]/((1 + [I]/K_1)/K_m + [S])$$

Enzyme Reaction Mechanism

In the previous discussion, a two-step mechanism is assumed. The first step is the formation of ES complex, and the second the formation of products from ES. The energy profile of each step (reaction) would be described by a curve like Figure I.17. In other words, the first step goes over a barrier of $\Delta H_1{}^*$, and the second step has to go over a barrier of $\Delta H_2{}^*$. If $\Delta H_2{}^* \gg \Delta H_1{}^*$ (rather, rate of step 2 $\ll$ rate of step 1), the second step would be much slower than the first step, and the overall rate would be determined by the slower step rate; the slowest step is called the rate-determining step. Many such cases where the second step is rate-determining have been found, but it is not always so. In other cases, the first step can be rate-determining.

In reality, many enzyme reactions may consist of more than two steps. However, often only one step may be much slower than all the other steps. If so, the overall rate is determined by that slowest step. Enzyme kinetic studies can then identify such a step, though other steps can only be hypothesized. Many physicochemical methods have been applied to identify intermediate entities, and hence shed light on possible mechanism (ensemble of stepwise reactions).

REVIEW QUESTIONS

1.   Define *biosphere*.

2.   How different are eukaryotic cells from prokaryotic cells?

3.   What are the major differences between animal cells and plant cells?

4.   Give the basic chemical formula for carbohydrate (monosaccharide).

5. As mentioned in the text, the hydrolysis of a protein is relatively simple, whereas the reverse reaction, formation of polypeptide, is an elaborate and complex process. Discuss why from a structural point of view.

6. In general, condensation reactions (such as formation of pyrophosphate bond in ATP and peptide bond) are usually endothermic, whereas the reverse processes are exothermic. Discuss why and the implications of this fact for biochemical processes.

7. Derive the Michaelis-Menten equation assuming the steady state of ES complex.

8. Characterize each of the following reactions either as acid-base type or oxidation-reduction type.

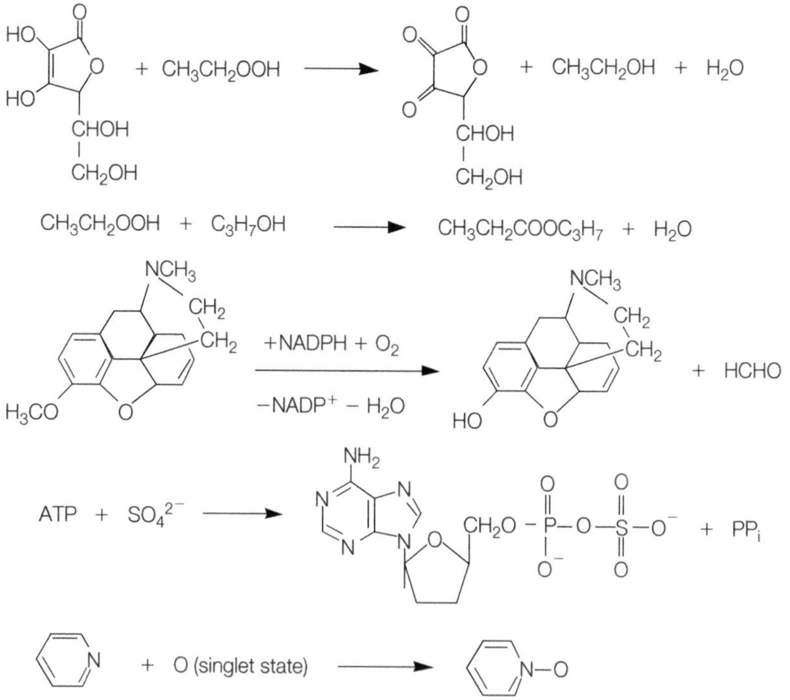

PROBLEMS TO EXPLORE

1. Why are *autotrophs* and *heterotrophs* in the biosphere, and how might they have evolved?

2. Both cellulose and starch (amylose) are made of the same monosaccharide glucose. But their physical and chemical properties are quite different. Explore the reasons for the difference.

3. The reaction enthalpies are relatively small in reactions of acid-base type (except for neutralization reaction), whereas those of oxidation reduction reactions are relatively large. Discuss.

4. Fat is more energetic (releases more energy) than the corresponding carbohydrate on the weight basis (or carbon basis). For example, compare sugar $C_{12}H_{22}O_{11}$ with $C_{12}H_{26}$ or $C_{12}H_{22}$ (i.e., the energy released per carbon atom or unit weight). Explain why. This is related to the fact that the hummingbird, for example, stores energy in the form of fat rather than carbohydrates before taking off on migration.

5. Give examples of the primary, secondary, tertiary, and quarternary structure of a protein other than that shown in Figure I.9.

6. An idea has been expressed that RNAs are the working gene whereas DNAs are the long-lasting storage of the RNA gene copies. Discuss the idea.

The Distribution of Elements

1.1. THE DISTRIBUTION OF ELEMENTS IN THE EARTH'S CRUST, SEAWATER, AND ORGANISMS

The material in the universe is composed of about 100 elements, which have been created through different nuclear reactions since the time of the Big Bang. The cosmic distribution of elements hence reflects these processes, and also the relative nuclear stability of different nuclides. The distribution of elements on Earth is significantly different from that in the overall universe. This fact reflects the way the Earth was formed, and the constraints due to its size. Against this background, organisms originated on Earth, and evolved through its long history.

It is convenient to regard the established Earth as consisting of several components: the core, mantle, crust, atmosphere, hydrosphere, and biosphere. The magnitudes of these components are shown in terms of their masses in Figure 1.1. The biosphere is tiny compared with the nonorganic components. The quantity of living organisms presented in this diagram may well be an underestimation, because the deep ocean is now recognized to harbor a significant proportion of living organisms, which have not been evaluated. In addition, the ecology of these organisms may be quite different from those on land and in shallow water.

From a thermodynamic standpoint, an organism, whether unicellular or multicellular, is an open system. Therefore, it will exchange energy and material with its surroundings. An implication is that organisms will ingest, utilize, and hence contain all the elements present in their surroundings.

37

■ **Figure 1.1.** The magnitudes (in terms of mass) of components of the Earth.

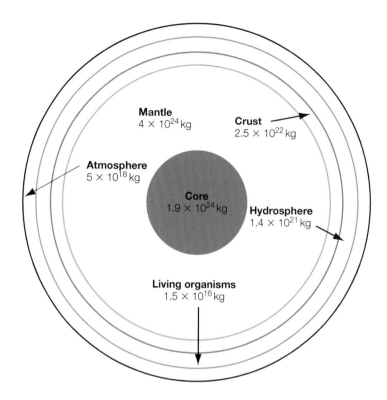

A set of elemental compositions of the human body, alfalfa (plant) and copepod (seawater crustacean), is given in Figure 1.2. Of course, the elements are not distributed uniformly in the body. The data for humans in Figure 1.2 were obtained for an organ (the liver), collected from a number of different literature sources. It is also not expected that different individuals would show a numerically similar distribution of elements. Hence, such a set of data as presented in Figure 1.2 can be considered to represent average or "ballpark" figures. It is obvious that living organisms contain all kinds of elements in addition to the four elements that constitute the bulk of the organic compounds. Many of these elements are essential to the organisms, and their behaviors in the organisms are essentially the subject of bioinorganic chemical studies.

As an organism is an open system, its elemental composition may reflect that of its surroundings. It is expected that some elements are actively taken up by the body and others may simply enter inadvertently. Hence there is not necessarily a very high correlation between

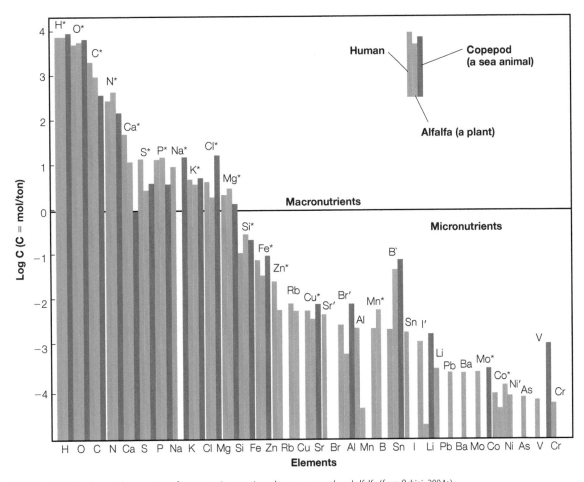

■ **Figure 1.2.** The elemental composition of representative organisms: human, copepod, and alfalfa (from Ochiai, 2004a).

the elemental composition in humans and that of the surroundings. The correlation between the elemental composition of the human body and that of the upper crust is shown in Figure 1.3, and that between the human body and the seawater on the current Earth is shown in Figure 1.4. The elemental composition of humans in these diagrams is that in the liver except for C, H, N, O, P, S, and Cl; the figures for the latter elements are those in the total body. In Figure 1.4, the estimated level of iron (as Fe(II)) in ancient seawater also is indicated. The correlation appears to be better with this corrected distribution in seawater. No matter how they are looked at, these figures imply that the living organisms are open systems that interact intimately with their environments.

■ **Figure 1.3.** The correlation of the elemental compositions of human tissue (liver) and of the crust (from Ochiai, 2004a).

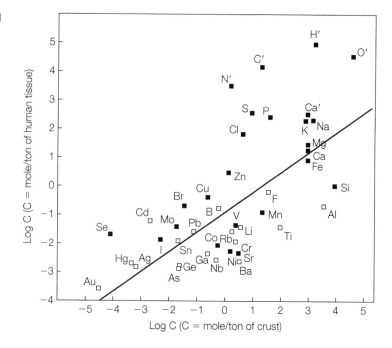

■ **Figure 1.4.** The correlation of the elemental compositions of human tissue (liver) and of seawater (on the present Earth); the position shown by an arrow represents an estimated concentration of iron in seawater on the anoxic Earth (from Ochiai, 2004a).

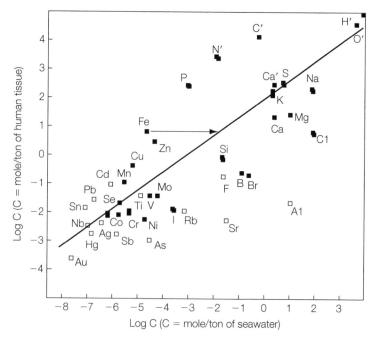

Whether iron was in the form of Fe(II) or Fe(III) in ancient seawater is not known, but it can be related to the atmospheric oxygen content at the time. This issue has not been settled among the geochemists. The most recent account of the controversy is found in *Science* (2005; **308**, 1730–1732). The prevailing notion championed by Holland (1984) is that the atmosphere was quite low in free oxygen content from the beginning of the Earth (i.e., about 4.6 billion years ago) until about 2.2 to 2.4 billion years ago and then rose rapidly to the current level (i.e., 0.2). An alternative idea proposed (by Ohmoto and Felder, 1987; Ohmoto *et al.*, 2006) asserts that the oxygen content of the ancient atmosphere went up quickly in the early stage (4 billion years ago) to reach the current level, where it has remained throughout the rest of the Earth's history. If the atmosphere was anoxic as the prevailing hypothesis asserts, the iron in the ancient seawater was in the form of Fe(II), which is soluble, and consequently the iron content was much higher, perhaps as much as a thousand-fold more than that in today's seawater. This seems also to be consistent with the formation of the so-called BIF (banded iron formation) that is the predominant source of today's iron ores, and is believed to have formed in the period about 3.0 billion to 2.0 billion years ago, peaking at 2.2 to 2.0 billion years ago. If all these hypotheses are reasonable, then the iron content in living organisms is commensurate with the iron content in the ancient seawater (see Fig. 1.4). This issue will be discussed again later.

1.2. THE ENGINES THAT DRIVE THE BIOCHEMICAL CYCLING OF THE ELEMENTS

Living systems are open and exchange material and energy with their environments. The major life processes are: (a) production of reduced carbon compounds (mostly carbohydrates) from carbon dioxide and water (and hydrogen sulfide in some organisms) through photosynthesis and chemosynthesis, (b) oxidative metabolism of reduced carbon compounds to extract their chemical energy (production of ATP), and (c) metabolic processes that produce all the other necessary compounds; many such chemical reactions require negative free energy in the form of hydrolysis of ATP. In accomplishing these processes, living organisms make use of a number of other elements; these are partially shown in Figure 1.2. An element is taken up by an organism, incorporated, and eventually released back to its surroundings. That same element may be directly incorporated into another organism, in the predator–prey relationship. This includes decaying processes,

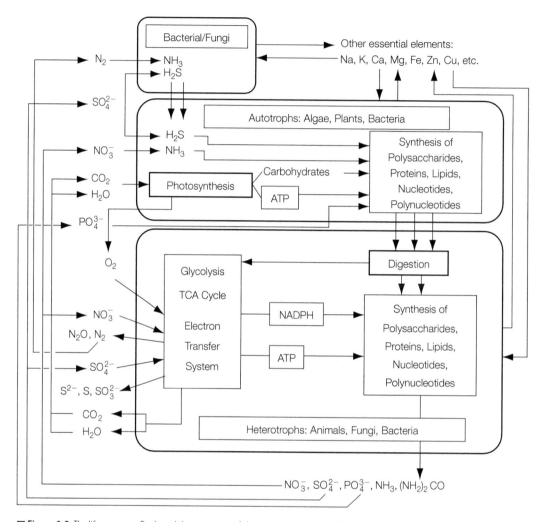

■ **Figure 1.5.** The life process on Earth, and the movement of elements associated with life (from Ochiai, 1992).

and is the generalized food chain. The life processes, being dynamic, thus move the elements. An outline of life processes on Earth and the movement of associated elements are given in Figure 1.5.

1.3. **FLOW OF THE ELEMENTS—BIOGEOCHEMICAL CYCLING**

The elements and their compounds are constantly being moved by natural nonbiological forces as well as biological systems. The major

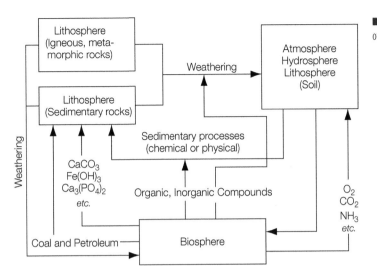

processes are summarized in Figure 1.6. Weathering erodes rocks through physical and chemical processes. Weathered material is carried by streams to the oceans, and is laid down there, eventually being turned into sedimentary rocks. The eroded material can also be carried through the atmosphere by wind. Living organisms and the biosphere as a whole contribute to these processes. As suggested in Figure 1.6, some minerals are produced by organisms. Examples include calcite ($CaCO_3$) and limonite ($Fe(OH)_3 \cdot xH_2O$). The latter is formed as a result of chemoautrophic process of such microorganisms as *Arthrobacter siderocapsulatus* and *Gallionella*. Some microorganisms help weathering processes; for example, some sulfur bacteria oxidize copper sulfide to leach out copper in a soluble form.

Hence elements are constantly being cycled throughout the entire Earth. This process partially involves the biosphere, and is called *biogeochemical cycling*. Figure 1.7 is a schematic representation of the biogeochemical cycling of the basic element, carbon (Ochiai, 2004). Carbon, in the form of carbon dioxide, is incorporated into organic compounds such as carbohydrates through photosynthesis in plants; they are then cycled among organisms through the food chain in the biosphere. When living organisms die, carbon will be deposited in the hydrosphere and lithosphere. It is also partially turned back to carbon dioxide in the atmosphere. Overall, it cycles among these four spheres of the Earth. The quantities underlined in the figure

Figure 1.7. The biogeochemical cycling of element carbon; the figure underlined represents the quantity (kg) of C currently present in each of the atmosphere, hydrosphere, lithosphere, and biosphere. The figure (kg/y) along a solid line (arrowed) is the flow of C; the figure (kg/y) along a dotted line (arrowed) represents the flow caused by anthropogenic activities (from Ochiai, 2004a).

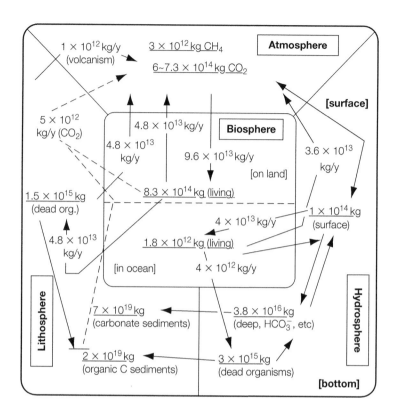

give the steady state quantities of carbon in various segments of the planet. Figure 1.7 also gives some quantitative estimates of the quantity and the flow (among different spheres) rate in units of kg/year. The broken lines indicate the anthropogenic flow of carbon and its compounds, particularly carbon dioxide. The combustion of fossil fuel (organic carbon sediment) is seen to be contributing significantly to the increase in the atmospheric carbon dioxide content.

The quantity and the flow rate of carbon in living and dead organisms (i.e., the biosphere) have been estimated (see, for example, Chameides and Perdue, 1997), though no very accurate evaluation is possible. The quantity and the flow rate of other elements in the biosphere are more difficult to estimate. Nonetheless, some estimates have been attempted. Similar diagrams of the biogeochemical cycling of nitrogen and sulfur are found in the Appendix (A.1 and A.2).

Similar estimates have been made for other elements (Ochiai, 1997, 2004b). Figure 1.8 depicts the biogeochemical cycling of iron.

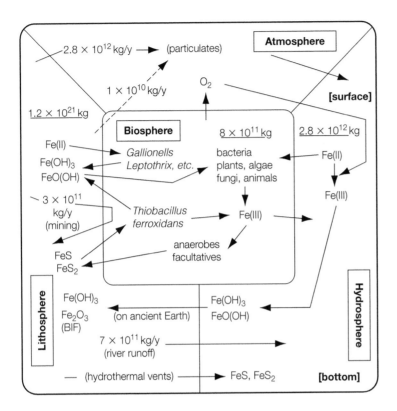

■ **Figure 1.8.** The biogeochemical cycling of iron; figures underlined and figures along arrows represent quantities and flows as in Figure 1.7 (from Ochiai, 2004b).

Some other diagrams of the biogeochemical cycling are found in the Appendix (A.3 for Ca and A.4 for Hg), and more extensively in Ochiai (2004a, 2004b).

1.4. HISTORICAL CHANGE IN THE BIOGEOCHEMICAL CYCLING OF ELEMENTS

The biogeochemical cycling described so far is that on the present Earth. The conditions on Earth have changed drastically since its formation. The major changes have been brought about by the emergence of life and also by the production of free oxygen by water-decomposing photosynthetic organisms. In addition, the geological changes involving the surface (plate tectonics, glaciation, etc.) and the interior of the Earth have influenced profoundly the geochemical cycling of elements over geological time. A brief survey will be given here on the historical change over geological time regarding the biogeochemical cycling of elements.

The elements were cycled by physical and chemical forces alone before the emergence of life. Sunlight and the internal energy source (such as heat created by radioactivity, gravity, and mantle movement) were the driving forces (as on the present Earth). It is believed that lightning and the bombardment of meteors and asteroids occurred more frequently and more severely on the ancient Earth, and that these events would also have contributed to the cycling of elements. As on the present Earth, water flow could dissolve material, and dislodge and carry particulate matter, and the wind transported material.

In the earliest period, the atmosphere contained a significant level of carbon dioxide (from volcanoes) but it was devoid of free oxygen (dioxygen). It is assumed that oxygen, being reactive, was consumed by reacting with oxidizable substances in the process of the formation of the Earth and the atmosphere, and that, as a result, little oxygen remained in free form in the atmosphere, according to the standard model (Holland, 1984).

Life is believed to have emerged under such an anoxic condition, and is now widely believed to have originated in hydrothermal vents at the ocean floor (see Ochiai, 1995b for several ideas on origin of life). Wächterhäuser and his coworkers have produced some experimental evidence for the formation of some crucial compounds such as pyruvate involving the Fe-Ni sulfide type of metallic compounds (Wächterhäuser, 2000). The original organisms thus are believed to be chemoautotrophs. There is some indication that such chemoautotrophs evolved as early as 3.5 billion years ago, based on S-isotopes fractionation data (Thamdrup, 2007; Philippot *et al.*, 2007).

The first water-decomposing photosynthetic organisms are believed to have been cyanobacteria (blue-green algae), and it has been suggested that they appeared about 3.0 to 2.5 billion years ago. Free oxygen (O_2) is produced as a byproduct of the photosynthetic process: $CO_2 + H_2O \longrightarrow CH_2O$ (carbohydrates) $+ O_2$. Photosynthetic oxygen caused no significant increase in the overall oxygen level in the atmosphere until about 2.2 to 2.4 billion years ago, because the oxygen was consumed by oxidizable material. A rise of the atmospheric oxygen now is believed to have started at 2.45 to 2.22 billion years ago (Bekker *et al.*, 2004; Kaufman *et al.*, 2007). The largest of such oxygen sinks was iron (Fe(II)) in the ocean. Iron is oxidized readily by oxygen and changes to Fe(III), which precipitates as iron hydroxide in seawater. Perhaps this was how the vast amount of iron oxide sediments known as the Banded Iron Formation (BIF) formed. The formation of BIF sediments started about 3 billion years ago and ended about 1.8 billion years ago (see Fig. 1.9). The iron oxide ores formed thereafter are of different types. This suggests that the iron in the ocean was exhausted by then, and also a sufficient level of atmospheric oxygen had formed. This changed the conditions of formation of the iron oxide minerals.

Nitrogen is another essential element. Almost all nitrogen is present as the inert form N_2 in the atmosphere. Ammonia is not and was not present at any significant level in the atmosphere at any time in the history of the Earth. Other usable forms (i.e., usable by organisms) are nitrite and nitrate, NO_2^- and NO_3^-. These compounds can be produced in lightning and other natural processes, but the quantity was never sufficient for organisms. Hence organisms must have developed their *nitrogen fixation* capability at earlier stages, perhaps from the very beginning of life. Nitrogen fixation (formation of ammonia) requires an enzyme, nitrogenase, which requires molybdenum. Molybdenum is and was relatively abundant in seawater in the form of molybdate MoO_4^{2-} even when the atmospheric oxygen level was very low. This assertion can be made on the basis of the reduction potential of MoO_4^{2-} and the potential (E_h) of seawater as related to oxygen pressure (Ochiai, 1978b). This suggests that molybdenum was amply available to organisms from the earliest stage of their evolution.

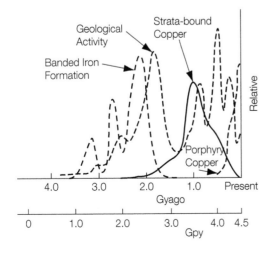

■ **Figure 1.9.** The age profile of (a) geological (magmatic) activity, (b) banded iron formation (BIF), and (c) strata-bound (sedimentary) and porphyritic (igneous) copper ore (production in 1974) (from Ochiai, 1983, 2004b). ((a) and (b): adapted from Eichler, 1976; (c) adapted from Bowen and Gunatilaka, 1977.)

The free oxygen level in the atmosphere governs the oxidation-reduction potential in seawater and the potential governs the oxidation states of different elements. As in the case of iron, a given element can be in different oxidation states and hence will have different solubility in seawater. Therefore, the oxygen level in the atmosphere may determine the organisms' accessibility to the elements in seawater, as mentioned earlier with regard to molybdenum. If we knew the evolution of the atmospheric oxygen level throughout the history of the Earth, we would be able to deduce the change in the availability of different elements over geological time. The availability (i.e., solubility in seawater) influences the geochemical cycles involving the biosphere.

It may even have dictated biological evolution, because different species require different sets of elements. For example, an organism that requires copper for its survival would not have emerged when copper was not available to it. Copper has much higher reduction potentials (for Cu(II)/Cu(I), Cu(II)/Cu(0)) than iron. This means that the metallic state of copper is much more stable relative to Cu(II) than that of metallic iron relative to Fe(II) or Fe(III). In fact, copper is found often as copper metal in nature, whereas no significant amount of iron has been found in the metallic state in the Earth's upper crust. Copper, therefore, would not have become available to organisms in the soluble form of Cu(II) until the oxygen level in the atmosphere had become sufficiently high. Figure 1.9 shows the age profile of the quantity of strata-bound sedimentary copper ores. This suggests that organisms could not have been able to use copper before about 1.8 billion years ago; therefore those organisms that had appeared before that time did not use copper. Proteins and enzymes that depend on copper have been identified mostly in eukaryotic organisms, though they are found in a very few prokaryotes as well. Eukaryotes (i.e., higher organisms with a separate nucleus in cells) are believed to have emerged sometime between 2.0 and 1.5 billion years ago. This is consistent with the hypothesized evolution of Cu(II) availability. As far as copper geochemical cycling is concerned, the biosphere was not involved before about 1.8 billion years ago.

The same idea can be applied to all other elements, and employed to deduce the historical change of the biogeochemical cycles of elements over geological time (Ochiai, 1978a, 1983). To obtain a meaningful insight on this issue from such an analysis, however, more detailed and definitive data on atmospheric oxygen and its change and geological activities are required. Unfortunately, these critical data are still lacking (see earlier).

A little more data are available for the more recent past; that is, the last 600 million years since the end of the Precambrian period. At the beginning of this period, atmospheric oxygen is estimated to have reached a few percent of the present level. It increased rapidly, reaching the present level 550 million years or so ago, and it seems to have remained steady since then. The data were isotope ratios of C^{13}/C^{12} and S^{34}/S^{32} of the rocks over this period of time. The significance of such data cannot be discussed here, but it suffices to note that they can indicate biological activities. From the isotopic data, Garrels and Lerman (1981) attempted to deduce the change in the rate of organic carbon burial and that of the formation of gypsum $CaSO_4$ over a period of time, from 600 million years ago to the present. Their results are shown in Figure 1.10.

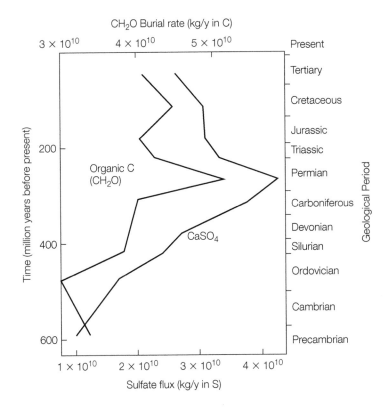

■ **Figure 1.10.** The change in the organic carbon burial rate and the gypsum formation rate over the last 600 million years. (After Garrels, R. M. and Lerman, A. 1981. Phanerozoic cycles of sedimentary carbon and sulfur. *Proc. Natl. Acad. Sci., USA* **78**, 4652–4656.) (Modified from Fig. 1.2 (p. 6) of Schlesinger, W. H. 1997. *Biogeochemistry*. Academic Press, San Diego.) (Ochiai, 2004b)

The figure clearly shows that the fluxes of organic carbon and gypsum formation were interrelated over this time period. That is, when more organic carbon was buried, more gypsum formed. When organic carbon is buried without being subjected to oxidation (respiration), the oxygen is not consumed and its level in the atmosphere would go up. The oxygen (unused by respiration) would then oxidize sulfide, particularly pyrite FeS_2, producing sulfate that would end up in gypsum. Hence the oxygen content would not go up as high as the burial of organic material would allow. These two processes, if coordinated as indicated by the figure, would control the level of the atmospheric oxygen. It is inferred that the net production rate of the organic carbon had not been quite matched, as on today's Earth, by the rate of consumption by the heterotrophs and its removal rate by burial, until about 200 million years ago.

It is interesting to note that a very large quantity of organic carbon was buried in the period about 300 to 200 million years ago (see Fig. 1.10). This is the period from which most of the coal originates. This suggests that the biosphere might have consisted of more producers (plants) than consumers (animals, fungi, and bacteria) in this period, compared to the present biosphere. Besides, somehow the environment was such that plants were rapidly buried in marsh during this period. A more recent data on the atmospheric O_2 content (Berner *et al.*, 2007) shows a curve similar to that of Figure 1.10, with a maximum of about 30% of O_2 content (as compared to about 22% in the current one) at around 280 million years ago.

REVIEW QUESTIONS

1.   Convert the masses of various portions on the Earth (core, mantle, crust, hydrosphere, atmosphere, and biosphere) to ratio (e.g., set the core to be 10^6 million and express others accordingly). See how small the biosphere is.

2.   Why is iron (Fe) a prominent element on the Earth as well as in the universe? This requires a little knowledge of nuclear chemistry, which you can obtain from a basic chemistry textbook.

3.   Figures 1.3 and 1.4 indicate that the elemental composition of the human body correlates better with that of seawater than with that of the crust. What is the implication of this fact? Discuss.

4. Pick one or two elements and elaborate the biochemical and biogeochemical cycling, referring to Figures 1.5, 1.7, and 1.8, identifying, if possible, special (meaning other than humans, mammals, common plants, etc.) organisms that deal (absorb, utilize, or otherwise) with it.

5. A greenhouse gas CO_2 is believed to be contributing to global warming. Figure 1.7 gives a rough estimate of quantities of C in several segments of the Earth and the rate of cycling among segments. It also gives an estimate of the contribution of anthropological activities to the C cycling (in the form of CO_2). Calculate, using the numbers in Figures 1.1 and 1.7, (1) the current content of CO_2 in the atmosphere in units of ppm (molar) and (2) the rate of annual increase of CO_2 (per the current quantity) due to human activities.

6. Oxygen (O_2) is essentially toxic, and therefore, it caused very severe pollution problems for those organisms that had lived in an essentially anoxic atmosphere, when the atmospheric oxygen became significant. What was the response of the biosphere for this emergence of toxic atmospheric oxygen?

PROBLEMS TO EXPLORE

1. Investigate how an element (pick one from among Fe, Cu, Zn, Mo, Ca) was made into ores, and how organisms contributed to their formation, if any.

2. Figure 1.5 contains a lot of information other than that given in the text. It is intended, for example, to show that the biochemistry of the entire life system can be divided into two groups. In the figure they are separated into the left half and the right half. It must be pointed out, though, that such separation cannot be done perfectly. The left half can then be divided into upper half and lower half. Try to extract as much information as possible from the figure and discuss.

3. An interesting element in terms of abundance is uranium. The natural abundance of uranium in the Earth's crust is on average similar to those of molybdenum and tungsten, and even higher than that of mercury. All the isotopes of uranium are alpha-particle emitters. What is the implication of this fact for health effects of uranium? Investigate and discuss.

4. Many elements, particularly those of transition metals, readily change the oxidation states. As presumed, the atmospheric oxygen content varied very much over the course of the Earth's history, and that would have affected the oxidation states of the elements present in the ocean and the upper part of crust. That in turn would affect the availability of the elements to organisms, because organisms would utilize elements readily available in dissolved state. This has an implication that the variation of the atmospheric oxygen, hence that of the readily available elements, might have had significant effects on the evolution of various organisms. Explore the idea.

Biological Needs for and the Behaviors of Inorganic Elements

2.1. INTRODUCTION

Any biosystem, as an evolving entity, has tried and still tries to increase its survivability. In this process, the biosystem has utilized a number of inorganic elements in addition to the essential organic compounds. Alternatively, the bare minimum requirement for the emergence of living cells may have included some inorganic elements such as calcium and iron. Calcium is required to form a cell bounded by a membrane. Iron as iron sulfides may have been involved in the original energy metabolism. Whatever the case, the fact is that living organisms require a number of inorganic elements in addition to the bioorganic compounds for their proper functioning. In fact, the living systems that we know were unlikely to emerge without at least calcium, magnesium, sodium, potassium, iron, and perhaps zinc as well.

There are several levels at which inorganic elements play roles (see Fig. 2.1):

(1) At the lowest level, that is, the molecular level, inorganic elements' major functions are (a) catalytic, (b) receptors and carriers for electrons and small biomolecules such as dioxygen and ethylene, and (c) structural—to maintain the structural integrity of biocompounds such as proteins and nucleic acids.

(2) At the cellular levels, Ca(II) plays the roles of maintaining the integrity of cell membranes and an intracellular second messenger. Some metalloproteins play the roles of sensor for certain compounds such as dioxygen, NO, and CO, though mechanistically this is a matter of molecular level.

(3) At the physiological level, alkali metal cations (Na(I) and K(I)) create the across-membrane electric potential gradients. In

53

■ **Figure 2.1.** Inorganic elements at work at molecular level (shown in red), cellular level (blue), and physiological level (green).

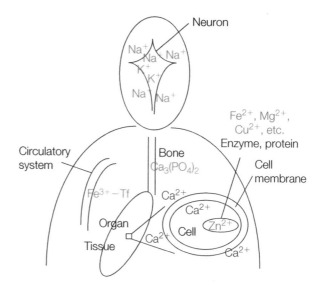

addition, they play a major role in maintaining the osmotic balance, along with anions such as Cl^- and CO_3^{2-}. Solid macroscopic minerals such as $CaCO_3$, $Ca_3(PO_4)_2$, and SiO_2(hydrate) provide mechanical strength to organisms like bones, dentin, egg shell, for example. Some inorganic solids function as the cellular structural elements and small magnetite (Fe_3O_4) crystals play the role of orientation-indicator in some organisms.

In addition, there are a number of ways in which these elements are dealt with in biological systems. They are:

(1) Systems and mechanisms to store, supply, transport, and regulate the level of these necessary elements.

(2) Mechanisms to defend themselves against the adverse effects of the elements.

Needless to say, the functions of an element at any level (cellular, physiological, or otherwise) ultimately are attributable to the interactions at the molecular level.

2.2. INORGANIC ELEMENTS IN THE BIOLOGICAL SYSTEMS

2.2.1. Inorganic Elements Involved at the Molecular Level

Chemical reactions can be classified into several categories. The organic compounds' repertoires for chemical reactions are limited, as they consist of only four elements, C, H, N, and O; plus S and P in a smaller number of compounds. Hence there is a necessity for other types of elements that would provide more functions at the molecular levels.

The 3D structure of a protein is determined by the overall energy minimum of interaction among all the atoms present. That energy-minimum structure is usually the biologically active one, but it may not be so in some cases. Another specific conformation/structure may be required to be active. In many such cases, a metal cation, particularly Zn(II), plays the role of maintaining the specific structure by strategically binding specific amino acid residues.

The functions provided by inorganic elements are based on their Lewis acidity, electron transfer capacity (multivalency), and formation of relatively weak carbon-metal bonds. A variety of roles emerge from these basic characteristics. The major one is to constitute the catalytic site of enzymes (Mg, Mn, Fe, Co, Ni, Cu, Zn, Mo, W). They also provide the active site for the functional proteins such as oxygen carriers (hemoglobin (Fe), hemocyanin (Cu), etc.), electron carriers (cytochromes (Fe), iron-sulfur proteins (Fe-S), blue copper proteins (Cu), etc.), and ethylene-receptor proteins (Cu) in plants.

2.2.2. Inorganic Elements Involved at the Cellular Level

Calcium (II) seems to play a whole variety of roles in the functioning of cells: maintenance of the cell shape and integrity of membrane, endo- and exocytosis, mitosis, muscle contraction, causing changes in some proteins/enzymes to modify their functions (second messenger), and so on. Cells may even be impossible without Ca(II). These functions rely on the ability of Ca(II) to bind O-ligands (such as that of carboxylate or phosphate) and the lability of such binding (i.e., binding is fast and reversible).

Some metalloproteins play the role of sensor of the cellular levels of small molecules such as O_2, CO, N_2, and H_2. Certain metal ions, including both essential and toxic ones, can function as inducers for expression for certain functional proteins. These proteins can be carriers, chaperons or storages, or detoxifiers for the metal ions.

2.2.3. Inorganic Elements Involved at the Physiological Level

The electric potential across a cell membrane is caused by the uneven distribution of Na^+ and K^+ in and out of a cell, particularly in the nerve cells. The electric potential changes as these ions go in and out of a cell. The mechanisms involve several enzymes including Na^+/K^+-ATPase.

The cellular effects of Ca(II) would manifest at the physiological level; for example, hormonal actions, muscle contraction, and stoppage of hemorrhage. Solid calcium compounds provide the mechanical strength for biological systems. The exoskeletons and shells of many organisms are made of calcium carbonate. On the other hand, the endoskeletons usually are made of calcium phosphate; for example, bones and teeth. In the vertebrates, a function of the bones is also the storage of calcium and phosphate.

A tiny magnetic crystal made of magnetite Fe_3O_4 functions in some bacteria as an orientation indicator against the Earth's magnetic field. Similar mechanisms exist in migratory birds. Biological systems have not yet found a way to make a magnetic material entirely of organic compounds.

2.2.4. Biological Systems Involved in the Metabolism of Inorganic Elements

Inorganic elements thus are utilized, so the biological systems have developed mechanisms to deal with them. They are ingestion, absorption, and incorporation of one into a specific protein; transport; converting the elements to metabolizable forms; detoxifying them; and so on.

Let's take an example of iron in the human body. Iron compounds, either Fe(II) or Fe(III), associated with various ligands will be mostly absorbed in the upper portion of duodenum in humans. Iron in an artificial cereal often is added in the form of metallic iron. This iron will be dissolved in stomach acid (1M HCl), and stays as Fe(II)-aqua complex. Some Fe(III) compounds will be reduced in the stomach by reducing agents such as ascorbic acid in the food stuff. However some Fe(III) compounds such as oxidized hemoglobin (in the form of Fe(III)-heme) in beef, for example, may remain as Fe(III). The major iron absorption mechanism seems to accept iron in the form of Fe(II) in the surface of cells lining the duodenum (mucosal cells). There are other mechanisms; for example, an absorption mechanism absorbs Fe-heme as a whole. Once Fe(II) is absorbed in the cell, it has to be converted to Fe(III) in order for it to be transported to the other side (blood side) of the lining cells, and further transported through blood vessels. That is, Fe(III) has to bind to a protein transferrin in order to be transported. Oxidation of Fe(II) to Fe(III) is carried out by an enzyme called ferroxidase or ceruloplasmin. Iron will be unloaded to cells from transferrin. Most of the iron in the human body is utilized in blood-producing cells in the bone marrow. Iron is then incorporated into heme by an enzyme, iron-chelatase.

The level of iron in body fluid is monitored by an iron-sulfur protein, aconitase. This protein acts as an inducer or repressor for transferrin by acting on a messenger RNA. Hemoglobin is destroyed once its life is spent, when the blood cells become brittle. The iron thus released is incorporated in a protein, ferritin.

Readily accessible iron is rather scarce in the environment, and the organisms have developed means to secure iron from it. Bacteria and other microorganisms have developed a variety of compounds collectively termed *siderophores*. These compounds are excreted into the surroundings and sequester iron. Some of these compounds effectively function as antibiotics. That is, by sequestering iron in the surroundings, the organism may be able to prevent proliferation of other microorganisms present in the vicinity. Plant roots secrete iron-capturing compounds. For example, barley and oat secrete such compounds called mugineic acid and avenic acid. The details of iron metabolism will be given in Chapter 10.

Organisms have developed especially elaborate mechanisms to deal with iron, as it is one of the crucial elements and rather scarce in readily accessible forms in the environment on today's Earth. Organisms have also developed a variety of mechanisms to deal with other necessary elements including Ca, P, S, Zn, Mg, Na, K, Cu, and Ni.

Toxic elements may also inadvertently enter into organisms, and have to be removed or detoxified in order for the organisms to survive. Hence, organisms have developed mechanisms for this purpose. Let's take a typical example of a toxic element, Hg. Single-cell microorganisms convert mercury compounds to either metallic mercury or dimethyl mercury. Both of these entities are neutral and relatively volatile, and hence will be removed by diffusion through the cell membrane. An excessive presence of Hg(II) seems to induce mercury reductase, which reduces Hg(II) compounds to Hg(0). Methyl-cobalamain is a methylating agent. In many animal species, including the human, the presence of a high level of mercury induces the production of a small protein metallothionein. Metallothionein and its analogs bind strongly to such metallic ions as mercury, cadmium, zinc, and copper.

These are but a few examples of the metabolism of inorganic elements. More details will be given in Chapter 10.

2.3. WHY HAS A SPECIFIC ORGANISM CHOSEN SPECIFIC ELEMENTS FOR ITS SPECIFIC NEEDS?

The next question asks why organisms have chosen the elements they did. Four general basic reasons were proposed to explain why a specific element has been chosen by an organism for a specific purpose (Ochiai, 1978; 1987b). They are:

(1) Basic fitness (chemical suitability)

(2) Abundance

(3) Efficiency

(4) Evolutionary adaptation

These factors combine to make a certain element be chosen for a specific biological function. In other words, an element is intrinsically (chemically) suitable for a certain function (1). If there are more than two such elements for a specific function, organisms prefer a more abundantly present and more readily available one (2). Figures 1.2 and 1.3 show that the correlation between the elemental distribution in the human body and that in the surroundings is consistent with this notion (2). If, again, there are more than two such elements that are more or less equally available, a more efficient (in the general sense) one is preferred (3). Once an element (or its compound) was employed, the organisms try to improve its functionality through evolution. In this process, specificity (for an element) becomes more prominent.

2.4. BEHAVIORS OF INORGANIC ELEMENTS-I: FUNDAMENTALS OF COORDINATION CHEMISTRY

As previously stated, the functions provided by inorganic elements are based on their Lewis acidity, electron transfer capacity (multivalency), and formation of relatively weak carbon-metal bonds. A very minimum of the basics of coordination and organometallic chemistry that are necessary for understanding the behaviors of metallic ions in biological systems will be provided here.

2.4.1. Coordination Compounds or Metal Complexes

A metallic cation (M(II) for example) is a Lewis acid, and will bind with an entity L, which has a Lewis base character. Thus:

$$M(II) + L \Leftrightarrow ML$$

L in such a compound as ML is called a ligand, and the bond between M and L is called a coordination bond. The coordination bond is formed by sharing the lone pair on the base with the acid; it can vary from virtually ionic to virtually covalent. A metal cation can coordinate a number of L, from 1 to 12 or so. The resulting compound is called a coordination compound or a metal complex. The number of the coordinated ligand atoms is defined as coordination number (CON). Most common CONs are 4 and 6. When no extraneous constraint applies, the coordination compound of CON = 4 takes either a tetrahedral (more common) or a square planar structure. Likewise, that of CON = 6 is most commonly octahedral. Other structures of course are known; for example, a trigonal prism in which the ligands occupy the apex positions of a prism and the metal is at the center.

L (ligand) can be a single atomic entity such as O or Cl^-, a molecular entity with a single coordinating atom with or without electric charge such as H_2O, NH_3, and CN^-, or a molecular entity with two or more coordinating atoms such as $NH_2CH_2CH_2NH_2$ (ethylenediamine, often abbreviated as "en"), $NH_2CH_2COO^-$ (glycinate), CO_3^{2-} and EDTA $(=(^-OOCCH_2)_2NCH_2CH_2N(CH_2COO^-)_2)$. Molecular ligands with two or more coordinating atoms often are called chelating agents. The number of atoms on ligand entities that can coordinate is expressed in dentate terms; that is, mono-dentate, bi-dentate, hexa-dentate, and so on. The structures of some typical coordination compounds are shown in Figure 2.2.

The structure of a coordination compound will be determined as a minimum energy state. The major factors involved in determining the structure are: (a) the metal–ligand bond energy, (b) ligand field energy, (c) ligand–ligand interaction energy, and (d) conformational energy of ligand. As a metal–ligand bond formation is exothermic

■ **Figure 2.2.** Examples of typical metal complexes (coordination compounds); (a) $[Ni^{II}(NH_3)_4(en)]^{2+}$, where en $= NH_2CH_2CH_2NH_2$ and the six coordinating N-atoms constitute an octahedral structure; (b) tetrahedral $[Fe^{III}Cl_4]^-$; (c) $[Cu^{II}(gly)_2]$, where gly $=$ glycinate and two O atoms and two N atoms coordinate to Cu(II) in a square planar manner.

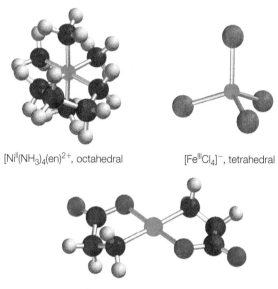

$[Ni^{II}(NH_3)_4(en)^{2+},$ octahedral $[Fe^{III}Cl_4]^-,$ tetrahedral

$[Cu^{II}(gly)_2]$ (gly=glycinate), square planar

($\Delta H < 0$), the higher the CON the better. However, as the CON increases, the negative effect of ligand–ligand interaction that is repulsive sterically or electrostatically will increase. For example, Cl^- ligands would not form an octahedral structure with 3d-transition metal ions under usual conditions, because they repel each other as they become crowded; the typical Cl^- complex is tetrahedral, such as $[Fe^{III}Cl_4]^-$. If the size of a cation is sufficiently large, even Cl^- may form an octahedral shape with CON $= 6$, as the crowding of ligands is lessened. F^-, being significantly smaller than Cl^-, forms complexes whose CON is even higher than 6, if the cations are relatively large. However, CN^- ions usually form octahedral complexes as in $[Fe^{III}(CN)_6]^{3-}$, very likely because the ion CN^- is narrower than Cl^-, and also because the negative charge on CN^- is more diffuse than that on Cl^-.

2.4.2. Ligand Field Theory—How the Predominant Structure Is Determined

The transition elements of the first series, Sc to Zn, in cationic forms, have 3d orbitals as their valence shell. The electronic configurations in divalent ions are d^1 in Sc(II), d^2 Ti(II), d^3 V(II), (Cr(III)), d^4 Cr(II), d^5 Mn(II), (Fe(III)), d^6 Fe(II), (Co(III)), d^7 Co(II), d^8 Ni(II),

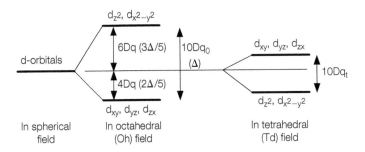

d^9 Cu(II), and d^{10} in Zn(II). There are five d-orbitals: $d_{x^2-y^2}$, d_{z^2}, d_{xy}, d_{yz}, d_{zx}. All five d-orbitals are equivalent in energy when a cation is placed in a spherical environment. However, when they are placed in a nonspherical environment such as the case of an octahedral complex, they are not equivalent anymore, because the electrons in different d-orbitals would interact with the ligands in different manners (to a different degree). As a result, d-orbitals would split into two groups: $(d_{x^2-y^2}, d_{z^2})$ and (d_{xy}, d_{yz}, d_{zx}) in an octahedral structure. The former has a higher energy than the latter, as shown in Figure 2.3. The energy difference between the two groups is defined as Δ or 10Dq, and is known as ligand field strength. In the case of a tetrahedral complex (ligand field), the d-orbitals will split into the same two groups as in the octahedral case, but the order of energy is reversed; now the set (d_{xy}, d_{yz}, d_{zx}) is higher than the set $(d_{x^2-y^2}, d_{z^2})$ (see Fig. 2.3). The energy difference between the lower set and that of d-orbitals in an equivalent spherical field is 4Dq, whereas the difference between the higher set and that of the equivalent spherical one is 6Dq in the case of an octahedral structure, as indicated in the figure. The d-orbital splitting scheme can be derived for other structures but is omitted here.

Let us concentrate on the octahedral structure, which is most common. When Sc(II) with d^1 forms an octahedral complex, the d-electron will occupy one of the lower sets; as a result, the Sc(II)-octahedral complex will be lower by 4Dq in energy than a corresponding spherical complex. This energy is called ligand field stabilization energy (LFSE). In an octahedral d^2-complex the two d-electrons will occupy two orbitals of the lower set, where the two electrons have parallel electron spins (due to the Hund rule), hence the total spin $S = 1$. The LFSE will be 8Dq. Likewise, in the d^3 case three electrons occupy the three d-orbitals of the lower set with parallel spin ($S = 3/2$), and LFSE = 12Dq.

■ **Figure 2.4.** Low and high spin complexes: examples of (a) d^5 and (b) d^7 in octahedral structure. The up arrow represents $+1/2$ spin and the down arrow represents $-1/2$ spin.

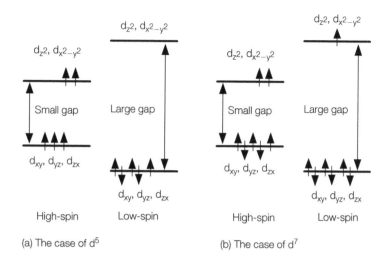

(a) The case of d^5 　　　　 (b) The case of d^7

There are two possibilities when it comes to d^4. The first is that all four occupy the lower set where two electrons have to occupy the same orbital with anti-parallel spins. The second is that three electrons occupy the lower set and the remaining one occupies one of the upper set with all parallel spins. The former complex would have $S = 1$ whereas the latter would have $S = 2$; the complex of the former type is designated as Low-Spin and the latter as High-Spin. Which case is realized depends on two counteracting factors: the energy gap between the two sets and the electron-electron interaction energy. If the gap (called ligand field strength) is large, a low-spin complex will be preferred. On the other hand, if the gap is not large, a high-spin complex will form. Likewise, for d^5–d^7, two types of complexes, high-spin and low-spin, will form depending on the ligand field strength (see Fig. 2.4). The low- or high-spin distinction is meaningless for octahedral complexes of d^1, d^2, d^3, d^8, d^9, and d^{10}.

The ligand field strength, which determines the energy gap, depends on the metal ion, its oxidation state, and the ligand. For a metal ion, the ligand field strength of ligands follows this general order, which is called a *spectrochemical* series:

$$I^- < Br^- < Cl^- < OH^- < F^- < H_2O <$$
$$NCS^- < NH_3 < en < phen < CN^-, CO$$

(en = ethylene diamine; phen = *o*-phenanthroline)

Halogen anions, I^-, Br^-, and Cl^-, are among the weakest ligands, whereas CN^- is one of the strongest ligands. H_2O and OH^- are relatively weak, and NH_3 and other amines are relatively strong ligands. CN^- is sometimes called pseudo-halogen; that is, it will behave like a halogen anion. However, CN^- is quite different from halogen anions as far as the ligand field strength is concerned. Halogen anion(s), H_2O, OH^-, and NH_3 all bind to a metal ion (Lewis acid) through a σ-donative bond; a pair of electrons for the bonding is supplied from the lone pair on the ligand. CN^- indeed also binds metal ion through such a bond, but it has also a relatively low-lying empty π^*–antibonding orbital. Electrons on the metal ion can then be accommodated into this low-lying π^*–antibonding orbital. In this bond, electrons are donated from the metal onto the ligand; often this is called π-back donation. When this type of bonding contributes to the coordination bond, the energy gap becomes larger than otherwise; hence such a ligand behaves as a strong ligand. This is because the energy level of the lower set (d_{xy}, d_{yz}, d_{zx}) is lowered by the π-back donative interaction. Another strong ligand is carbon monoxide, CO. It is isoelectronic with CN^- and turns out to be even better in attracting metal electrons into its empty π^*–antibonding orbital. This fact that CO is quite unique in forming organometallic compounds will be considered shortly.

The typical structure for d^8 (Ni(II)) is octahedral. However, if the ligand field is particularly strong, the structure tends to change to square planar, because the overall energy will be lower in square-planar shape. For example, $[NiCl_4]^{2-}$ is tetrahedral with S = 1, and is blue in color, whereas $[Ni(CN)_4]^{2-}$ is colorless with S = 0 and square-planar. In the case of d^8 of lower transition series (i.e., Pd(II) and Pt(II)), the predominant structure is square-planar.

Cu(II), a d^9 ion, often takes square planar structures as well. The reason for this is not the same as for that of Ni(II). Suppose that it takes an octahedral structure; the nine electrons occupy all but one of the two—$d_{x^2-y^2}$, d_{z^2}. In other words, either d_{z^2} or $d_{x^2-y^2}$ would be empty. This situation is termed degenerate (possibility of two or more state of equivalent energy). When this is the case, the entire system (Cu(II) complex) would distort itself so as to lower the energy further. This phenomenon is called the Jahn-Teller effect. It turns out that if it distorts to a tetragonally distorted structure whose extreme case is square-planar, the energy would become lower than the original octahedral structure. Hence, the square-planar structure is preferred.

The ligand field strength of a tetrahedral structure is relatively weak. Theoretically it is about half of the equivalent octahedral one. Hence, the tetrahedral complexes are almost always high-spin. The reason that five d-orbitals would have different energy in a nonspherical environment is that the electron distribution in a d-orbital is not spherical; it is distributed in certain directions and spaces. Therefore, if the overall d-electron distribution itself is spherical, there is no effect of the nonspherical (octahedral or otherwise) environment. This is the case for d^0, d^5 (high-spin) (Mn(II) and Fe(III)), and d^{10} (Zn(II)). In these cases, no ligand effect is exerted, and the structure of the complex is determined by factors (a), (c), and (d) mentioned earlier ((a) = metal–ligand bond energy, (c) = ligand–ligand interaction energy, and (d) = conformational energy of ligand).

It is interesting to note that Co(II) in a tetrahedral structure will have a spherical electron distribution, because four out of its seven d-electrons will distribute in the two lower sets ($d_{x^2-y^2}$, d_{z^2} in this case) and the other three in the higher set. This will make Co(II) a good substitute of Zn(II). It has indeed been demonstrated that Co(II) can substitute for Zn(II) ions in many Zn(II)-enzymes without losing catalytic activity.

2.4.3. Thermodynamic Tendency to Form Coordination Compounds

A metallic cation M exists as an aquo complex $[M(H_2O)_n]$ (n = most often 6 but this varies) in an aqueous medium, where the water molecules act as ligands. In aqueous medium, the aquo complex will bind to another ligand L, but the reaction is in fact a substitution reaction:

$$[M(H_2O)_n] + L \rightleftharpoons [ML(H_2O)_{n-1}] + H_2O \quad : K_1$$
$$[ML(H_2O)_{n-1}] + L \rightleftharpoons [ML_2(H_2O)_{n-2}] + H_2O \quad : K_2, \text{ etc.}$$

The equilibrium constants for reactions of this type often are called the stability constant K or formation constant. The equilibrium constant is related to the associated thermodynamic factors in the following manner:

$$\Delta G^0 = -RT\ln K = \Delta H^0 - T\Delta S^0.$$

ΔH^0 represents the binding enthalpy (of L to M or rather the difference of binding energies of M-L and M-OH$_2$), and ΔS^0 is the accompanying entropy change.

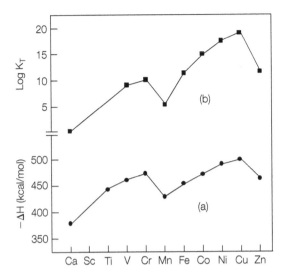

■ **Figure 2.5.** The trends in (a) hydration energy (ΔH) of divalent cations and (b) Log K_T where K_T is the formation constant ($K_1K_2K_3$) of [$M^{II}(en)_3$].

There is an interesting trend in K values or ΔH^0 values among the 3d-transition metal coordination compounds. A set of typical data is presented in Figure 2.5. Log K_T is proportional to ΔG^0, and the ΔS^0 value for the formation of [$M(en)_3$] from M(II) and 3 "en" molecules would be about the same, regardless of the metal ion. Therefore, both of these represent the trend in ΔH^0 for the formation of octahedral coordination compounds: [$M(II)(H_2O)_6$] and [$M(II)(en)_3$]. The trend is a double-humped curve with a dip at Mn(II) and an upward slope; this trend often is called the Irving-Williams series. This trend is satisfactorily explained in terms of two factors. One is the general monotonous increase, which is due to the increase in the effective nuclear charge and the decrease in the ionic radius. The second factor is LFSE, ligand field stabilization energy, which was mentioned earlier. In the case of a high-spin d^5 complex, the five electrons would be distributed equally in the two sets; the result is LFSE = 0, because the three electrons in the lower set contribute $3 \times (-4Dq) = -12Dq$, but two electrons in the upper set contribute $2 \times (+6Dq) = +12Dq$ and they cancel each other. That is, an octahedral d^5-coordination compound (in the case of high-spin) would not enjoy the LFSE. d^6 and above would follow the suit of d^1–d^5. Hence, the octahedral compounds of d^1–d^4 (but not d^5) and d^6–d^9 (but not d^{10}) would gain an extra stability due to LFSE. This causes the two-hump shape in the trend in ΔH^0 values of the coordination compound formation.

Ligands can bind to a metal cation stepwise. For example, starting with $n = 1$:

$$[Co^{II}(H_2O)_{7-n}(NH_3)_{n-1}] + NH_3 \rightleftharpoons [Co^{II}(H_2O)_{6-n}(NH_3)_n]$$
$$+ H_2O \ : K_n(n = 1\sim6)$$

In other words, NH_3 replaces H_2O one by one. K_n values have been determined to be $K_1 = 10^{2.1}$, $K_2 = 10^{1.6}$, $K_3 = 10^{1.0}$, $K_4 = 10^{0.8}$, $K_5 = 10^{-0.2}$, and $K_6 = 10^{-0.6}$. A statistical factor is partially responsible for the decrease, because the next NH_3 has less choice for its binding sites as n increases. If this statistical factor is taken into account, the K_n values would be $10^{2.1}$, $10^{1.7}$, $10^{1.2}$, $10^{1.1}$, $10^{0.3}$, and $10^{0.2}$. Since NH_3 is a stronger base, the replacement of a water molecule by NH_3 reduces the Lewis acidity of Co(II) and hence the binding strength for the next NH_3 is reduced. Overall, a difference is by a factor of 10^2 over K_1 to K_6. Hence on average, the binding strength will be reduced by a factor of $10^{0.4}$ every time a H_2O is replaced by a NH_3. In the case of formation of $[Cd^{II}(CN)_n(H_2O)_{4-n}]$, the corrected (for the statistical factor) K values are $K_1 = 10^{5.5}$, $K_2 = 10^{5.2}$, $K_3 = 10^{4.9}$, and $K_4 = 10^{4.2}$. The average change in K_n value is approximately by a factor of $10^{0.4}$ here. For another octahedral complex $[Fe^{III}(NCS)_n(H_2O)_{6-n}]$, the K values range from $K_1 = 10^{2.9}$ to $K_6 = 10^{-0.3}$; the average decrement factor is about $10^{0.6}$.

2.4.4. **Chelate Effect**

NH_3 binds a metal ion via a nitrogen donor atom, and so does $NH_2CH_2CH_2NH_2$ (en). Only the latter is bidentate; that is, it has two nitrogen donor atoms in a molecule. Both Ns (i.e., of NH_3 and en) would bind a metal cation with approximately the same strength; that is, ΔH^0 would be about the same. However, there is a significant difference between the two K values as shown here:

$$[Cu(H_2O)_6]^{2+} + 2NH_3 \rightleftharpoons [Cu(NH_3)_2(H_2O)_4]^{2+} + 2H_2O : K = 10^{7.7}$$
$$[Cu(H_2O)_6]^{2+} + en \rightleftharpoons [Cu(en)(H_2O)_4]^{2+} + 2H_2O : K = 10^{10.5}$$

That is, there is a difference of three orders of magnitude. Yet, the ΔH^0 values are approximately the same; $\Delta H_1^0 = -46\,kJ/mol$ and $\Delta H_2^0 = -54\,kJ/mol$. This phenomenon is known as the chelate effect.

From these values and K values, ΔS values can be estimated as $\Delta S_1^0 = -8.4\,J/K/mol$ and $\Delta S_2^0 = +23\,J/K/mol$, respectively. As evident from the entropy values, the difference in K value is due mainly to the difference in the entropy values. In the case of a monodentate ligand such as NH_3, the number of water molecules that is released would be the same as the number of ligands that bind. Hence the magnitude of entropy change is relatively small. On the other hand, one molecule of the chelating agent "en" replaces two water molecules, resulting in a large increase in entropy, and hence in a larger K value.

A metal ion is coordinated by a number of residues in a protein. The protein functions as a multidentate ligand and the resulting binding constant (of the metal to the protein) is usually large, partly due to the chelate effect.

2.4.5. **Ligand Substitution Reactions— Kinetic Factors**

A ligand in a coordination compound; for example, $[ML_5X]$ can be replaced by another ligand. Reactions of this type are called ligand substitution reactions, for example,

$$[ML_5X] + Y \longrightarrow [ML_5Y] + X$$

As discussed earlier, the formation reaction of a coordination compound from an aquo complex is an example of ligand substitution reaction. If X and Y are of the same chemical entity, the reaction can be regarded as ligand exchange. Two intrinsically different mechanisms are recognized for ligand substitution reactions; one is associative and the other dissociative.

In the associative mechanism, the entering ligand Y would associate with the $[ML_5X]$ forming $[ML_5X]-Y$; Y may occupy the seventh coordination site resulting in a seven-coordinated complex $[ML_5XY]$ or Y may be associated in the second coordination sphere of $[ML_5X]$. Y replaces X and then X leaves from the complex; the result is the formation of $[ML_5Y]$. Distinctions can further be made regarding the details of mechanism, depending on which step is rate-determining.

In the dissociative mechanism, the leaving ligand X dissociates first, and the entering ligand enters the vacated coordination site, though the position may not necessarily be the same since ligands can rearrange themselves while in the 5-coordination state. Based on a simple ligand-field theory outlined earlier, the activation energy for ligand substitution reactions may be predicted (roughly) for metal complexes of different electron configuration (d^n). Such an argument predicts high activation energy (slow rate) for ligand substitution reactions in the case of 6-coordinate octahedral complexes with d^3, d^6 (low-spin complexes), and d^8 electron configurations. Activation energy of ligand substitution reactions is predicted to be low (fast rate) for the metal complexes of other electron configurations. Indeed ligand substitution has been found to be very slow with Cr(III) (d^3), and the low-spin complexes of d^6 (Co(III), Rh(III), Ru(II)). See Chapter 4 for further discussion.

4-coordinate complexes, either tetrahedral or square planar, usually have a capacity to allow binding of the incoming ligand before exchange of ligands, and hence the ligand substitution reactions of these complexes proceed usually in an associated mechanism.

A metal cation catalytic center in acid-base reactions often has to accommodate two entities, a substrate and an attacking entity such as OH^- in the same coordination sphere. The most widely used cations in such enzymes are Mg(II) and Zn(II); their electron configurations are d^0 and d^{10}, respectively. That is, their valence shell electron configuration is spherical, implying that the number of ligands that can bind to them and the manner in which such ligands arrange themselves on Mg(II) or Zn(II) are relatively flexible, unlike those of incomplete d-shell. This allows Mg(II) or Zn(II) to accommodate readily the required number of ligands binding to it in a manner required by the surrounding amino acid residues. In other words, this makes Mg(II) or Zn(II) a good acid catalytic center. In the same sense, Ca(II)'s coordination number and coordination structure are rather flexible, and this makes Ca(II) a widely usable cation (second messenger). Ca(II) has other characters that make it useful for its use as second messenger (see Chapter 10).

2.4.6. Oxidation–Reduction and Reduction Potential

One of the characteristics of transition elements compared to main group elements is that different oxidation states do not differ very much in their energy; in other words, different oxidation states are relatively easily accessible, so many of them can take different oxidation states, and interconversion between different oxidation state is relatively facile.

This is contrary to most of the nontransition metallic elements; each usually takes a single specific oxidation state, and the other oxidation states are inaccessible under ordinary circumstances. Examples include Na(I), K(I), Mg(II), and Ca(II). Exceptional cases do exist; for example, Tl(I) and Tl(III), Sn(II) and Sn(IV), and Pb(II) and Pb(IV). Nonmetallic elements are multivalent; for example, CH_4 (nominally $-IV$) to CO_2 ($+IV$), NH_3 ($-III$) to NO_3^- (V), H_2S ($-II$) to SO_4^{2-} (VI). Of course, these different oxidation states are the basis for the energy metabolism of organic compounds. For example, carbohydrate $C_6H_{12}O_6$ (the oxidation state of C is nominally "0" here) is oxidized to H_2O and CO_2 ($+IV$). The change of oxidation states in carbon compounds is not facile, and thus such reactions require catalysts.

Two types of oxidation are involved; one is dehydrogenation (removal of H, H^- or H^+ plus (2)e), and the other is removal of electron(s). Dehydrogenation involves organic compounds NAD^+ or FAD and its analogues as the hydrogen acceptor, and requires a Lewis acid catalyst, usually Zn(II). Electron transfer is where transition metals play essential roles, as they are capable of facile change of oxidation states.

Two important factors are thermodynamic and kinetic. The thermodynamic factor for the change of oxidation state is expressed in terms of reduction potential, since typically it is determined electrochemically. It represents the tendency for an entity to be reduced. For example, the standard reduction potential E^0 of $[Mn^{IV}L_6]/[Mn^{II}L_6]$ is defined for:

$$[MN^{IV}L_6] + 2e \rightleftharpoons [Mn^{II}L_6] : E^0; \quad \Delta G^0 = -2FE^0$$

The relationship between the Gibbs free energy and the reduction is given in general by:

$$\Delta G = -nFE$$

where n is the number of moles of electrons involved in reducing one mole of the species of interest, and F is the Faraday constant. The superscript "0" implies standard state, which is defined as a state where P (pressure) = 1 atm (or 101.3 kPa) and the concentrations (activity for more rigor) of all the species involved are 1 mol/L. The relationship between the standard potential and that under an arbitrary condition is given by Nernst equation:

$$E = E^0 - (RT/nF)\ln([\text{oxidized form}]/[\text{reduced form}])$$

When a reduction reaction involves H^+ as in the following reaction, the reduction potential depends on the pH of medium.

$$MnO_4^- + 5e + 8H^+ \Leftrightarrow Mn^{II} + 4H_2O$$

The Nernst equation can then be written (where m is the coefficient of H^+) as:

$$E = E^0 - (RT/nF)\ln([O]/[R]) + (RT/nF)\ln([H^+]^m)$$
$$= E^0 - (RT/nF)\ln([O]/[R]) - 2.303(RTm/nF)pH$$
$$= E^0 - (RT/nF)\ln([O]/[R]) - (0.0592m/n)pH \text{ (at T = 298 K)}$$

In chemistry, the standard state is defined as [any of the species involved] = 1 mol/L; that is, pH 0 ($[H^+]$ = 1 M) is the standard. (Again, note that activity rather than molar concentration should be used throughout for rigorous treatment.) However, it is not convenient to set it as the standard in dealing with biosystems, an aqueous medium. The convention in biochemistry is then that the standard state is defined for pH 7 and is expressed as $E^{0'}$.

The medium water itself can undergo oxidation and reduction. It can be oxidized to O_2, or be reduced to H_2. In terms of reduction process:

$$O_2 + 4H^+ + 4e \rightleftharpoons 2H_2O : E^0 = +1.23 \text{ V}, E^{0'} = +0.816 \text{ V}$$
$$2H_2O + 2e \rightleftharpoons H_2 + 2OH^- : E^0 = 0.0 \text{ V}, E^{0'} = -0.414 \text{ V}$$

The oxidation of water is the reverse of the first reaction. A chemical species that has a reduction potential higher than the reduction potential of O_2 in thermodynamic sense will oxidize water to O_2, and hence cannot exist in water as such. On the other hand, a species whose reduction potential is lower than that of water will reduce water spontaneously and would not exist as such in water. For example, the reduction potential of Na(I) is -2.71 V, which is much lower than that of water. Hence, if you add sodium metal, Na, to water, a vigorous reaction takes place in which sodium reduces water:

$$2Na + 2H_2O \rightarrow 2Na(I) + 2OH^- + H_2$$

On the other hand, Co(III), whose reduction (to Co(II)) potential is $+1.81$ V, oxidizes water:

$$4Co(III) + 2H_2O \rightarrow 4H^+ + O_2 + 4Co(II)$$

These thermodynamic data suggest that neither Na metal nor Co(III) would be able to remain as such in water; indeed they cannot.

However, there is another factor that governs whether such oxidation–reduction will take place in practice—the kinetic factor. This is dependent on the activation barrier (activation energy); the activation barriers of water oxidation and reduction reaction are quite high. Hence even if the reduction potential is higher than $+1.23\,V$ (in a medium of pH 0), often the oxidation of water may not proceed at an appreciable rate. For example, MnO_4^-, whose reduction potential (to Mn(II)) is $+1.51\,V$, can remain intact almost indefinitely in water without oxidizing water. However, $+1.81\,V$ of Co(III) is high enough to overcome this barrier. Kinetic issues will be discussed in the next section.

The most important issue regarding the reduction potential is the effect of ligand. The reduction potential represents the relative stability of two different oxidation states.

$$[M^{III}(H_2O)_6] + e \rightleftharpoons [M^{II}(H_2O)_6] : \Delta G_h^0 = -FE_h^0$$

$$[M^{III}(H_2O)_6] + 6L \rightleftharpoons [M^{III}L_6] + 6H_2O : \Delta G_{III}^0 = -RT \ln K_{III}$$

$$[M^{II}(H_2O)_6] + 6L \rightleftharpoons [M^{II}L_6] + 6H_2O : \Delta G_{II}^0 = -RT \ln K_{II}$$

$$[M^{III}L_6] + e \rightleftharpoons [M^{II}L_6] : \Delta G_L^0 = -FE_L^0$$

The second and third equations represent the stability of the metal complex in different oxidation states. These factors are related to each other in the following way:

$$\Delta G_L^0 = \Delta G_h^0 - \Delta G_{III}^0 + \Delta G_{II}^0 \text{ or}$$
$$- E_L^0 F = -E_h^0 F + RT \ln K_{III} - RT \ln K_{II}$$

Hence,

$$E_L^0 = E_h^0 + (RT/F)\ln(K_{II}/K_{III})$$

It is clear that the reduction potential will increase when the ligand tends to stabilize the lower oxidation state more than the higher oxidation state. A set of data of reduction potentials of Fe-complexes is given in Table 2.1. It shows that the reduction potential of Fe^{III}/Fe^{II} varies widely depending on the ligands. Among the strong ligands that cause the iron complexes to be low-spin, CN^- tends to stabilize the higher oxidation state more than the lower oxidation state; hence CN^- decreases the reduction potential in the case of iron. It seems that negatively charged ligands tend to stabilize the higher oxidation state, as expected; this is regardless of the spin-state. Phen (phenanthroline) and bipy (bipyridyl) are both planar aromatic systems with coordinating N atoms. The electrons on the metal ion will be

Table 2.1. The Reduction Potential Fe^{III}/Fe^{II} of Representative Iron Complexes

Metal complex (Fe^{III}/Fe^{II})	High or low spin	Reduction potential (V)
[Fe(nitro-phen)$_3$]	Low	+1.25
[Fe(phen)$_3$]	Low	+1.14
[Fe(bipy)$_3$]	Low	+1.10
[Fe(H$_2$O)$_6$]	High	+0.77
[Fe(PO$_4$)(H$_2$O)$_5$]	High	+0.61
[FeF(H$_2$O)$_5$]	High	+0.40
[Fe(CN)$_6$]	Low	+0.36

spread onto the empty π–antibonding molecular orbitals. This will relieve the charge concentration on the Fe-atom, and hence tends to stabilize the lower oxidation state. This tendency is exaggerated when the ligand has an electron withdrawing group such as nitro-group. Because CN^- is isoelectronic with CO and hence is a π-acid, CN^- may stabilize the lower oxidation state when the metal is in a sufficiently low oxidation state because this electron withdrawing effect may only then become significant compared to the σ-donative effect of CN^-.

The difference in spin-state (i.e., high or low spin) also would have a significant effect on the reduction potential. A high-spin Co(III) complex [Co^{III}(H$_2$O)$_6$] has a very high reduction potential, +1.81 V. This simply reflects a high electron affinity of the high positive charge (+III), as in the case of Fe^{III}(aq) seen earlier. However, with strong ligands, Co(III) forms low-spin octahedral complexes, in which all the six electrons are accommodated in the lower set of d-orbitals. Hence, the extra electron has to be accommodated in the high-lying set. This is energetically unfavorable. Accordingly the reduction potential of low-spin Co(III) complex is lowered; for example, E^0 for [Co^{III}(NH$_3$)$_6$] (to [Co^{II}(NH$_3$)$_6$]) is +0.10 V and that of [Co^{III}(CN)$_6$] is −0.83 V.

Kinetic issues of oxidation-reduction or electron transfer reactions including the long-distance electron transfer in proteins will be discussed in Chapter 5.

2.5. BEHAVIORS OF INORGANIC ELEMENTS-II: BASICS OF ORGANOMETALLIC CHEMISTRY

Organometallic compounds are those that have metal-to-carbon bonds. This bond does not necessarily have to be a single σ-bond; it can be between a metal and the π-electron cloud. Metal atoms may also bind to the π-conjugated double-bond system of an aromatic ring.

The classic organometallic compounds are represented by the Grignard compound (e.g., C_2H_5MgBr). The carbon atom (of C_2H_5) binds to Mg through a typical σ-bond, though the bond is quite ionic in this case; C is virtually a carbanion. On the other hand, the Hg-C bonds in H_3CHgCH_3 are quite covalent. One of the earliest organometallic compounds to be prepared was a Pt-ethylene compound (see Fig. 2.7(f)). When a mixture of ethane and ethene was bubbled through an aqueous medium containing $[PtCl_4]^{2-}$, ethene was trapped, and hence only ethane was recovered. It turned out that ethene binds to Pt through its π-bond, forming $[PtCl_3(\eta^2\text{-}C_2H_4)]^-$.

A number of organometallic compounds have been discovered in recent years among metalloenzymes. These enzymes contain metal-carbon bonds such as Co—C(alkyl) in vitamin B_{12} coenzyme and Ni—CO in some hydrogenases. The latter is a metal carbonyl, which can provide some of the basic characteristics of organometallic compounds and their chemistry.

2.5.1. Metal Carbonyls and the 18-Electron (18 e⁻) Rule

First, let's look at the simplest so-called organometallic compounds, metal carbonyls (i.e., those of carbon monoxide). For example, carbon monoxide readily reacts with metallic nickel and forms a compound $[Ni(CO)_4]$. The oxidation state of Ni in this compound is nominally zero. The simplest metal carbonyls that have been synthesized include $[Cr(CO)_6]$, $[Mn_2(CO)_{10}]$, $[Fe(CO)_5]$, $[Co_2(CO)_8]$, and $[Ni(CO)_4]$. The structures of these compounds are shown in Figure 2.6. The oxidation states of metals in them are all zero; hence the electron configuration of the metal is d^6 for Cr, d^7 for Mn, d^8 for Fe, d^9 for Co, and d^{10} for Ni. The ligand CO has a lone pair on the C-atom, which is used to bind with a metal. Therefore, each CO that binds to a metal through this interaction is taken to contribute two electrons to the metal. If we count the number of electrons that surround a metal in this manner,

■ **Figure 2.6.**
Simpler
metallocarbonyls.

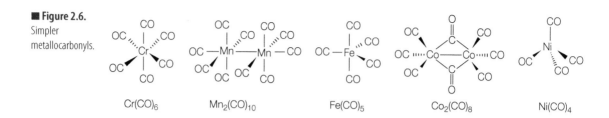

$Cr(CO)_6$ $Mn_2(CO)_{10}$ $Fe(CO)_5$ $Co_2(CO)_8$ $Ni(CO)_4$

we find that all three mononuclear carbonyls have 18 electrons; 6 (of d) + 6 × 2 = 18 for $[Cr(CO)_6]$, and so on. In the case of dinuclear carbonyls, Mn and Co both have an odd number of d-electrons, and there are intermetallic single bonds as shown. Therefore, one of the d-electrons on each is shared by two metals, and if this is taken into account, the total number of electrons surrounding each metal is also 18 in the dinuclear metal carbonyls.

Indeed most of stable organo (transition) metal compounds have 18 electrons; this is called the 18-electron rule. An element that has completely filled 3d/4s/4p orbitals (i.e., 18 electrons altogether) is Kr, one of the inert gas elements that is thermodynamically stable and inert. Hence compounds that have this electron configuration would have similar stability, and this is the basis of the 18 electron rule. However, as the oxidation state of a metal in these compounds (and others as will be discussed later) tends to be lower, it can readily be subject to reactions. Important types of reaction will be discussed later.

In counting the number of electrons, CO is assumed to donate two electrons toward a metal. However, another type of interaction, π-back donative bond, is important in the binding of CO. This interaction is between a filled d_π orbital (d_{xy}, d_{zx}, d_{yz}) and π*-empty orbital on the CO. In other words, the electrons on the metal would be shifted into this empty orbital of CO. In this sense, such an entity as CO is often called π-acid. This electron shift is supposed to reduce the relatively high electron density around the metal and hence enhance the stability of the low oxidation state. The π-back donative bond has been found to be important in organometallic compounds. CO, with relatively low-lying π*-empty orbital is particularly effective. The isoelectronic CN^- does have a π-acid character but it is less effective as such than CO due to the negative charge.

One exception to the $18e^-$ rule is that a stable compound often is obtained with 16 electrons when the metal has d^8 configuration.

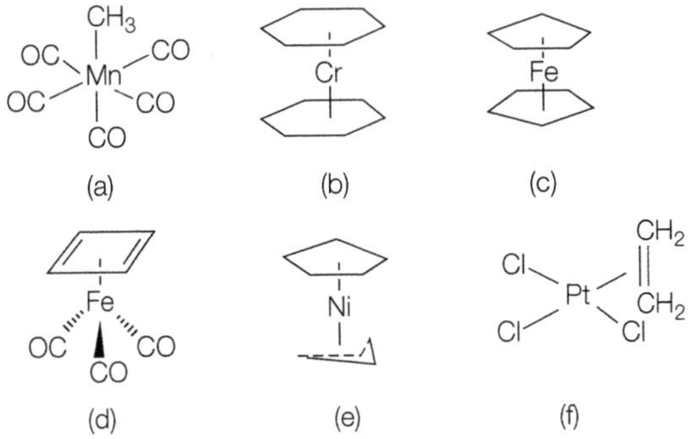

■ **Figure 2.7.** Typical examples of organometallic compounds.

Such a compound takes a square planar structure; this phenomenon was explained in Section 2.4.2. This configuration of 16 electrons is said to be coordinatively unsaturated, and is not sufficiently stable, and tends somewhat to be subject to a change in such a way to attain the coordinatively saturated state (18 electrons).

2.5.2. **Other Organometallic Compounds**

Since the discovery of ferrocene $(Fe(\eta^5\text{-}C_5H_5)_2)$ in the mid 1950s, organometallic chemistry has flourished; an enormous number of organometallic compounds have been synthesized. It turned out that the majority of such stable compounds obey the 18 electron rule, but that the compounds of square planar structure with 16 electrons with d^8-metals are also sufficiently stable.

A few examples of typical organometallic compounds are shown in Figure 2.7. Compound (a) has Mn(I) at the center, methyl group (regarded as a carbanion) and five COs as ligands. Since Mn(I) has a d^6 electron configuration and each of six ligands contributes two electrons, the 18-electron rule holds true. In compound (b), dibenzene chromium, each of benzene rings provide its six π-electrons for binding; and hence 18 electrons surround the Cr. This compound was synthesized during World War II in Germany, from acetylene. Compound (c) is the famous ferrocene, dicyclopentadienyl iron. The cyclopentadienyl ring is assumed to carry a negative charge, and hence contains six p-electrons to bind to the metal. The Fe is assumed to be Fe(II) d^6; hence the 18-electron rule holds. All four p-electrons in the butadiene ring are involved in binding to the Fe along with three CO ligands in compound (d). Here Fe should nominally be in the zero oxidation state (d^8), and the butadiene ring is square. In compound (e), Ni is Ni(II) (d^8) and cyclopentadienyl carries a negative charge, and the propenyl (π-allyl) also carries a negative charge and contributes four electrons to the bonding; hence 18 electrons altogether. Compound (f) carries a negative electric charge as a whole, and is an example of d^8 square planar complexes with 16 electrons. In this compound, the ethene provides its two bonding π-electrons for binding to Pt (II), and a Cl^- provides a pair of electrons.

One of the biologically important types of organometallic compounds is exemplified by compound (a); that is, the one containing σ-bonded alkyl group. This type of organometallic compound is found in vitamin B_{12} coenzyme, NiFe-hydrogenase, and acetyl-CoA synthase. The σ-bond is essentially between the hybrid orbital (of 3d, 4s, and 4p with σ character) of the metal such as Co, Fe, or Ni and the 2s orbital of a carbon atom. The overlap between those orbitals and hence the binding is not very effective. This is reflected in a relatively small bond energy in such compounds. For example the Co—C bond energy in vitamin B_{12} coenzyme and its model compounds is about 100 kJ/mol. This should be compared to the bond energies of a typical bond such as C—C, C—H, or O—H, which range around 400 kJ/mol. It seems that the metal needs to be in lower oxidation states in order for this metal-carbon σ–bonding to be effective, because the metal atomic orbitals will spread better in lower oxidation states. And the electron withdrawing π–acid ligand such as CO and an extended π-conjugate system such as porphyrin or the like are effective in stabilizing a lower oxidation state and hence a metal-carbon σ-bond.

Another type of organometallic compound may also be involved in a biological process. Ethene (ethylene) is a plant hormone, and its receptor protein may contain a copper ion. It appears that a Cu(I) in the protein binds ethene in the manner of Figure 2.7(f).

2.5.3. Some Special Types of Reactions Involving Organometallic Compounds

Organometallic compounds undergo several interesting types of reaction. The important ones are: (1) oxidative addition, (2) reductive elimination, and (3) *cis*-migration.

Since the metallic entity in most of organometallic compounds is in a low oxidation state, the metal would be able to readily donate electron(s) to an electron acceptor. However, it may not be a simple oxidation. If the metal simply loses electron(s), the 18-electron rule will be broken. Therefore, in order for the metal to maintain the 18-electron rule it would not give away the electron(s) entirely but share them with the acceptor. For example,

$$[Mn^{-I}(CO)_5]^- + CH_3I \rightarrow [(H_3C^-)Mn^I(CO)_5] + I^-$$

Because the Mn is in a low oxidation state, a pair of electrons on Mn^{-I} would easily be donated toward the electrophilic carbon of CH_3I; that is, Mn^{-I} acts as a Lewis base, a nucleophile. The result would be described as the carbanion $^-CH_3$ coordinated to the now oxidized Mn^{+I}. Therefore, as the CH_3 entity binds (adds as a carbanion), the Mn entity is formally oxidized. Hence this is called oxidative addition. The Mn entities on both sides of the equation are found to obey the 18-electron rule. From the standpoint of CH_3I, this reaction can be said to be a nucleophilic substitution reaction. If the metal Lewis base is coordinatively unsaturated, alkylhalide (e.g., CH_3Br) adds to it as two separate entities. For example,

$$CH_3Br + [Ir^ICl(CO)_3](16\ e's) \rightarrow [(H_3C^-)(Br^-)Ir^{III}Cl(CO)_3](18\ e's)$$

H_2, ordinarily a reducing agent, can oxidatively add to a metal of an organometallic compound. For example,

$$H_2 + [(OC)_5Mn - Mn(CO)_5] \rightarrow 2[(H^-)Mn^I(CO)_5]$$
$$H_2 + [Rh^ICl(PPh_3)_3] \rightarrow [(H^-)_2Rh^{III}Cl(PPh_3)_3]$$

Note that the Rh entity in the second reaction is not an organometallic compound in its true sense. That is, oxidative addition reactions can take place to any transition metal in a low oxidation state. In each of these reactions H_2 formally oxidizes the metal and binds to it as hydride.

Vitamin B_{12} coenzyme or adenosyl cobalamin is an example of biological organometallic compound with a Co—C bond. Another derivative, methyl cobalamin, can be synthesized by adding CH_3I to the reduced (Co^I) cobalamin. This is an example of oxidative addition; that is:

$$CH_3I + Co^I(corrin)(cobalamin) \rightarrow H_3C—Co^{III}(corrin) + I^-$$

Another important type of reaction is *cis*-migration, in which an alkyl or hydride ligand migrates to an unsaturated ligand on the adjacent position. This is illustrated by the following reactions:

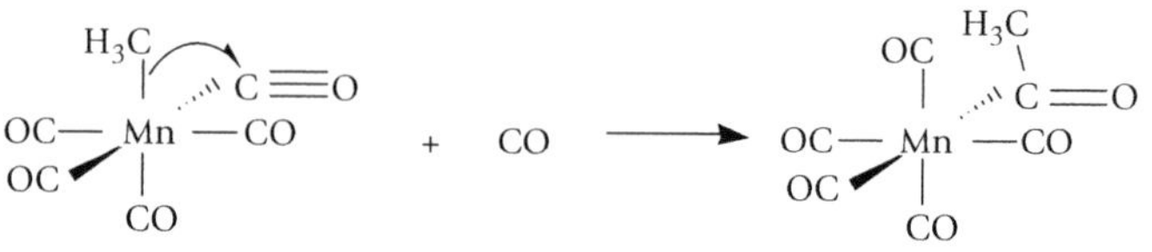

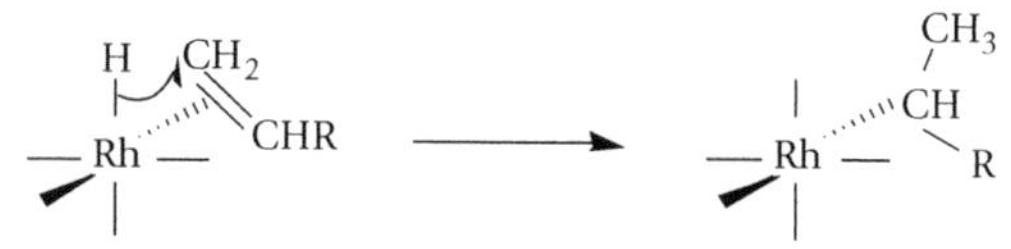

In these reactions, the unsaturated ligand, either CO or alkene, inserts itself between a M—H or M—C bond. In this sense these reactions are also called insertion reaction.

The following reaction involves the bond formation between two adjacent ligands, and the formed compound (CH_3CH_2R in this case) separates leaving a reduced metal behind. A reaction of this type is called reductive elimination, and is the exact reverse of oxidative addition.

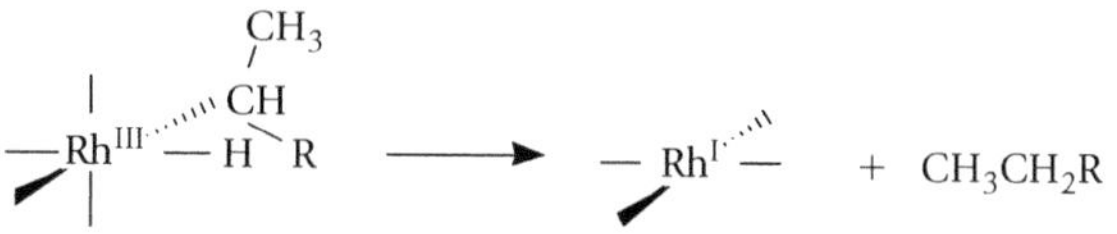

Yet, R—H would not oxidatively add to a metal directly in this case; in other words, the reverse of this reaction is not possible. The reason is that (1) the C—H bond energy is large, and (2) R—H is not sufficiently electrophilic to allure the two electrons from the metal. In other cases, the oxidative addition of R—H is possible, though, if the metal is a sufficiently strong nucleophile.

REVIEW QUESTIONS

1. Why is the reduction potential of $[Fe^{III}(phenanthroline)_3]^{3+}$ higher than that of $[Fe^{III}(CN)_6]^{3-}$?

2. Give the LFSE value for each of d^2, d^4, and d^7 (both high-spin and low-spin case) octahedral complexes.

3. *Chelate effect* is recognized in terms of the formation constant (or stability constant, a thermodynamic factor). A part of the reason for the chelate effect is entropy effect. Explain.

4. An octahedral (CON = 6) structure usually is preferred to a tetrahedral one (CON = 4) in transition metal complexes.

Why? However, although $[Fe(CN)_6]^{3-}$ exists, $[FeCl_6]^{3-}$ does not; it is $[FeCl_4]^-$ instead. Why?

5. A metal ion with d^8 tends to prefer a square planar structure to an octahedral on when binding strong ligands. Explain using a ligand-field diagram; this is a case of tetragonal distortion.

6. $[NiCl_4]^{2-}$ is blue and paramagnetic (with two unpaired electrons) whereas $[Ni(CN)_4]^{2-}$ is white and diamagnetic (no unpaired electron). Explain.

7. State your understanding of Jahn-Teller effects.

8. The Ni(II) complex with ammonia takes an octahedral structure $[Ni(NH_3)_6]^{2+}$, whereas Cu(II) forms a square planar $[Cu(NH_3)_4]^{2+}$. Explain.

9. Co(II) has been shown to be a good substitute for Zn(II) in many of Zn(II) enzymes. That is, Co(II) can mimic the behavior(s) of Zn(II) in many Zn-enzymes. Explain.

10. The standard reduction potential of O_2 is $E^0 = +1.23$ V. The standard condition is $P(O_2) = 1$ atm and $[H^+] = 1$ M (i.e., pH $= 0$). The reaction is: $O_2 + 4H^+ + 4e \rightarrow 2H_2O$. This reaction indicates that the thermodynamic factors must be pH-dependent as the reaction involves H^+. Calculate numerically the standard reduction potential of O_2 at pH $= 7$ using Nernst equation.

11. Give the number of the unpaired electrons in each of the following coordination compounds: (a) $[Fe(phen)_3]^{2+}$ (phen = phenanthroline, strong field (SF)), (b) $[MnCl_4]^{2-}$, (c) $[Gd^{III}(glycine)_3]^0$, (d) $[Co(NH_3)_6]^{3+}$, (e) [cytochrom c (Fe^{III})].

12. Give the number of electrons surrounding the metal in each of the following electrically neutral organometallic compounds, as illustrated in Figure 2.7.

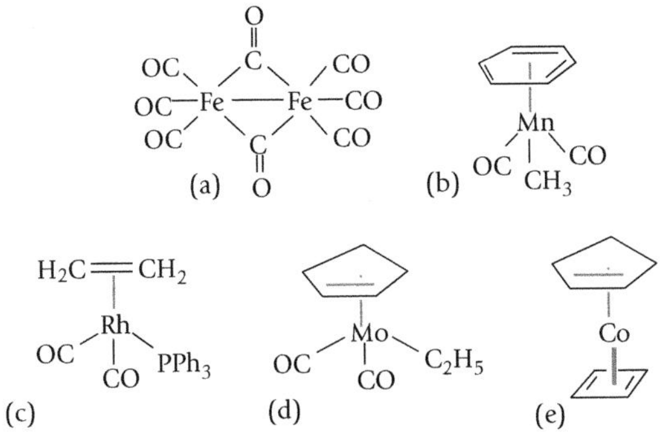

PROBLEMS TO EXPLORE

1. How and why has an element been adopted and used by organisms? Pick an element and investigate. Suggested elements: Mn, Mo, Cu, Se, Ca.

2. The main factors that determine the structure of a metal-coordination compound are discussed in this chapter; the main structures assumed by metal complexes include octahedral (with CON = 6), tetrahedral (CON = 4), and square-planar (CON = 4). However, these structures are rarely assumed strictly in those surrounding a metal ion in proteins. Discuss the factors other than those mentioned that contribute to determining the coordination structure in proteins.

3. Explore quantum mechanical explanations of the relatively small binding energy of a sigma bond between carbon and a 3D-transition element.

Chapter 3

How Do Enzymes Work?

This chapter explores the workings of enzymes at a deeper level than ordinarily attempted. Exposure to the material discussed in this chapter is not essential for understanding subsequent chapters. Therefore, it is suggested that the readers who are not familiar with the intricacies of enzyme reaction mechanisms postpone reading this chapter until they become familiar with the content of Chapters 4 through 8.

3.1. ENZYMATIC ENHANCEMENT OF REACTION RATE: GENERAL CONSIDERATIONS

An enzyme enhances enormously the reaction rate as compared to the corresponding uncatalyzed reaction, sometimes by as much as 10^{17}-fold. The reasons for this enormous enhancement have not been fully understood. Two contrasting basic ideas have been proposed. The first is the classic idea, emphasizing the electrostatic interaction between the protein and the substrate, based on the Transition State theory (Villa and Warshel, 2001). The second idea is that the dynamic motions of an enzyme (protein molecule) have something to do with the enzymatic power (see Wilson, 2000). Admittedly this idea is still fairly vague, lacking specific theories to connect the dynamics of the protein and its catalytic function.

In this author's opinion, the choice is not between either of these, as Warshel strongly suggests. In his examples, he finds by theoretical quantum calculations that the random fluctuation of a protein part does not significantly change the transmission coefficient, and hence he concludes that "dynamic effects" are insignificant. It may indeed be the case, assuming that all his assumptions and calculations are correct. Yet it may be so only for the enzyme systems studied.

It is more likely that (1) dynamic motions that affect the transition state and its temporal variation represents a whole picture, (2) there is a whole range of variation in the dynamic effects, and (3) in certain cases the classic, static picture represented by Transition State theory may give enough of an accurate picture. We will discuss these theories (ideas) separately, though the reality will be the composite of the three.

3.1.1. **Transition State Theory**

The simplest kinetic treatment of enzymatic reactions assumes to involve substrate (S) binding to enzyme (E) and its subsequent transition to product (P):

$$E \ (enzyme) + S \ (substrate) \rightleftharpoons ES \longrightarrow E + P \ (product)$$

The Michaelis-Menten equation can then be derived; thus

$$v = k_2[E]_0[S]/((k_1 + k_{-1})/k_1 + [S]) = V_{max}[S]/(K_m + [S])$$

where k_1, k_{-1}, and k_2 are, respectively, the rate constant of the forward reaction of the first step, of the reverse reaction, and of the second step; and $V_{max} = k_2[E]_T$ and $K_m = (k_{-1} + K_2)/k_1$. The Michaelis-Menten constant K_m can often be approximated as $K_m \sim K_1^{-1}$, where $K_1 \ (=k_1/k_{-1})$ is the association (binding) constant of the ES-complex. In terms of free energy of binding,

$$\Delta G_1 = -RT\ln K_1 = \Delta H_1 - T\Delta S_1$$

The rate constant k_2 can be expressed by a general form of Arrhenius equation:

$$k_2 = \kappa \exp(-\Delta G^*/RT) = \kappa \exp(-(\Delta H^* - T\Delta S^*)/RT)$$

The asterisks designate the activation parameters, and κ is the transmission coefficient.

The activation parameters can then be expressed as the difference in the parameter between the initial state (i) and the transition state (t); thus

$$\Delta G^* = G_t - G_i$$
$$\Delta H^* = H_t - H_i$$
$$\Delta S^* = S_t - S_i$$

A catalyst (enzyme) can affect one, some, or all of these parameters, κ, Hs, and Ss of the system, which consists of the protein (or RNA in the case of ribozymes) and the substrate(s). ΔH^* is typically a positive value, and its magnitude needs to be reduced for the rate to be enhanced. ΔS^*, typically a negative value, and its magnitude need to be reduced. To accomplish these effects, either H_t must be reduced or H_i increased, or both; and S_t must be increased or S_i reduced, or both (Ochiai, 1987). The binding free energy ΔG_1 would affect these enthalpy and entropy values. This is in the general sense what is termed as the Circe effect by Jencks (1975) or as orbital steering by Koshland (Mesecar *et al.*, 1997). A special case where the initial state is raised in anticipation of the final state in such a way that the ΔH^* be reduced is termed *entatic effect* (Vallee and William, 1968).

As the relationship between the energy value and the rate is given by the Arrhenius equation, a small change in the energy will result in a large change in the rate. At 300 K, a change in ΔG^* of 5.9, 11.7, 23, and 46 kJ/mole (1.4, 2.8, 5.5 and 11 kcal/mole) would bring about a change in k_2 of 10-, 100-, 10^4-, and 10^8-fold, respectively.

Hexokinase catalyzes the phosphorylation of glucose (and other hexoses) with ATP. This is the type of reaction:

$$ATP + R\!-\!OH \rightarrow R\!-\!O\!-\!PO_3^{2-} + ADP + H^+$$

If R is H, this is the hydrolysis of ATP. Indeed, the enzyme can catalyze the hydrolysis as well. But the rate of the hydrolysis reaction is 4×10^4 times smaller than that of phosphorylation. HOH is small enough to fit in the active site of the enzyme, but does not have the specific binding group, the pyranose ring of glucose. This group binds to the active site, and a part of the binding energy is supposed to cause a conformational change in the protein. The binding energy ΔG has been estimated to be -46 kJ/mole in this reaction, but the experimentally observed binding energy is -21 kJ/mole. The difference 25 kJ/mole is believed to be used for the catalytic effect (Jenkins, 1975).

Diol dehydratase binds adenosyl cobalamin (AdoCbl) and catalyzes the dehydration of diols. From the kinetic data, ΔG^0 of AdoCbl binding is estimated to be -36 kJ/mole. ΔH^0 can then be estimated to be -86 kJ/mole (assuming ΔS^0 to be about -40 eu). AdoCbl has 16 hydrogen-binding sites. Let us assume that all 16 bind through hydrogen bonds to the protein residues, as has been shown for similar enzymes, and that the hydrogen bond is conservatively assumed to be about -12 kJ/mole. Then the binding ΔH^0 of AdoCbl to the enzyme should be about -190 kJ/mole. The difference of approximately 100 kJ/mole may be used to effect the catalysis (Ochiai, 1989, 1994).

Quantum chemical calculations have increasingly been employed to estimate these energy values, including activation parameters $\Delta G^*(H^*)$ assuming reasonable transition states (for example, Villa and Warshel, 2001; Garcia-Viloca et al., 2004; Siegbahn, 2003 for metalloenzymes). Warshel and his coworkers studied a number of enzyme reactions using quantum chemical theories (plus molecular mechanics); examples include lysozyme, serine protease, acetylcholinesterase, alcohol dehydrogenase, ribonuclease, carbonic anhydrasse, DNA polymerase, aldose reductase, and orotidine 5′-monophosphate decarboxylase. Their overall conclusion from these studies is that the major contributor to the catalytic effect of a protein is electrostatic interaction, which in a way solvates the transition state more strongly in the case of the enzyme than in the case of mere solution (water).

In a more generalized treatment (Garcia-Viloca et al., 2004), the basic premise is stated as "the entire and sole source of the catalytic power of enzymes is due to the lowering of the free energy of activation and any increase in the generalized transmission coefficient, as compared to that of the uncatalyzed reaction." This is indeed the basic idea for catalysis in general, provided in any general chemistry textbook, though the treatment is not detailed at the general chemistry level.

Lowering the activation free energy in chorismate mutase has been studied theoretically (Ranagan et al., 2003 quoted in Garcia-Viloca et al., 2004). Two factors are involved. The electrostatic stabilization of the transition state lowers it (G_t in Eq. 3.5) by a few tens kJ/mole, and the binding by the enzyme of the substrate in an inactive form changes it to the active form; this raises G_i (as in Eq. 3.5) by about 20 kcal/mole. Overall, these two effects would lower the activation free energy ΔG^* by about 40 kJ/mole or so. This would account for about a 10^7-fold rate enhancement.

As stated before, the reaction system consists of the protein and substrate(s). Hence the effect can manifest not only on the substrate (its initial state and its transition state) but also on the protein portion. For orotidine monophosphate decarboxylase, it has been estimated (Garcia-Viloca et al., 2004) that the conformational energy of the protein is reduced by about 80 kJ/mole when it changes from the initial state to the transition state. This may be the major contribution to the lowering of the activation free energy in this case, and could account for as much as 10^{15}-fold rate enhancement.

Another example is xylose isomerase. The difference in the enzyme conformation between the initial state and the transition state makes a significant contribution to lowering the ΔG^*. In this case a Mg(II) ion's position shifts and it causes a change in the charge distribution on the two Os on the xylose, and that enhances the shift of the hydride. It was shown that the activation free energy was lowered by about 20 kJ/mole (Garcia-Viloca *et al.*, 2004).

3.1.2. **The Dynamic Effects**

This is a relatively new idea. Realization that detailed studies of the static structure of the enzyme, enzyme/substrate, or enzyme/transition state analogue and such have not produced convincing mechanisms for the rate enhancement has led some researchers to look for some other causes. As reported by Wilson (2000) on the Mesilla conference, a basic question was "Is 'Transition State Theory' an adequate description of what's going on in an enzyme?" (Benkovic, quoted in Wilson, 2000). They have proposed that dynamic effects are indeed responsible for the catalysis.

Villa and Warshel (2001) do concede that the fluctuations in the protein affect the transition state (as discussed later), but are of the opinion that the fundamental importance of random thermal motions does not imply that enzymes use particular vibrational modes to enhance reaction rates. Any dynamic effects on the rate are presumed to lie in the transmission coefficient, κ, in their opinion. Hence the authors attempted to estimate κ-values in some enzymatic reactions as well as those in the corresponding uncatalyzed reactions (Villa and Warshel, 2001). The results (autocorrelation of the electrostatic contribution to the energy gap) showed no significant difference between the protein-assisted case (enzyme reaction) and the reaction in water solution. These and other similar results obtained by them and their coworkers suggest to these authors that the rate of the enzymatic reaction is determined by the activation energy and that any dynamical effects are relatively minor.

In a refined version of Transition State theory (Garcia-Viloca *et al.*, 2004), the transmission coefficient κ can be expressed by:

$$\kappa = \Gamma(T)\kappa(T)g(T)$$

where $\Gamma(T)$ arises from dynamic recrossing, $\kappa(T)$ comes from quantum tunneling effect, and $g(T)$ arises from deviations of the equilibrium distribution in phase space. These parameters can be expressed in terms of appropriate ΔG value, such as,

$$\Delta G_{recross} = -RT\ln \Gamma(T)$$

And hence, the rate can be expressed as

$$\text{Rate constant } (k_2) = \exp(-\Delta G_{act}{}^*/RT)$$

where $\Delta G_{act}{}^* = \Delta G^* + \Delta G_{recross} + \Delta G_{tunneling} + \Delta G_{nonequil.}$

The recrossing factor $\Gamma(T)$ does not seem to change between the uncatalyzed reaction and the catalyzed one; changes are relatively small, contributing little in changing the reaction rate (Garcia-Viloca et al., 2004). Quantum chemical estimates of the tunneling effect ($\kappa(T)$) ranges from 1.5 for triose phosphate isomerase to 780 or so for soybean lipoxygenase (Garcia-Viloca et al., 2004). The third factor g(T) appears to be relatively small, at most a factor of 2 or 3. These results imply that the effect on the transmission coefficient by an enzyme in the rate enhancement seems to be relatively small. Therefore, if the dynamic effects should be reflected in the transmission coefficient, these authors have come to the same conclusion as Villa and Warshel (2001).

On the other hand, Benkovic and his coworkers modified amino acid residues far from the active site in dihydrofolate reductase and found that it had considerable influence on the enzymatic reaction rate. ^{15}N NMR studies by Wright and his group on this enzyme system found substantial changes in dynamics in various parts of the molecule. Petsko and his coworkers reduced the temperature below the glass transition temperature in the cytochrome P-450 associated enzymatic reaction, and then initiated the reaction by injecting electrons. The catalytic reaction did not proceed in any significant degree under this condition. The protein's dynamic motions are frozen below the glass transition temperature. These results imply that the dynamic motions of the remote site as well as the active site are important in the enzymatic reaction (quoted in Wilson, 2000).

Hydrogen transfer reactions often proceed through the tunneling effect (Liang and Klinman, 2004). The tunneling effect will diminish as the temperature increases. Yet, it has been found that the tunneling effect contribution did not diminish in the alcohol dehydrogenase reaction (Liang and Klinman, 2004) of a thermophile organism at 65°C as compared to that at 25°C. This result suggested to the researchers that the higher degree of dynamic motions reduced the barrier of tunneling and compensated the temperature effect.

Eisenmesser *et al.* (2002) studied ^{15}N NMR relaxation of amide nitrogens during the catalytic process of cyclophilin A. This enzyme catalyzes *cis-trans* isomerization of X-Pro peptide bond. In the presence of substrate (Suc-Ala-Phe-Pro-Phe-4-NA) at least three different states coexist; E (enzyme), ES_{trans} (S in the *trans* form), and ES_{cis}. Exchange among these different states increases the values of R_2 (transverse relaxation rate), as compared to that in the free enzyme where no such exchange takes place. Significant increases in R_2 have been observed at 10 amide nitrogen sites (out of 160 residues), and the increase was commensurate with the substrate concentration. The exchange rate was estimated to fall in the range of 10^{-5} to 10^{-3} s^{-1}. This time scale agrees with the microscopic rates of substrate turnover, suggesting that the fluctuations that manifest as the conformational exchange is associated with the catalytic process.

3.1.3. **A Composite Theory**

It is true that there is an energy barrier along the way the reaction proceeds; the peak of it is defined as Transition State, no matter how the reaction energy profile is looked at. However, it is also true that the residues in a protein fluctuate temporally in the time scale of picoseconds to milliseconds, and that these fluctuations affect the reaction energy profile. Another reality is that a large number of individual protein molecules as well as a large number of substrate molecules exist in a reaction system and that the temporal fluctuation is likely unsynchronized among individual protein molecules.

In the case of the hydride transfer from benzyl alcoholate to NAD^+ by alcohol dehydrogenase, Warshel and his coworkers (Villa and Warshel, 2001) showed by theoretical calculations that the fluctuations in the protein result in an electrostatic contribution to the energy difference of the reactants and the products system; the fluctuation centers around 60 kJ/mole and reaches zero (or below) value every two to three picoseconds. When the difference is zero, the energy barrier for the hydride transfer becomes insignificant, and presumably a reaction would take place instantaneously. According to their argument, the actual time scale is larger by a factor of 10^9 because the actual energy barrier is around 65 kJ/mole. If so, the hydride transfer would occur every two to three milliseconds. Yet, according to these authors, the fundamental importance of random thermal motions does not imply that enzymes use particular vibrational modes to enhance reaction rates.

It may be true that no specific vibrational modes are involved in enhancing the rate, but the fluctuations in protein do indeed bring about a situation momentarily that may allow the reaction to proceed instantaneously, and this could occur frequently at millisecond to maybe nanosecond (irregular) intervals. Very likely the evolutionary pressure on the enzyme system has brought about this situation.

The following is then a picture of enzyme reaction. Enzyme molecules are floating and are in constant molecular motions. Substrate molecules are also floating and are in random/molecular motions. A substrate molecule happens to enter an enzyme molecule, but soon comes out if it happens to enter a wrong portion of the enzyme, so that no reaction would take place. The evolutionary pressure would have made the overall structure in such a way that the following process is preferred. That is, if it enters the catalytic site and remains there for a short while, the molecular motions and fluctuations (of both the protein and the substrate) may bring this ensemble to a situation where the transition state can readily be overcome (e.g., as in the way described earlier for alcohol dehydrogenase). Now the reactant almost instantaneously would be converted to the product. If the product leaves the site in the next instant, a reaction has been accomplished. However, if it remains bound there for a short while more, there could be an opportunity to revert back to the reactant state. Again, the evolutionary pressure has made the first situation (fast release of product) a preferred process. On another molecule of the enzyme, similar events are taking place, though not necessarily synchronously with those on the first protein molecule. The kinetic results would be the apparent, temporal average of these events.

Obviously the dynamic effect just described would not be entirely random. If it is completely stochastic, the enzyme will never be efficient. The dynamic effect is an add-on to the situation provided by a snug fitting of a substrate to the catalytic site of an enzyme.

Another way of looking at the dynamics of enzyme reaction recently has been put forward; it is called the free energy channel model (Vendruscolo and Dobson, 2006; Boehr *et al.*, 2006). In the case of dihyrofolate (DHF) reductase, several free energy minima are recognized; the catalytic cycle goes in steps (Boehr *et al.*, 2006):

$$[E:NADPH]\text{-}(1) \rightarrow [E:NADPH:DHF]\text{-}(2) \rightarrow [E:NADP^+:THF]$$
$$(THF:tetrahydrofolate)\text{-}(3) \rightarrow [E:THF]\text{-}(4) \rightarrow$$
$$[E:NADPH:THF]\text{-}(5) \rightarrow [E:NADPH]$$

Step (1) is binding of the substrate, (2) the catalytic process, (3) release of the cofactor (NADP$^+$), (4) binding of the cofactor (NADPH), and (5) release of the product. The dynamic effect on steps (1) and then (2) is described by Vendruscolo and Dobson (2006) as follows. "Conformational fluctuations resulting from the concerted motions of many atoms can push the unbound states of enzymes into conformations closely resembling the bound states, thereby priming them to form complexes with specific ligands. Fluctuations take place preferentially in a way that prepares the protein to bind to its cofactors and substrates. The free-energy landscapes of the free and the bound states differ just enough to cause changes in relative populations of their principal states. After binding, the free-energy landscape is plastically deformed just enough to make a slightly different state of the protein become the most populated."

Note: Although it is true that the enzyme reactions are dynamic processes, it is not possible as yet to discuss the enzyme reactions as if the substrate and the enzyme are moving constantly. In this discourse, we heavily rely on the static picture obtained by x-ray crystallography in order to understand the reaction mechanism. Some dynamic nature of the protein/enzyme in solution can be and have been obtained by NMR method, but not enough studies have been conducted yet to elucidate the dynamic process of enzyme reactions.

3.2. METALLOENZYMES/PROTEINS AND METAL-ACTIVATED ENZYMES

A metalloenzyme consists of metallic ion(s) and a protein which could be multipolypeptides. Exceptions are ribozymes, which are RNA (instead of protein) and require metallic ion Mg(II) for catalytic effects. Hence they can be called metalloenzyme. Ribozymes are discussed in Chapter 4.

Typically the metallic ion(s) constitutes the main factor of the catalytic site in metalloenzymes. Strictly speaking, metalloenzymes are those with strongly bound metallic entity(ies), even in the absence of substrates. Some enzymes do not have bound metallic ion(s) but require one for full catalytic function. For example, hexokinase requires added Mg(II) for its catalytic activity and Mg(II) would bind to the catalytic site, but the native enzyme does not necessarily contain Mg(II). In this particular case, Mg(II) is not free, but associated with another substrate ATP; $[Mg(ATP)]^{2-}$ is the substrate. This is an example of metal-activated enzyme. No such strict distinction is employed in the present treatment.

As outlined in Chapter 2, metallic elements provide several functions that the organic entities (i.e., proteins/amino acids and ribonucleotides) cannot provide. They are Lewis acidity, oxidation/reduction center (electron transfer agent), and relatively weak organometallic bond. The Lewis acidity of metallic cations makes them acid catalytic entities beyond the organically available acid/base entities (OH, SH, COOH, NH$_2$, histidine N, etc.). This aspect will be discussed more fully in Chapter 4.

Organic compounds, in general, are poor electron-transfer agents; their oxidation-reduction by one electron is usually difficult to effect. One characteristic of transition metals is their multivalency; the different oxidation states are relatively easily accessible. This is true particularly with Mn, Fe, Cu, Mo, and W. These metals hence provide good electron-transfer capability and catalytic capability for oxidation-reduction reactions. This aspect is dealt with in Chapters 5 and 6. Chapter 8 concentrates on one particular reduction reaction, the reduction of N_2 to NH_3. Vitamin B_{12} coenzyme, or adenosyl cobalamin, is a naturally occurring organometallic compound with a Co—C bond. This Co—C bond is relatively weak, and is readily cleaved, and the resulting free radical (on the carbon) is the basis of the catalysis by adenosyl cobalamin. This is dealt with in Chapter 7.

Though the center of catalysis in a metalloenzyme is indeed a metallic entity, it is insufficient to provide enough enhancement effects on the reaction. It requires auxiliary, strategically located amino acid residues and/or other cofactors. The amino acid residues can assist the reaction in various ways, including provision of donor or acceptor of proton(s), electrostatic effects or hydrogen bonding to the substrate, and hydrophilic or hydrophobic environments. Subtle variations of these auxiliary entities due to dynamic motions can bring about the free-energy channel through which a reactant state converts itself to the corresponding product state, as outlined earlier.

REVIEW QUESTIONS

1. Show your understanding of *Circe Effect* by giving examples (from the original article).

2. What is *Entatic Effect*? Explain by giving example(s).

3. Find experimental data for the rate of a solution reaction of $CO_2 + H_2O \longrightarrow H^+ + HCO_3^-$, and that for the corresponding enzymatic reaction. Compare.

4. The highest reaction rate (of a second order rate law) in solution that can be attained is called diffusion-limited. Explain.

PROBLEMS TO EXPLORE

1. The rate of carbonic anhydrase reaction seems to be higher than diffusion-limited. Discuss.

2. Explore the hydrogen tunneling process (e.g., see Liang and Klinman, 2004). How might this effect shed light on the controversy of dynamic effect *vs* transition state theory?

3. Select a recent paper regarding the dynamic effects on the catalytic activity of an enzyme. Discuss how the temporal movements of amino acid residues of the protein enhance the reaction rate or reduce the activation energy.

Reactions of Acid-Base Type and the Functions of Metal Cations

4.1. GENERAL CONSIDERATIONS

4.1.1. Different Types (Definitions) of Acid-Base

General chemistry textbooks discuss several different definitions of acid-base. Two are essential in this treatment, Brønsted acid-base and Lewis acid-base. Brønsted acid is a chemical entity that has proton(s) to be donated to another entity in aqueous medium, and the entity that accepts proton(s) is defined as Brønsted base. Typical Brønsted acids are HCl (hydrochloric acid) and acetic acid (CH_3COOH); they dissociate proton(s). That is:

$$HCl\,(aq) + H_2O \rightleftharpoons H_3O^+ + Cl^-$$

$$CH_3COOH\,(aq) + H_2O \rightleftharpoons H_3O^+ + CH_3COO^-$$

Often H_3O^+ is abbreviated as H^+. The degree to which an entity dissociates a proton is measured as K_a (acid dissociation constant). K_a for HCl is known to be very large ($\gg 1$), whereas K_a for acetic acid is relatively small (1.8×10^{-5}). pK_a values often are used instead of K_a itself; pK_a is defined as $-\log K_a$ (pK_a of CH_3COOH is 4.85). A typical Brønsted base is OH^- (of NaOH, for example), which accepts H^+ readily. NH_3 can be defined as Brønsted base when it combines with H^+. However, in aqueous medium NH_3 may accept H^+ from a water molecule and partially turn into NH_4^+ and OH^-; it is a Brønsted base in this sense, too.

The Lewis definition of acid-base is more fundamental; an acid is defined as an entity that accepts a lone pair (of electrons) and a base is a donor of a lone pair (of electrons). H^+ can accept a lone pair of, for example, OH^-, and hence it can be defined as a Lewis acid, and OH^- is obviously a Lewis base. BF_3 has a strong tendency to accept a pair of electrons to attain an octet electron configuration and thus is a strong Lewis acid. NH_3 has a lone pair (of electrons) on its nitrogen atom ($|NH_3$; "|" denotes a lone pair) and this lone pair can relatively easily be donated. Hence:

$$F_3B + |NH_3 \rightleftharpoons F_3B^- - {}^+NH_3$$

A metal cation accommodates a lone pair of electrons and hence a Lewis acid. As discussed in Chapter 2, a metal coordination compound is formed as a result of Lewis acid-base interaction between a Lewis acid metal (cation) and lone pair donating ligand(s).

Organic residues found in proteins and nucleic acids can act as Brønsted acids and bases and Lewis bases. RCOOH in glutamic acid and aspartic acid and HS (of cysteine) are obviously a Brønsted acid, and the dissociated entities $RCOO^-$ and ^-S (cysteine) are Brønsted as well as Lewis bases. Alcoholic OH (of serine, threonine, and tyrosine) can be proton-donors (Brønsted acids) and the alcoholates are bases. An amine group in lysine and histidine (imidazole) and some bases in nucleotides and polynucleotides is a good proton acceptor (Brønsted base) but also can donate its lone pair (to a Lewis acid); hence it is a Lewis base, too. No Lewis acid, however, is found among organic compounds in the biological systems.

4.1.2. Reactions of Acid-Base Type Catalyzed by Enzymes

Reactions of acid-base type are numerous, but all involve transfer or sharing of a lone pair in a general sense (see the Introduction). Examples of enzymatic reactions of this type are found in the ensuing sections. The catalytic entity that can deal with such a reaction should be capable of accepting or donating lone pairs from or to reactants. That means that the catalytic entity should be acid or base. Such repertoire in the bioorganic compounds is limited, as mentioned earlier. No true Lewis acid exists among organic compounds except for carbocation. That necessitates a true Lewis acid, metal cation in certain cases.

Indeed enzymes that hydrolyze peptide bonds, variously called proteases, proteinases, or peptidases, use all these functional groups. They are grouped into serine proteases, cysteine proteases, aspartic proteases, and metallo-proteases. Why then is a metal cation required for certain cases even though many such reactions can be catalyzed by nonmetallic entities?

What metal cations are available and how are they used for the catalytic entities? Virtually any metal cation can be a Lewis acid; for example, Na(I), K(I), Mg(II), Ca(II), Mn(II), Fe(II), Fe(III), Co(II), Ni(II), Cu(II), and Zn(II) in the biological systems; even xenobiotic ions such as Cd(II) or Hg(II) in certain cases. The alkali metal cations are obviously too weak an acid and will not be useful for acid-base catalytic purposes. All the other cations listed are capable of performing acid-base catalytic functions, though their abilities vary.

The issue of measure of acidity (basicity) is discussed in the next section. Another issue is that the preference for the nature of ligand depends on the metal cation. This basically is related to the electron overlap; this issue can also be dealt with in terms of soft-hardness. For example, the elements in the left-hand side of the periodic table up to Mn(II) including Mg(II) and Ca(II) prefer to bind O-ligands, whereas Co(II) and those further on the right prefer N-ligands and S-ligands. The cations of the former category are harder than those in the second category. It is not, though, that Mn(II) would never bind an S- or N-ligand, nor that Co(II) would never bind O-ligands.

In terms of soft-hardness, Fe(II) is softer than Fe(III). It is generally true, by the way, that a metal cation in higher oxidation states is harder than the same cation in the lower oxidation states. Therefore, Fe(II) binds more preferentially with the softer N ligands than with the harder O ligands. Fe(III) and Fe of higher oxidation states prefer to bind with negatively charged ligands, particularly those of O, N, and S. This is likely due to an increased covalency caused by its highly electron withdrawing ability coupled with the more polarizable entities (ligand). This issue is, however, not so much relevant to acid-base catalysis as to oxidation-reduction.

Another factor that is important and common to any type of catalysis is that the catalytic entity should be able to bind a substrate but also should be able to release it rapidly. That is, it has to be kinetically labile.

Some types of RNA shown to exhibit catalytic activities are called *ribozymes*, which require Mg(II) for the activity. Besides, RNA and DNA interact with metal ions in a variety of ways; a later section will be devoted to a brief discussion of metal-nucleic acid interactions.

4.1.3. Acidity Scale and Acid Character of Metal Cations: Prominence of Zn(II) and Mg(II)

What factors determine the Lewis acidity is not a simple issue. A number of factors would contribute to it, including the extent of electron overlap and the ionic potential defined as z/r, where z is the electric charge and r the ionic radius. If the acid-base interaction is largely ionic, the second factor would be a good measure, where z can be regarded as the nominal electric charge. However, if the acid-base interaction involves a significant degree of electron overlap (covalency), the first factor and the second factor are equally important, and, besides, z should be more like the effective nuclear charge of the accepting orbitals.

There is no perfect, all-encompassing acidity scale as far as Lewis acids are concerned, though pK_a (pK_b) value is a good scale in the case of Brønsted acids and bases. pK_a values have been determined also for metal cations in reactions of the following type.

$$[M^{+II}(H_2O)_n] \rightleftharpoons [M^{+II}(H_2O)_{n-1}(OH)] + H^+$$

How realistically this scale represents the Lewis acid character of a cation can be disputed, but let us use this data for a measure of acidity of metal cations. Figure 4.1 presents the plotting of pK_a value against an ionic potential. The correlation is relatively good, except for the trivalent cations. The correlation coefficient between two sets of data (excluding Fe(III), Cr(III), and Al(III)) is ca 0.96. The ionic potential here is Z_{eff}/r, where r is the ionic radius for CON (coordination number) = 6 and the Z_{eff} is based on that of Z_{eff} for 4 s orbital (or an appropriate one) in the neutral atom plus 2 (or 3) for +II (or +III) cations. Z_{eff} values used for 4 s orbital in the neutral atom are

■ **Figure 4.1.** The correlation between pK_a of aquo complex of metallic ions and the ionic potential as defined by z_{eff}/r where z_{eff} is the effective nuclear charge and r the ionic radius (in 6 coordinate octahedral structure). The ionic potentials of Zn* and Cu* represent those of 4-coordinate structure, and these values suggest correspondingly higher pK_a if they form 4-coordinate aquo complex.

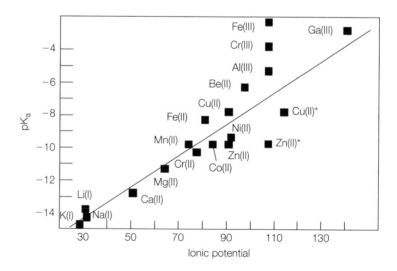

those calculated by Clementi and Raimondi. A reason for the use of 4 s orbital is that the ligand lone pair likely would be accepted into a mostly 4 s orbital of the metal cation.

It is obvious that the ionic potential as calculated here is not appropriate for the trivalent cations. The Z_{eff} values for these trivalent cations likely would be greater than those calculated based on merely adding the nominal electric charge. The ionic potential of Zn(II) in the figure is that for CON = 6, for the aquo complex of Zn(II) is known to be six-coordinated. However, the CON of Zn(II) in enzymes and proteins is usually 4, and hence the ionic potential of Zn(II) would be greater than that indicated in the figure (shown as Zn(II)* in the figure).

The Lewis acidity of a metal cation obviously is modified by the other ligands coordinated to it, as illustrated by the data given in Table 4.1. Therefore, it may not be very fruitful to pursue this line of arguments to characterize the Lewis acidity of a cation any further.

Zn(II) seems to be one of the strongest Lewis acids among the common divalent cations, judging from the Z_{eff}/r value and assuming CON = 4 for it. Cu(II) (for CON = 4) is better in this sense, but Cu(II) would not have been available to the organisms in the ancient seawater as discussed in Chapter 1, whereas Zn(II) was available in the seawater since the beginning of the Earth, and more abundant (by at least one order of magnitude) than copper. As a matter of fact, Zn is the most abundant transition element in seawater, except for iron (see Fig. 1.4). Hence Zn(II) was the best Lewis acid that the organisms could choose from ancient time on.

This argument would not necessarily preclude the use of other cations for Lewis acid catalyst in certain specific cases. Another cation

Table 4.1. Variation of pK_a (of the Coordinated H_2O) Caused by Ligands

Complex	pK_a	Complex	pK_a
$[Co^{III}(NH_3)_5(H_2O)]$	6.6	trans-$[Co^{III}(en)_2(NO_2)(H_2O)]$	6.4
cis-$[Co^{III}(NH_3)_4(H_2O)_2]$	6.0	$[Co^{III}(H_2O)_6]$	− 0.7
cis-$[Co^{III}(en)_2(H_2O)_2]$	6.1	$[Cr^{III}(NH_3)_5(H_2O)]$	5.3
trans-$[Co^{III}(en)_2(H_2O)_2]$	4.5	cis-$[Cr^{III}(en)_2(H_2O)_2]$	4.8
cis-$[Co^{III}(en)_2(NO_2)(H_2O)]$	6.3	trans-$[Cr^{III}(en)_2(H_2O)_2]$	4.1

such as Mn(II), Fe(II), or Ni(II) may be more appropriate than Zn(II) for certain enzymes in terms of the structural requirements and preference for the ligands. In yet another case, Cu(II) might be more useful. By the way, these metal ions can undergo oxidation-reduction, which may or may not be appropriate for the catalysis. Zn takes only a single oxidation state Zn(II); it could be another advantage.

The arguments made so far imply that the catalytic site containing Zn(II) would be often tetrahedral with CON = 4 and the coordinating ligands are N- or S-containing ones. However, the coordination structure of Zn(II) is rather flexible, readily taking other arrangements and other CONs, because of the spherical electron distribution (d^{10}) without any specific electronic preference as in the square planar structure of Cu(II) of d^9. This can also lend an advantage to Zn(II) over other cations. According to a study on the structures of Zn(II)-containing small coordination compounds (Bock *et al.*, 1995), the percentages of the coordination compounds of CON = 4 (tetrahedral), CON = 5, and CON = 6 (octahedral) are, respectively, 42, 19, and 35%. Co(II) in a tetrahedral environment takes a spherical electron distribution, and hence can mimic Zn(II) in many enzyme systems.

Another widely used cation is Mg(II). It is a hard acid, preferring a hard base, particularly O-ligands. That is, it has a good affinity toward the O^- of carboxylate or phosphate. In particular, Mg(II) is the preferred cation to bind ATP; Mg(II) would chelate two Os of the triphosphate part as well as an N of the adenine part. The Mg(II)-coordination compounds are predominantly octahedral with CON = 6 (76% of all compounds; Bock *et al.*, 1995). Mg(II) seems to play two roles; one is to reduce the overall electric charge of ATP, and the Mg(II) binds to amino acid residues too, hence bringing the Mg-ATP to the catalytic site. The Lewis acid character of Mg(II) can then enhance the reaction. Mn(II) has a similar character to Mg(II) both in terms of the electronic configuration (spherical electron distribution) and the effective ionic potential (Z_{eff}/r) (see Fig. 4.2). Hence, Mg(II) can often be interchanged with Mn(II) in Mg(II)-dependent enzymes without losing catalytic activity. However, the biological choice is usually Mg(II), as it is more abundantly available than Mn(II).

4.1.4. **Kinetic Factors**

The kinetic factors are also critical in catalysis; a substrate has to bind and the product has to leave the catalytic entity promptly. The kinetic

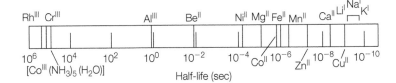

■ **Figure 4.2.** The half-life of water ligand exchange rate of aquo complexes; note the extremely slow rate of Cr(III), Co(III) (of low-spin complex), and Rh(III), whereas water exchange rate of all alkali metals, Ca(II), Cu(II), and Zn(II), are quite fast.

factors of interest here are: (a) rate of coordination and dissociation of a ligand (substrate on and off rates); and/or (b) rate of ligand substitution by another (substrate). Figure 4.2 gives the half-life of water exchange reaction of the aquo complexes of various cations. The half-life is inversely proportional to the rate of the exchange reaction. The figure gives an overall picture of ligand substitution (or exchange) rate of various metal cations. Cu(II), Ca(II), and alkali metal cations are among the fastest. Those for Ni(II), Co(II), Fe(II), and Mn(II) are quite fast; so are Mg(II) and Zn(II). On the other hand, the ligand exchange (substitution) of Cr(III) (d^3), and Rh(III) and low-spin Co(III) complexes (d^6) are very slow. This implies that when t_{2g} orbitals (d_{xy}, d_{yz}, d_{zx}) are either half-filled or fully filled (in the case of octahedral complexes), the ligand substitution reactions are extremely slow. This would preclude Cr(III) or low-spin Co(III) from utilization as a catalytic entity. An exception is cobalamin, which works in an entirely different way.

By the way, the lability of Ca(II) in ligand substitution reactions makes it useful for its second messenger role, and will be discussed in section 10.2.2.

4.1.5. **Enhancement of Reaction by Protein Residues**

As discussed previously (in Chapter 3), the Lewis acidity of a metal ion alone may not be sufficient to facilitate the reaction. Reactions of acid-base type involve transfers of proton and OH^- (or other basic entities such as ROH (RO^-) and phosphate O^-). This requires strategically placed proton acceptors and proton donors, such as carboxylic acid groups of Asp or Glu; OH of Ser, Thr or Tyr; SH of Cys, N/NH of His. In addition, a substrate may need to be secured in an appropriate conformation through hydrogen bonds and electrostatic interactions. The rate enhancement by an enzyme is the result of the cooperative actions of these factors, and of the dynamic effects.

In this chapter, the focus is on the effect of metal cations as the Lewis acid. Little attention will be paid to the associated factors (proton donors and acceptors, etc.), not because they are unimportant, but because this treatment is on the functions of metallic elements and, besides, the details of associated factors are yet to be delineated in most of the cases, particularly in view of the dynamic effects.

4.2. Mg(II)-DEPENDENT ENZYMES

Some representative examples of Mg(II)-dependent enzymes will be discussed. An emphasis is placed on how Mg(II) contributes to the catalytic effect. Ribozyme is also a Mg(II)-dependent enzyme, but will be treated in a later section (4.6).

4.2.1. Rubisco (Ribulose 1,5-Bisphosphate Carboxylase/Oxygenase)

The enzyme catalyzes formation of two molecules of 3-phosphoglyceric acid from carbon dioxide and ribulose 1,5-bisphosphate (RuBP):

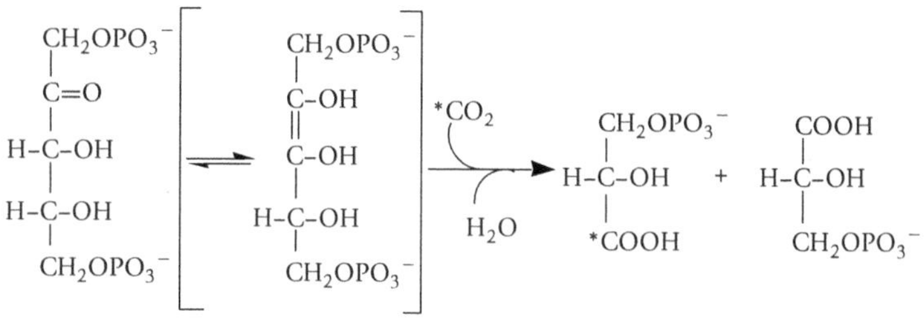

As the C atom of CO_2 binds to C2 of ribulose, the C3—OH turns to C=O releasing H^+. H_2O would then attack the C3 of C=O, resulting in a cleavage of C2—C3 bond. This enzyme is pivotal in the dark reaction of photosynthesis, and is believed to be the most abundant enzyme on Earth.

O_2 can compete with CO_2 for the active site of rubisco. So when the O_2 pressure is high, it produces one molecule of 3-phosphoglyceric acid and one molecule of phosphoglycolic acid. This process is called *photorespiration*. As this is an oxygenation reaction, the enzyme is called oxygenase as well as carboxylase. This oxygenation reaction is rather unusual and is not well understood.

The enzyme usually consists of L_8S_8, the large subunit L and the small subunit S. The smallest functional unit seems to be L_2S_2.

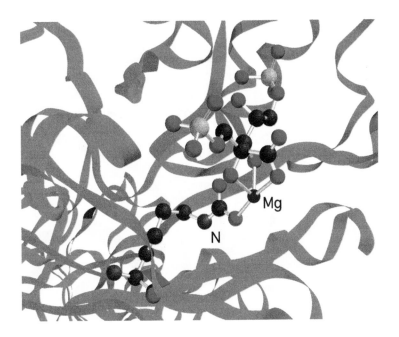

■ **Figure 4.3.** The catalytic Mg(II) site of Rubisco (ribulose 1,5-biphosphate carboxylase/oxygenase) bound with a substrate analogue with incorporated CO_2 and CO_2 of carbamate. (PDB ID:1BWV: Shibata, N., Inoue, T., Fukuhara, K., Nagara, Y., Kitagawa, R., Harada, S., Kasai, N., Uemura, K., Kato, K., Yokota, A., Kai, Y. 1996. Orderly disposition of heterogeneous small subunits in D-ribulose-1,5-bisphosphate carboxylase/oxygenase from spinach. *J. Biol. Chem.* **271**, 26449–26452.)

Mg(II) is essential for the carboxylation reaction. The enzyme binds tightly RuBP in its deactivated closed form. It would not allow the entrance of water molecules. It has to be activated by another enzyme, rubisco activase, that is a kind of ATPase; that is, it requires hydrolysis of ATP. It converts rubisco into an open active form, in which a CO_2 binds to a lysine NH_2 forming a carbamate group. Mg(II) binds to this carbamate group (see Fig. 4.3), and then binds to another CO_2 molecule that is to be incorporated into RuBP. The mechanism of this enzymatic reaction is far from well understood, but the critical function would be polarization of O=C bond (of CO_2) enhancing its attack onto the π-electron of the C2=C3 bond. It also assists the nucleophilic attack of C3 (C=O by now) by H_2O. Figure 4.3 shows binding of an analogue of the intermediate (after addition of CO_2 to the C2 carbon).

4.2.2. **Pyruvate Kinase**

Phosphoryl transferases, enzymes to transfer a phosphate group of ATP to an alcoholic entity as well as ATPases, are all Mg(II)-dependent. A typical example is pyruvate kinase, which catalyzes the following reaction:

$$ATP + CH_3(C=O)COO^- \rightarrow ADP + CH_2=C(OPO_3^-)(COO^-)$$

■ **Figure 4.4.** The catalytic site of pyruvate kinase with one Mg(II) bound with oxalate (an analogue of pyruvate) and the γ-PO$_3^-$ of ATP and another Mg(II) bound with ATP(3O$^-$PO$_2$S). (PDB 1 A49: Larsen, T.M., Benning, M.M., Rayment, I., Reed, G.H. 1998. Structure of the bis(Mg^{2+})-ATP-oxalate complex of the rabbit muscle pyruvate kinase at 2.1 Å resolution: ATP binding over a barrel. *Biochemistry* **37**, 6247–6255.)

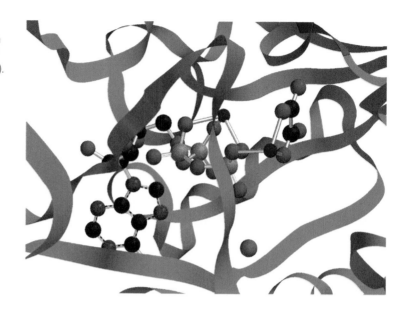

Presumably the enol form of pyruvate will react with ATP; the reaction is then a nucleophilic attack by OH of the enol on the P of the terminal phosphate of ATP. It requires Mg(II) (or Mn(II)) and K(I). An x-ray crystallographic structure of pyruvate plus ATP and oxalate (nonreactive substrate analogue) is shown in Figure 4.4 (PDB 1A49; Larsen *et al.*, 1998). It shows that all three Os of the triphosphate portion bind to Mg(II) and another Mg(II) ion nearby binds the other substrate, perhaps pyruvate in the enzymatic reaction. The second Mg(II) likely stabilizes the O$^-$ form of the enol OH of pyruvate (and there must be a proton receptor nearby), resulting in an increased nucleophilicity of the oxygen atom. The other Mg(II) increases the electrophilicity of the P atom of the terminal phosphate. Together the two Mg(II)s enhance the reaction.

4.3. Zn(II)-DEPENDENT ENZYMES

Zn(II)-dependent enzymes are widespread among all the classes of enzymes. A list of a few representative Zn(II)-enzymes is given in Table 4.2. A more comprehensive list will be found in Lipscomb and Sträter (1996). A few better-understood enzymes will be discussed, illustrating the main catalytic functions of Zn(II).

Table 4.2. Examples of Zn(II)-Enzymes (from Lipscomb and Sträter, 1996)

Enzyme	Ligands binding Zn(II) (enzyme source)
Mononuclear Zn-enzymes	
Adenosine deaminase	His,His,His,Asp,H_2O (mouse)
Alcohol dehydrogenase	Cyc,His,Cys,H_2O (horse liver)
Astacin	His,His,His,Tyr,H_2O (*Astacus astacus*, crayfish)
Carbonic anhydrase	His,His,His,H_2O (human)
Carboxypeptidase	His,Glu,His,H_2O (cow)
Fibroblast collagenase	His,His,His (human)
Lysozyme	His,His,Cys,H_2O (bacteriophage T7)
Thermolysin	His,His,Glu,H_2O (*Bacillus thermoproteolyticus*)
Polynuclear Zn-enzymes	
Acid phosphatase	Zn:Asp,Asn,His,His; Fe:Asp,Asp,Tyr. His (kidney bean)
Alkaline phosphatase	Asp,Asp,His; Asp,Asp,His (*E. coli*)
Bovine leucine aminopeptidase	Asp,Asp,Asp,CO,Glu, H_2O; Lys,Asp,Asp,Glu, H_2O

4.3.1. **Carbonic Anhydrase**

This enzyme catalyzed the following simple reactions and is among the fastest known enzymes:

$$CO_2 + H_2O \rightleftharpoons HCO_3^- + H^+$$

Three different types of carbonic anhydrase are known: α (animals), β (plants), and γ (bacteria). They have low sequence homology, but have similar catalytic sites with Zn(II). The ligands in the resting enzyme are three histidine residues and a water molecule and the coordination structure is approximately tetrahedral. This active site is sitting at the end of a groove lined by hydrophobic amino acid residues. The hydrophobic groove is filled with water molecules bound to each other through hydrogen-bond (ice-structure). The details of the structure, kinetic, and other data have been discussed and reviewed since the 1970s, including Ochiai (1977), Lipscomb and Sträter (1996), and more recently Christianson and Cox (1999).

Ochiai (1977) suggested the mechanism shown in Figure 4.5, based on the data then available, which seems to be virtually the same as

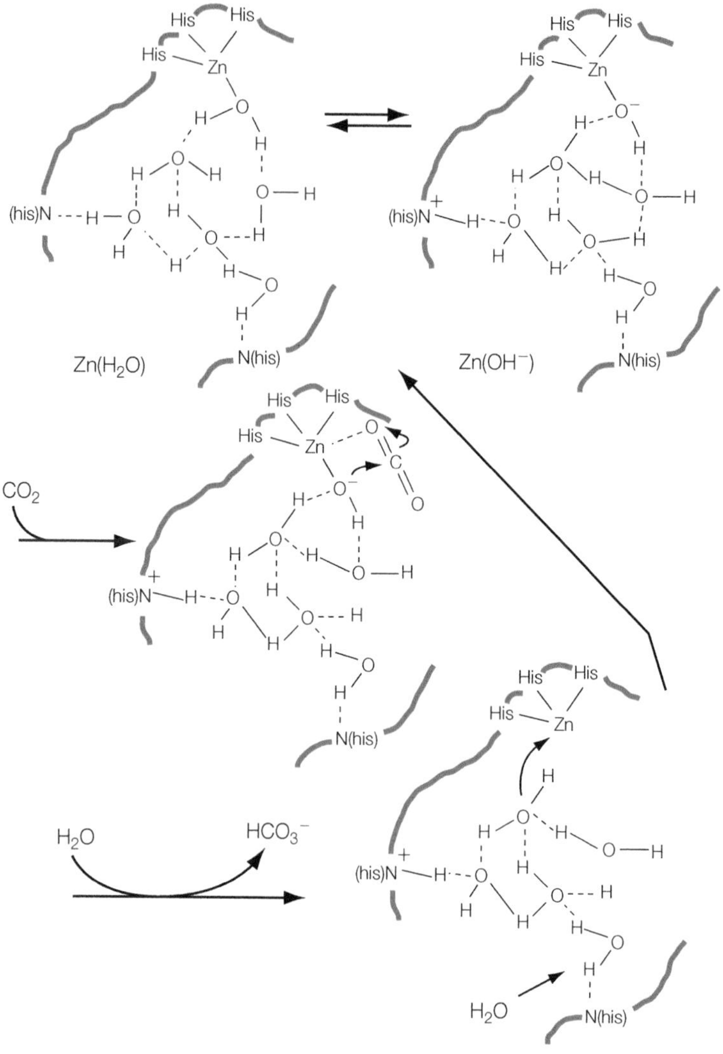

Figure 4.5. A mechanism of carbonic anhydrase (after Ochiai, 1977).

that shown in Christianson and Cox's review (1999). Newer data seems to indicate that a Thr-OH binds to CO_2 through hydrogen bonds. The details may not be exactly right, but the main points seem to be portrayed by the mechanism. The mechanism shows (a) that Zn(II) coordinates a H_2O molecule, removes H^+ from it with assistance by other moieties turning it into OH^-, and (b) that Zn(II) provides assistance to polarize $O{=}C$ (of CO_2) bond enhancing the attack of OH^- to this carbon atom (of CO_2).

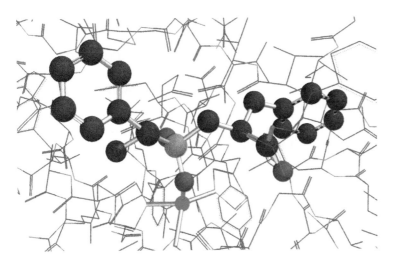

■ **Figure 4.6.** The catalytic site of thermolysin where Zn(II) is bound in an approximately tetrahedral manner by two histidine residues, one carboxylate (glu) and O of phosphate of a diester. This entity is meant to be a structural analogue of a tetrahedral transition state involving the carbonyl C of a peptide that is bound with the attacking OH. (PDB 1OSO: Selkti, M., Tomas, A., Gaucher, J.F., Prange, T., Fournie-Zaluski, M.C., Chen, H., Roques, B.P. 2003. Interactions of a new alpha-aminophosphinic derivative inside the active site of TLN (thermolysin): A model for zinc-metalloendopeptidase inhibition. *Acta Crystallogr., Sect. D* **59**, 1200–1205.)

Marine diatom *Thalassiosira weissflogii* produces a Cd-specific carbonic anhydrase when it is grown in a Zn-deficient environment so that it cannot produce enough of Zn-carbonic anhydrate. The Cd-protein is apparently different from the Zn-protein (Lane and Morel, 2000).

4.3.2. **Thermolysin, Carboxypeptidase A, and Others**

Thermolysin is a representative of mononuclear Zn(II)-peptidases with the conserved His-Glu-X-X-His-His motif. In addition to Zn(II), *Bacillus thermolyticus* thermolysin contains Ca(II), which is required for the thermal stability. The catalytic Zn(II) is coordinated by His-142, His-146, H_2O, and Glu-166, whose carboxylate group coordinates the Zn(II) in the bidentate mode. That is, the coordination number for the Zn(II) is formally five. The C=O of the peptide bond of a substrate peptide coordinates to Zn(II), and the OH^- group that forms from the coordinated H_2O molecule attacks the carbon of the C=O, forming an sp^3-C intermediate state (C(OH)(O—)(NH—)(C—)), which will cleave spontaneously between the carbon C and N (see Fig. 4.6). The H^+-transfers involved are assisted by His (231), Tyr, Asp, and Glu residues strategically located (Lipscomb and Sträter, 1996).

Carboxypeptidase A (and B) has also a single Zn(II), which is coordinated by two His's, one Glu, and one H_2O molecule. It catalyzes the hydrolytic cleavage of C-terminal amino acid residue, especially of an aromatic and hydrophobic nature. Carboxypeptidase B, on the other hand, requires the presence of a positively charged side chain.

The mechanism includes (a) coordination of C=O of the peptide bond to be cleaved to Zn(II), though the direct binding has not been fully established; and (b) attack by OH$^-$ bound to Zn(II) to the C of the C=O, resulting in the formation of a tetrahedral intermediate. In other words, the mechanism is essentially the same as that for thermolysin.

There are a large number of peptidases, proteases, or the like that contain a single catalytic Zn(II) (Lipscomb and Sträter, 1996). Their hydrolytic mechanisms seem to be of the same nature as before, though the Zn(II) ligands and their local structures may not necessarily be the same. To cite another example, lysozyme of T7 bacteriophage hydrolyzes the amide bond between N-acetylneuramic acid and L-alanine in the peptidoglycan of the bacterial cell wall. The ligands of the single Zn(II) are two His, one Cys residue, and one water molecule.

4.3.3. Leucine Aminopeptidase

This is a typical di-zinc peptidase, and catalyzes the hydrolytic cleavage of the N-terminal amino acid of an oligopeptide. The catalytic site consists of a dimeric Zn(II) bridged by an OH$^-$. One of the Zn(II)s is coordinated by NH$_2$ of a Lys and a monodentate COO$^-$ of an Asp, and the other by two monodentate Asp residues; and the two Zn(II) are bridged by a bidentate Glu, in addition to the OH group. The coordination structure of both of these Zn(II) ions is approximately square pyramidal with coordination number = 5. The N-terminal amino group of an oligopeptide binds to one of the Zn(II)s and the carbonyl group to the other Zn(II), and its reaction intermediate, a tetrahedral *gem*-diolate binds between the two Zn(II)s. A single Zn(II) may not be enough to make the OH$^-$ sufficiently strong nucleophile; two Zn(II)s would make it a better nucleophile. An alternative explanation for the necessity of two Zn(II)s would be that all the substrates for the enzymes of this type (including aminopeptidase A) have a terminal amino group, which is utilized to be a hook to the catalytic site, one of the Zn(II)s.

4.3.4. Alkaline Phosphatase and Purple Acid Phosphatase

Alkaline phosphatase has a dinuclear Zn(II) catalytic site and also has Mg(II) or Zn(II) to bind the substrate phosphate. It catalyzes the cleavage of phosphate from phosphomonoester or transfer of the

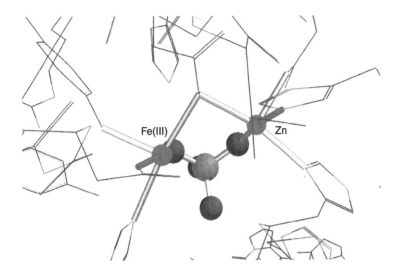

■ **Figure 4.7.** The catalytic site (Fe(III)-Zn(II)) of purple acid phosphatase. (PDB 4KBP: Klabunde, T., Strater, N., Frohlich, R., Witzel, H., Krebs, B. 1996. Mechanism of Fe(III)-Zn(II) purple acid phosphatase based on crystal structures. *J. Mol. Biol.* **259**, 737–748.)

phosphoryl group to an alcohol. The mechanism involves the formation of phosphate ester of a serine residue; the phosphate group is then transferred to another alcohol (substrate). The ester O and one of the phosphate Os bridge the two Zn(II)s. The electrophilicity of the P of the phosphate is enhanced, and hence the attack of the serine O(H) is facilitated. The proton of the serine OH is removed by a nearby proton acceptor. The remaining alcoholate on one of the Zn(II)s is then replaced by a water molecule or another alcoholate, which then attacks the P of the serine phosphate; this completes the enzymatic reaction (Lipscomb and Sträter, 1996).

The catalytic site of purple acid phosphatase consists of a dinuclear Fe(III)—M(II) unit. One of the ligands for the Fe(III) is a tyrosyl OH, and this bond is responsible for the purple color. The M(II) is Zn(II) in kidney bean's enzyme, and either Fe(II) or Mn(II) in mammalian enzymes. Fe(III), being highly acidic, can maintain the attacking OH^- even at an acidic pH, whereas a divalent cation Mn(II), Zn(II), or even Fe(II) cannot maintain or create OH^- (from H_2O). Then the OH^- group on Fe(III) attacks the P center of the phosphate ester substrate, cleaving the ester bond. A result is the formation of a dinuclear (Fe(III)—M(II)) entity bridged by a PO_4^{3-} group (see Fig. 4.7).

4.3.5. **Alcohol Dehydrogenase**

Horse liver alcohol dehydrogenase contains two types of Zn(II); one is structural and the other catalytic. The catalytic Zn(II) is coordinated by two cysteine thiols and one histidine, and the fourth ligand is water, which is to be replaced by OH of the substrate alcohol. The function of Zn(II) is to help remove the H^+ from the OH group. A hydride H^- will then be transferred to NAD^+ from α-carbon of the resulting alcoholate $R—CH_2—O^-$, forming an aldehyde $R—CH=O$. The hydride transfer here may take place through hydrogen tunneling (Liang and Klinman, 2004).

4.4. OTHER METAL CATION-DEPENDENT ACID-BASE ENZYMES

4.4.1. Aconitase, an Iron-Sulfur Enzyme, and Others

Proteins containing iron-sulfur clusters such as $[Fe_2S_2]$ and $[Fe_4S_4]$ are typically electron-transfer agents. However, a number of hydrolyase or other types of enzymes have been discovered to contain iron-sulfur proteins (Flint and Allen, 1996). A typical example is aconitase, which was reviewed in detail by Beinert *et al.* (1996).

An active aconitase contains a $[Fe_4S_4]$ cluster; each of the three iron atoms of the cluster are bound to a cysteine residue as in most ferreodoxins, but the remaining one has a water molecule instead of cysteine. The native enzyme has the cluster in the form of $[Fe_4S_4]^{2+}$; that is, nominally 2Fe(II) + 2Fe(III). The special iron is labile and relatively easily lost, and the substrate citrate binds to this labile iron, which is believed to be Fe(II) in order to be active. Aconitase catalyzes the conversion of citrate to isocitrate. It appears to be a 1,2-rearrangement of an OH group. The removal of OH^- from C-2 position is affected by the Fe(II), resulting in the formation of *cis*-aconitate. Rehydration of *cis*-aconitate under a different conformation will result in the product isocitrate. The change of conformation of *cis*-aconitate may require a special coordination of the carboxylate groups, and this may require an octahedral (CON = 6) coordination site. This is provided by Fe(II), though other cations such as Ni(II) can also provide such a coordination site. Fe(II) may be most favorable for this situation, because it prefers an octahedral structure and O-ligands. If this requirement is paramount, other choices such as Zn(II) and Cu(II) may not be suitable (Ochiai, 1987).

Another reason why a $[Fe_4S_4]$ cluster is used for this purpose could be evolutionary. A genetic modification of some iron-sulfur protein in such a way to lose the crucial cysteine residue may have resulted in opening up an opportunity for the iron without cysteine to be used as Lewis acid catalytic site. Besides, the other part of the cluster may function to keep the special iron in Fe(II) state.

Other similar enzymes include homoaconitase, methylcitrate dehydratase, and 2-methylisocitrate dehydratase, and fumarase (malate dehydratase). They catalyze reactions similar to that of aconitase. Evidences suggest that they contain Fe-S clusters (Flint and Allen, 1996).

4.4.2. **Arginase: Mn Enzyme**

Manganese enzymes as Lewis acid catalyst include arginase, inorganic pyrophosphatase, serine/threonine protein phosphatase, and proline-specific aminopeptidase (of *E. coli*). Arginase's catalytic metal center is a dimeric Mn(II) unit bridged by a hydroxide. A typical enzyme arginase consists of three homologous polypeptide subunits. Each of the Mn(II) in the dimeric catalytic site is coordinated by one His and a bidentate Asp residue and bridged by one OH^-, one bidentate Asp and another Asp that seems to function as a monodentate ligand. The OH^--bridged dimeric Mn unit is also present in the other hydrolases mentioned earlier (Christianson and Cox, 1999; Dismuke, 1996).

Arginase catalyzes the hydrolysis of arginine into ornithine and urea, as shown here:

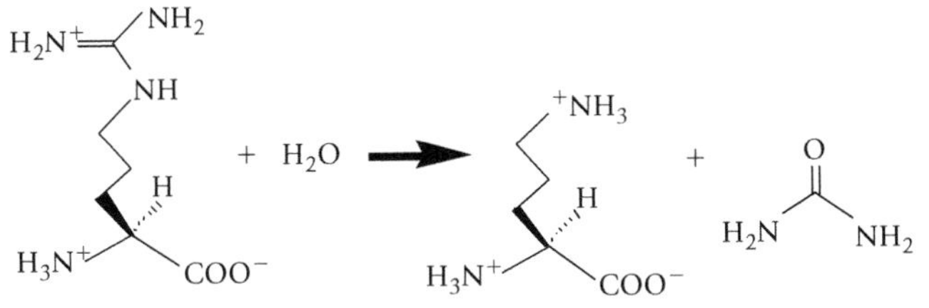

The transition state is believed to be a tetrahedral $[C(OH)(NH_2)_2 (NH-)]$, whose OH comes from the bridging OH. This transition state would spontaneously decompose into urea and ornithine, assisted by proton transfer (through His-141 present in the catalytic cavity) to the ω-NH_2 to be formed.

4.4.3. **Urease and Other Ni Enzymes**

Urease catalyzes the hydrolysis of urea to form two molecules of ammonia and one molecule of carbon dioxide (($NH_2)_2$ C$=$O + $H_2O \rightarrow$ 2NH_3 + CO_2). The active site of urease from *Klebsiella aerogenes* consists of two Ni(II)s bridged by OH^- and a carbamate (Ermler *et al.*,

1998). The carbamate is to be formed by binding CO_2 to a Lys residue in the protein, as in ribulose 1,5-bisphosphate carboxylase/oxygenase (see earlier). This has been indicated by the fact that this urease is inactive in the absence of CO_2. In addition to the bridging ligands, one of the Ni(II)s is bound with two histidines and one water molecule, and the other with two histidines, one monodentate carboxylate (of Asp), and one water molecule. The carbonyl oxygen atom of urea binds to the first Ni(II), which polarizes the carbonyl bond making the carbon atom more electrophilic. The water molecule loses H^+ to form OH^- on the second Ni(II). This nucleophile OH^- then attacks the carbon atom of urea, forming a tetrahedral geminal diol intermediate. It will spontaneously split between C and N of urea, forming ammonia and carbamate (NH_2COOH). The final step is the decomposition of the carbamate to CO_2 and NH_3, which is likely facilitated by H^+-donation by a strategically located proton donor (maybe a cysteine residue).

The functions in catalysis of Ni(II) seem to be of the same nature as those of Zn(II) and Mg(II) as outlined earlier. Hence an interesting question is, why Ni(II)? Before we consider this question, it might be interesting to see how selective the use of Ni might be in similar enzymes.

Glyoxalase converts α-keto aldehyde to 2-hydroxycarboxylic acid, and consists of two components: glyoxalase I and glyoxalase II (Maroney, 1999). Glyoxalase I of *E. coli* contains a single Ni(II). The activity of this enzyme is recovered partially when Ni(II) is replaced by Co(II), Mn(II), and Cd(II), but not by Zn(II). On the other hand, human and yeast glyoxalase I contains a single Zn(II) at its catalytic center. Another enzyme peptide deformylase turns out to be a Ni-enzyme, though Zn-enzyme is active to a smaller extent (about 30% of the Ni-version).

These facts may imply that there is no definite chemical reason for either Ni(II) or Zn(II) in these particular enzymes. Frausto da Silva and Williams (2001) suggests that these are historical relics. Perhaps on the ancient Earth, organisms tried to utilize all the available elements. Several different elements, say, Ni(II) and Zn(II), were utilized for the same chemical reactions, and they may have turned out to be approximately effective. Some organisms have since evolved to make Ni(II) more effective by modifying the associated protein. Once this particular enzyme had been perfected for Ni(II), the same enzyme with Zn(II) replacing Ni(II) may not be as effective. However, the same enzyme could have perfected for Zn(II), which is more abundant than nickel on Earth. This process is an example of one of the rules of selection of elements by organisms outlined in Chapter 2—evolutionary adaptation.

4.5. **STRUCTURAL EFFECTS OF METAL IONS**

Proteins and RNAs contain a number of self-structuring entities interacting electrostatistically through hydrogen bonding, and/or S—S covalent bonding. The result is a certain 3D structure, which is obtained by minimizing all the interaction energies including inter-actions with the medium (water molecules, etc.). The self-structuring components in a protein or RNA, however, may not be sufficient to produce a specific structure required for the role in some cases.

A metal cation, when bound to a protein or RNA, would change its conformation, due to the electrostatic and acid-base (coordinating) interactions. The biosystem utilizes this function to produce a nec-essary specific structure in proteins and RNA. An example of such effects on RNA structure is seen in Figure I.16. A further discussion on DNA and RNA will be given in the next section. A few examples that will be given next would suffice to show such effects on proteins.

Ca(II) binds to a number of proteins, and very likely changes their conformations upon binding. This character manifests in Ca(II) being a second messenger (see Chapter 10). A well-characterized example is calmodulin. Binding of four Ca(II) ions to a calmodulin molecule brings about a specific conformation to it. The conformation is appro-priate for binding, for example, a peptide on which the calmodulin has its effect (see Fig. 10.9). Another example of the conformation-changing effects of Ca(II) is seen in synaptotagmin (see Chapter 10).

Alcohol dehydrogenase contains two different types of Zn(II). One is the catalytic Zn(II), which was discussed earlier. Another Zn(II) is tightly coordinated by cysteine residues in a tetrahedral manner, solidifying a specific local structure, which in this case helps to main-tain a dimeric (two subunits bound) structure.

A transcription factor turns on and off certain specific segments of a DNA. Some of these factors contain Zn(II) as a structural factor; they are coined "zinc finger" protein. Zn(II) in a transcription factor is used to maintain a specific conformation of the protein in such a way that it can bind a specific segment of a DNA. Zif268, a zinc finger pro-tein, nicely fits to the groove of a DNA as shown in Figure 4.8. Each of the three mononuclear Zn(II) ions is strategically located, and bound

■ **Figure 4.8.** Zif268-DNA complex; the protein is shown as a ribbon, and the DNA is shown in the forms of sticks. The entities in orange color are phosphate P, and their arrangement shows the double helix structure of the DNA. (PDB 1 AAY: Elrod-Erickson, M., Rould, M.A., Nekludova, L., Pabo, C.O. 1996. Zif268 protein-DNA complex refined at 1.6 Å: A model system for understanding zinc finger-DNA interactions. *Structure* **4**, 1171–1180.)

with two His and two Cys residues to maintain this overall structure. GAL4, another transcription factor, contains dinuclear Zn(II)s. The two Zn(II) ions are doubly bridged by Cys residues and bound to two separate short helical segments of the protein, keeping the relative direction of the helices approximately perpendicular to each other.

Other metallic ions can also and do indeed play structural roles. For example, a subunit of a Cu-enzyme ascorbate oxidase contains the four catalytic Cus (see Fig. 5.7), and the two subunits are connected by a single Cu ion. The catalytic metal ion can be regarded to play also a structural role in the sense that it maintains that specific conformation of the enzyme in such a way to bring necessary amino acid residues to the catalytic site.

4.6. METAL IONS AND POLYNUCLEIC ACIDS (DNA AND RNA)

Metal ions, especially Mg(II) are widely involved in maintaining structure and providing catalytic character to RNA. First a general survey is presented regarding metal ion interactions with DNA and RNA.

4.6.1. **General Characteristics of Interactions of Metal Ions with Polynucleotides**

Polynucleotides, RNA and DNA, contain potentially ligating atoms, as well as some specific spatial structures. The possibly metal-ligating atoms in DNA and RNA include phosphate oxygen on the backbone; oxygen atoms of hydroxyl groups on ribose, deoxyribose, as well as on some bases; and nitrogen atoms in the bases. Hence there could be a rich chemistry of metal ion interactions with DNA and RNA.

The specific spatial structures of DNA invite some proteins to interact with them. The specific DNA binding proteins often are involved in controlling the transcription (expression) of DNA. An example is found in Figure 4.8. Some metals are involved in the regulation of transcription often through binding to the DNA-binding proteins. This is essentially metal-protein interaction, as no direct binding of the metal to the DNA is involved. However, this issue (i.e., metal-protein-DNA ternary complex) will be included in here.

RNAs occur in many different types of secondary and tertiary structures, and bind metal ion Mg(II) (or Zn(II), Pb(II) and others in some cases). A strategic coordination of Mg(II)s to maintain a t-RNA-specific structure is illustrated in Figure I.16. Ribozymes, the catalytic RNA, employ metal ions, particularly Mg(II) for their activities.

4.6.1.1. *Effects on Structures*

The double helix structure of DNA is maintained by two cohesive forces: van der Waals (London dispersion) force, particularly among the stacking bases, and the hydrogen bonds between base pairs. These cohesive forces are sufficiently strong to overcome the repulsive force that exists between the negative charges on the bridging phosphate groups on the backbone at normal temperatures. Raising the temperature of a DNA solution will cause the double helix to become random coils; the transition temperature is defined as melting temperature T_m. T_m is dependent on its G-C pair content (which is stronger in hydrogen bond than the A-T pair) as well as the presence of positively charged species including polyamines, and metal cations including Na(I). Divalent cations such as Mg(II) are most effective in stabilizing the double helix structure and raising T_m, by neutralizing the negative charges on the phosphates (Eichhorn, 1981). Zn(II) seems to stabilize the double helix structure not only by neutralizing the negative charges but also by bridging some base pairs. Cu(II), on the other hand, destabilizes the double helix structure perhaps by disrupting the interbase hydrogen bonds, thus lowering T_m (Eichhorn, 1981).

The tertiary structure of tRNA (as seen in Fig. I.16) apparently requires some Mg(II) ions in addition to the interbase hydrogen bonds in order to be maintained. The four Mg(II) ions are strategically located to assist the specific conformation (see Fig. I.16). Nucleic acid preparations from various sources have been found to usually contain metal ions of several kinds, including Mg(II), Ca(II), Cr(III), Mn(II), Fe(II/III), Ni(II), and Zn(II) (Wacker and Vallee, 1959; Bryan, 1981). Metal ions in RNA are often difficult to remove with chelating agents or dialysis. They, at least some of them if not all, are considered to play a major role in stabilizing the specific conformations of RNAs and maintaining the integrity of ribosomal RNAs.

As will be discussed later, ribozymes (catalytic RNA) use Mg(II) as a catalytic entity. However, Mg(II) may also be said to function as a center to bring together appropriate portions of RNA; that is, a structural function.

Other types of metal-DNA binding are found in the binding of anti-cancer agent Pt-complexes; Pt(II) would cross-link G—G or A—G in an intrastrand manner, and also in other modes including an interstrand fashion (see Chapter 12 for further details).

4.6.1.2. *Catalytic Metal Ions in DNA Polymerases and Nucleases*

E. coli's DNA polymerase I (Klenow fragment) and other DNA polymerases, and HIV-I's reverse transcriptase all turned out to have similar structures, consisting of the so-called finger domain, thumb domain, and palm region, as shown in Figure 4.9 (Klenow fragment of *E. coli* DNA polymerase; Teplove *et al.*, 1999) and Figure 4.10 (DNA polymerase of T7 phage; Li *et al.*, 2004). The palm domain is further connected to the exonuclease domain in some, as in the case of the Klenow fragment. The polymerization active center is believed to be located at the bottom of the thumb domain, which harbors two essential metal ions (see Figs. 4.9 and 4.10).

Since the polymerization is an esterification of the phosphate that can be catalyzed by general acid-base, typical metal cations such as Mg(II) and/or Zn(II) seem to play the catalytic role. The metal ions are strategically located to facilitate the formation of the bipyramidal transition state around the phosphorus atom. Probably this effect is more important than their Lewis acid character.

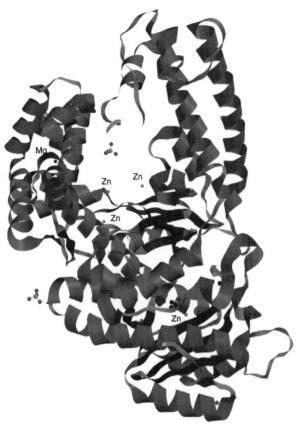

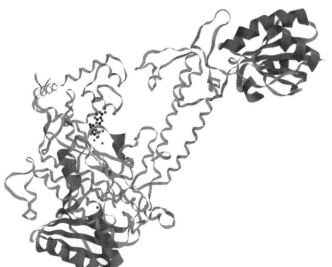

■ **Figure 4.9.** Klenow fragment of *E. coli* DNA polymerase. (PDB 1D8Y: Teplova, M., Wallace, S.T., Tereshko, V., Minasov, G., Symons, A.M., Cook, P.D., Manoharan, M., Egli, M. 1999. Structural origins of the exonuclease resistance of a zwitterionic RNA. *Proc. Natl. Acad. Sci. USA* **96**, 14240–14245.)

■ **Figure 4.10.** DNA polymerase of T7 phage. (PDB 1SKR: Li, Y., Dutta, S., Doublie, S., Bdour, H.M., Taylor, J.S., Ellenberger, T. 2004. Nucleotide insertion opposite a cis-syn thymine dimer by a replicative DNA polymerase from bacteriophage T7. *Nat. Struct. Mol. Biol.* **11**, 784–790.)

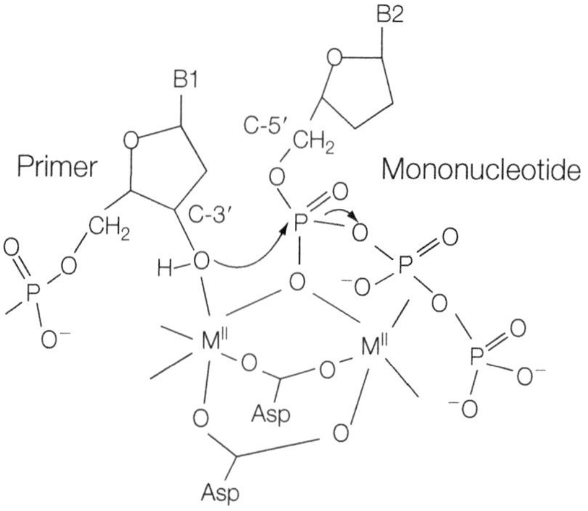

■ **Figure 4.11.** A mechanism of DNA polymerization (after Fig. 27.12 in Berg *et al.*, 2002).

The exonuclease activity (polymerization) of the Klenow fragment of *E. coli* DNA polymerase is also dependent on a pair of metal cations (Joyce and Steiz, 1994). It appears that the physiological cations are Zn(II) and Mg(II), as indicated in Figure 4.11. Zn(II) coordinates OH (at C-3′) and facilitates its attack on the phosphate P, as in Zn(II)-dependent hydrolases such as alkaline phosphatase. DNA polymerase of T7 phage seem to require Mg(II) specifically (see Fig. 4.10). A mechanism proposed of binding of the terminal of a primer DNA to the next mononucleotide (C-3′ to C-5′) is shown schematically in Figure 4.11.

The hydrolysis and transphosphorylation reactions of RNAs are also conducted by some RNA themselves (ribozymes) (see later). Pb(II) has been found to specifically cleave the U17-G18 of tRNA^{phe}. Pb(II) was shown by x-ray to bind directly to C60 and U59 and to be located near the 2′-OH of U17 (Pan *et al.*, 1993). Pb(II) seems to bind a water molecule that is hydrogen-bonded to the 2′-OH. Perhaps the Pb(II) changes the local conformation in such a way that the 2′-oxygen comes close to the phosphate between U17 and G18, thus facilitating the formation of the transition state.

4.6.2. **Gene Regulation and Metal Ions**

There is a wide variety of gene transcription factors, proteins that bind specifically to DNAs and regulate their expressions. A group of such proteins contain Zn(II) ions, whose function is to maintain the necessary configuration of the proteins that specifically bind

DNAs; one such protein, Zif268, was discussed in an earlier section. Another kind of transcription factor responds to the level of a metal, which needs to be dealt with a specific protein. A protein binds a metal ion, and becomes an active transcription factor, resulting in the expression of the gene of another protein that sequesters the metal ion present in excess. Several illustrative examples are presented in other chapters.

First, iron-responsive element binding protein (IRE-BP) is cytosolic aconitase, and is discussed in Chapter 10. Depending on the iron level in cytoplasm, IRE-BP binds a specific location of m-RNA of ferritin or transferrin.

A detoxifying mechanism of mercury in prokaryotes is MR (mercury reductase). The expression of this enzyme involves gene (*mer*) regulating proteins MerR and MerD. This is discussed in Chapter 11.

Metallothioneins (MTs) are induced by Hg, Cd, Zn, Cu, and Ag, and play a metal-level controlling factor as well as detoxifying role for these heavy metals; it is treated in Chapter 11. It is also induced by oxidative stress (see Chapter 11). How a heavy metal and oxidative stress induce MT expression has not been well understood. A protein MTF-1 (metal-responsive transcription factor) seems to be involved in expression of MT (Andrews, 2000). This protein is a kind of zinc-finger protein, and binds to the metal response element (MRE) on the promoter region of MT-1 gene. This element seems to be involved in activation of MT-1 gene by Zn(II) (and also *t*-butyl hydroquinone; Andrews, 2000). On the other hand, an increase of m-RNA of MT-1 by Cd(II) and H_2O_2 seems to be mediated by other elements of the MT-1 gene: ARE (antioxidant response element) and USF (upstream stimulating factor) (Andrews, 2000).

4.6.3. **Ribozymes**

Metal ions, K(I), Na(I), and Mg(II) are playing essential roles in maintaining the specific structure/conformation in a number of RNA molecules (Pyle, 2002). There are three known types of ribozyme (i.e., catalytic RNA):

(a) Large phosphoryl transfer ribozymes

(b) Small phosphoryl transfer ribozymes

(c) Aminoacylesterase ribozymes (Pyle, 2002)

The group (a) includes the self-splicing group I and group II introns, ribonuclease P and spliceosome (RNA plus protein). These ribozymes cleave or ligate the phosphodiester linkages of RNA and DNA involving an exogenous nucleophile such as water or alcohol moiety. They require Mg(II) for the catalytic effect. As shown in Figure 4.12, the intron I catalytically splices itself, making a single-stretch exon (m-RNA), and removing the intron portion. The exon region of m-RNA encodes a protein, but the intron segment does not. The qualification "catalytically" suggests that it can splice other polynucleotides, including DNA as well.

■ **Figure 4.12.** A schematic of self-splicing of exon (m-RNA) from intron (RNA) (after Fig. 28.36 on p. 805 of Berg *et al.*, 2002).

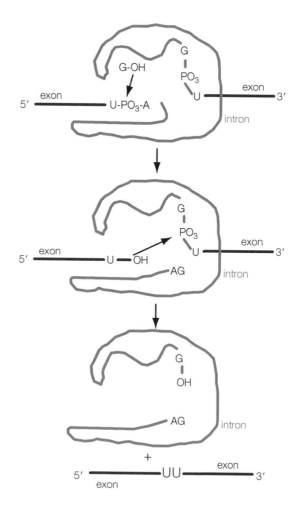

Stahley and Stroble (2005) determined crystallographically the structure of a group I intron with the G (with ribose 3'-phosphate) attached at the end of the intron. This is to represent the second step of the mechanism shown in Figure 4.12. The structure (at 3.4Å resolution) indicates, according to the authors, that two Mg(II) ions act as the catalyst, as shown in Figure 4.13. Other researchers proposed three-metal-ions mechanism, in which the additional M(II) binds the 3'-O of the leaving G (see Pyle, 2002; Stahley and Stroble, 2005). Pyle (2002) reviewed the effect of metal ions on the other types of ribozymes.

Whichever the case may be, the functions of the metal ion (mostly Mg(II)) seem to be (a) bringing the two nucleotides side-by-side, (b) enhancing electrophilicity of 3'-O of the intron's end G, (c) increasing the electrophilic character of the P of the phosphate group, and (d) stabilizing O$^-$-character of 3'-O of U (the end of 5'-exon side). The hydrogen of the 3'-OH of U needs to be removed; this implies the presence of a strong nucleophile nearby. This is yet to be investigated.

Protein synthesis system on ribosomal RNA can be regarded as a ribozyme. It is briefly discussed in Introduction's "formation and hydrolysis of proteins" section (p17) and Figure I.10 (Nissen, et al, 200).

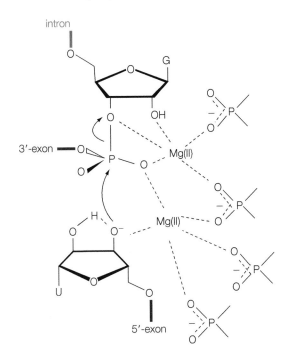

■ **Figure 4.13.** The coordination of RNA to 2 Mg(II) ions in self-splicing process (after Stahley and Stroble, 2005).

REVIEW QUESTIONS

1. Name Lewis basic amino acid residues, and Lewis acidic amino acid residue(s).

2. According to Figure 4.1, among common elements, Cu(II) and Al(III) are superior to Zn(II) in acidity. Why have organisms not adopted them for their needs (i.e., acid-base catalyst)?

3. A ligand X in [ML$_5$X] has a high formation constant (large K value for [ML$_5$] + X $\longrightarrow$ [ML$_5$X]) but can readily be substituted by another ligand Y (assume that the formation constant of [ML$_5$Y] is about the same as that of [ML$_5$X]). Can these apparently opposing facts (i.e., high formation constant and being readily substituted) be compatible? Find some examples, and explain your answer.

4. The water exchange rate of [Cr^{III}(H$_2$O)$_6$]$^{3+}$ is extremely slow (inert) compared to [Fe^{III}(H$_2$O)$_6$]$^{3+}$ for example. Show your understanding of why.

5. Draw your understanding of the reaction mechanism of Rubisco in detail. Does it make sense?

6. As seen in the text, Mg(II) in many Mg(II)-dependent enzyme reactions can be substituted by Mn(II) *in vitro* but often *in vivo* as well. Discuss the reasons for the fact.

7. The reaction catalyzed by alcohol dehydrogenase is a dehyrogenation, an oxidation reaction. Why is such an oxidation-reduction reaction catalyzed by an acid catalyst Zn(II)?

8. Draw the reaction mechanism for arginase reaction. How might Mn(II) ion(s) intervene in the reaction?

PROBLEMS TO EXPLORE

1. Mg(II) is not a very strong acid (see Fig. 4.1.), but is used as an acid catalyst in a wide variety of enzymatic reactions. Discuss why.

2. Some peptidases (proteases) depend on Zn(II) whereas a large number of peptidases do not require metal cations (i.e., can function with only some amino acid residues). Discuss why certain peptidases require Zn(II) but others do not.

3. Some literature indicated that Co(II) can activate arginase in place of Mn(II). Find the literature, and discuss why.

4. Explore PDB for glyoxalase structures with various divalent cations. Discuss the structural basis for effectiveness of various cations, particularly Ni(II) *vs* Zn(II).

5. Alkaline phosphatase uses Zn(II) as its catalyst, whereas acid phosphatase uses Fe(III) (and Zn(II) in some) for its catalyst. Discuss why.

6. Cations such as Ca(II) often are used to change a conformation of a protein by binding to its specific site. How can a conformation change be brought about? What about anion instead of cation? Any example?

7. When catalytic activities of RNA were discovered, it was a revelation. In other words, a dogma had been established by then that enzymes had to be proteins. However, RNAs have residues similar to those in proteins, such as basic N and OH. This suggests that some catalytic activities may be expected with RNAs, which turned out to be true. How about DNA?

Reactions of Oxidation–Reduction Type Including Electron Transfer Processes

5.1. **GENERAL CONSIDERATION**

Several different types of enzymatic reactions are grouped as oxidation–reduction reactions. They include:

(a) Reactions of oxidoreductases, in which the oxidizing agent is typically O_2

(b) Reactions of dehydrogenases, in which a hydride (or $2e/2\,H^+$) is accepted by NAD^+ (or FAD)

(c) Oxidative decomposition of water; this can be an oxidase

(d) Reactions of hydroperoxidases and catalase in which hydroperoxide or hydrogen peroxide is the oxidizing agent

(e) Reactions of oxygenases (mono- and di-oxygenases) in which the O-atom(s) of O_2 is (are) incorporated in the products

(f) Ni-containing hydrogenase and other hydrogenases

Oxygenases and several reductases such as ribonucleotide reductase and nitrogenase are excluded from the present chapter, as they are discussed in Chapters 6, 7, and 8, respectively.

Some of the oxidoreductases contain in themselves an oxidation site and a reduction site, and electrons are transferred between them. There are also other electron transfer processes that do not involve chemical reactions. The examples are the electron transport process involved in the mitochondrial oxidative phosphorylation and the electron transport process connecting photosystem I and photosystem II in chloroplast. These processes involve oxidation–reduction of the components, and hence will be included in the current chapter.

5.1.1. **Reduction Potential**

Oxidoreductases catalyze reactions of the following type:

$$S_1 \text{ (substrate 1)} + M(+n) \rightarrow S_1{}' \text{ (oxidized } S_1) + M(+n-m)$$

$$S_2 \text{ (substrate 2, often } O_2) + M(+n-m) \rightarrow S_2{}' \text{ (reduced } S_2) + M(+n)$$

For this set of reactions to take place at least thermodynamically, the reduction potential of the $M(+n)/M(+n-m)$ needs to be between those of S_2 and S_1 entities; that is, reduction potential $E(S_2) \geqslant E(M) \geqslant E(S_1{}')$. The second step may not necessarily involve S_2 directly as indicated. S_2 may oxidize yet another M' entity, and then oxidized M' oxidize $M(+n-m)$ back to $M(+n)$.

In the case of simple electron transfer process, electron(s) will move from a site of lower reduction potential to another of higher reduction potential. Regarded as an enzyme, an electron transfer agent such as cytochrome c acts on the substrate electron. Hence, the reduction potential, how it is regulated and modified, is of paramount importance.

The basic concepts of reduction potential are discussed in Chapter 2. Here we will see the reduction potential data of metalloenzymes and proteins, and those of important bioorganic/inorganic compounds. Such a set of data is shown in Figure 5.1. Figure 5.2 lists the reduction potential values of the components in the mitochondrial electron transfer system, and shows that the reduction potential is the major factor in electron transfer.

5.1.1.1. *Heme Proteins and Enzymes*

A large number of heme-containing enzymes and proteins are found in all the biological systems. Their reduction potentials vary from about $-0.4\,V$ to about $+0.5\,V$. They are modified (through evolution) so as to be commensurate with their functions. How is this done chemically?

The structures of representative porphyrin groups are shown in Figure 5.3. Heme A constitutes proteins of cytochrome a type, and its reduction potential is in the highest range. Heme C is found in proteins of cytochrome c type, whose reduction potentials range from about 0 to $+0.3\,V$. Protoporphyrin-heme is the heme group found in type b cytochrome, hemoglobin, cytochrome P-450, peroxidases, and catalase, and covers the lower range.

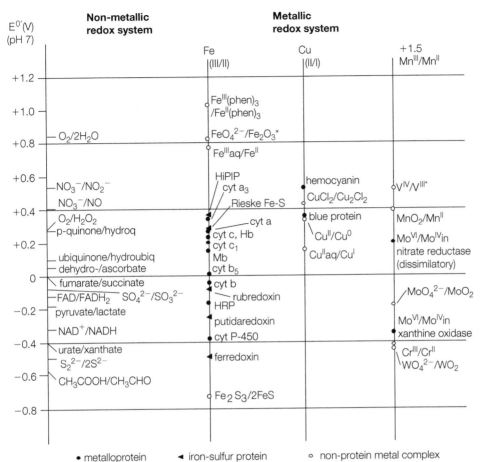

■ **Figure 5.1.**
Reduction potentials (at pH 7) of relevant systems.

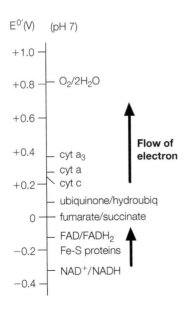

■ **Figure 5.2.** The flow of electrons in the respiratory systems and the reduction potentials of the components involved.

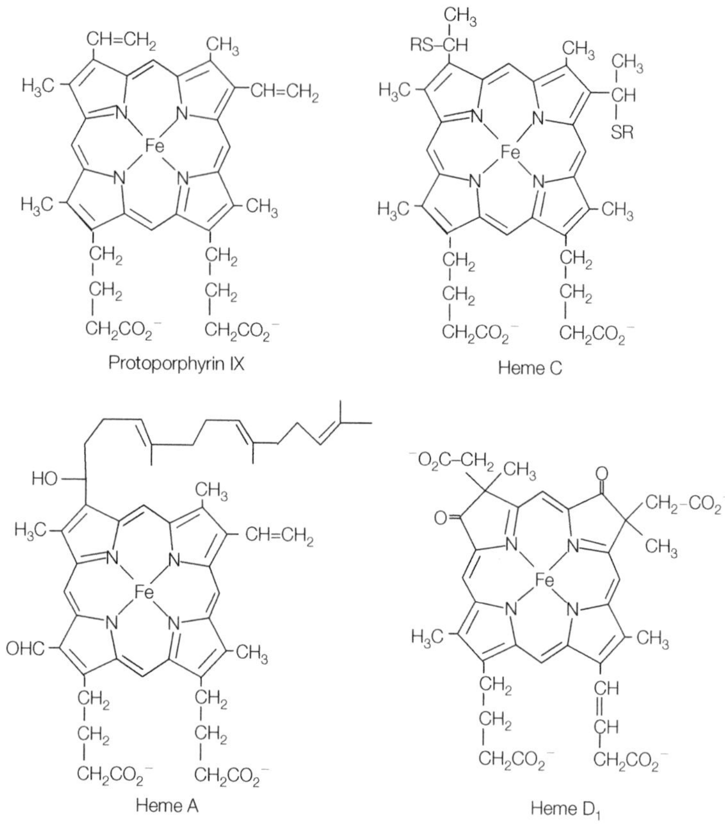

■ **Figure 5.3.** Major hemes (Fe-porphyrins).

Protoporphyrin IX

Heme C

Heme A

Heme D₁

The reduction potential is affected by a number of structural and chemical factors:

(a) Peripheral groups

(b) The presence and the nature of the fifth and/or sixth ligand(s)

(c) The position of Fe-atom

(d) The amino acid residues present in the vicinity of the heme group

One of the reasons for a high reduction potential of cytochrome a and a_3 is the presence of an aldehyde group on the periphery. The aldehyde group, being an electron-withdrawing one, tends to stabilize the lower oxidation state by spreading more extensively the extra electron deposited on the Fe-atom (effect (a)). However, even proteins with the same protoporphyrin IX have a wide range of reduction potentials, from low ($-0.42\,$V of catalase) to high ($+0.17\,$V of hemo-globin). Factors (b) through (d) must be operating.

A component in an electron transfer process does not require a chemical substrate binding site; hence the heme of the electron transfer agent is usually fully occupied; fifth and sixth coordination sites are coordinated by ligands. In the case of cytochrome c, the fifth and sixth ligands are a histidine N and the S of the thioether in a methionine. In this structure, the Fe-atom usually is located at the center of the porphyrin ring and subjects to a strong ligand field effect. Hence the Fe is in the lower-spin state in both Fe(III) and Fe(II). This raises the reduction potential, because the extra electron added (upon reduction) would not significantly affect the entire system ((b) and (c) effect). In the case of cytochrome b (bovine), the two ligands are histidine nitrogen atoms, and the two ligands in cytochrome f are histidine and NH of a peptide bond.

Hemes that are involved in chemical reactions usually have the fifth coordination site occupied, but the sixth unoccupied or occupied with a water molecule that can readily be replaced by a substrate. For example, Fe(II) in hemoglobin has a histidine at the fifth site, but none in the sixth site. In addition, the Fe atom in this state is not in the porphyrin plane; it is drawn toward the fifth ligand. Under this condition, the ligand field on the Fe(II) is relatively weak, and hence the Fe(II) is in a high-spin state (total spin $S = 4$). This will destabilize the Fe(II) state, and thus tends to reduce the reduction potential ((c) effect). This mild destabilization of Fe(II) is necessary for the function of hemoglobin.

Negatively charged ligands such as S^- (of cysteine), O^- (of tyrosine), and N^- (of histidine) will bind more strongly with Fe of higher oxidation state. Thus, binding of one of these ligands to the Fe of heme will result in lower reduction potential, as seen for cyt P-450 (the fifth ligand is S^-), catalase (O^-), and horseradish peroxidase (N^-) in Figure 5.1. This in turn means that the higher oxidation state(s) is stabilized in these enzymes, and high oxidation states such as Fe(IV) and Fe(V) seem to become accessible in these proteins ((b) effect).

5.1.1.2. *Iron-Sulfur Proteins*

Several different types of iron-sulfur clusters have been identified in iron-sulfur proteins (Beinert *et al.*, 1997). Important ones are:

- Rubredoxin [Fe(cys)$_4$]

- Green plant ferredoxin [Fe$_2$S$_2$(cys)$_4$]

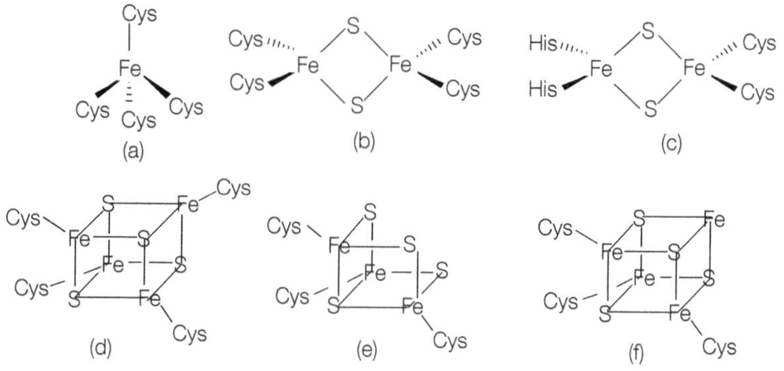

■ **Figure 5.4.** Major types of iron-sulfur clusters.

■ **Figure 5.4.** Major types of iron-sulfur clusters.

- Rieske type $[Fe_2S_2(cys)_2(his)_2]$
- Bacterial ferredoxin and other electron transfer agents $[Fe_4S_4(cys)_4]$
- Nonfunctional $[Fe_3S_4]$
- Nonelectron transfer type $[Fe_4S_4(cys)_3]$
- Part of FeMo-cluster of nitrogenase

These are sketched in Figure 5.4 (except for the FeMo, which will be discussed in Chapter 8).

In these clusters, the bridging S is nominally S^{2-}, which, along with the cysteinyl S^-, stabilizes the higher oxidation state of iron, Fe(III). Hence the iron-sulfur clusters typically have fairly low reduction potential (see Fig. 5.1). This is in agreement with the very low reduction potential of mineral $Fe_2S_3/2FeS$. An exception is a protein called HiPIP (high-potential iron sulfur protein), which will be discussed later. Rubredoxin $[Fe(cys)_4]$ has only cysteinyl S^-, and its reduction potential is relatively high. In the Rieske cluster, the two cysteinyl S^-s are replaced by neutral histidine Ns. A result of this replacement is an increase in the reduction potential (see Fig. 5.1).

The dinuclear cluster $[Fe_2S_2]$ can take three oxidation states: nominally $[Fe^{III}_2S_2]^{2+}$, $[Fe^{III}Fe^{II}S_2]^{1+}$, and $[Fe^{II}_2S_2]^0$. Under the physiological conditions, this cluster works with the pair of the first and the second states, $Fe^{III}Fe^{III}/Fe^{III}Fe^{II}$. In addition to the fact that S^{2-} stabilizes Fe^{III} (more so than Fe^{II}), an electronic factor seems to contribute to decreasing the reduction potential. The ground state of $[Fe^{III}Fe^{II}S_2]^{1+}$ has been found to be in $S = 1/2$ ($S = $ total spin), suggesting an antiferromagnetic coupling between the two Fes (Beinert *et al.*, 1997; Noodleman *et al.*, 1995). A bonding molecular orbital formed would accommodate a pair of antiparallel electrons in Fe^{III}_2 state. When an extra electron is added to this system, it has to be accommodated in the corresponding antibonding orbital (likely that between d_z2; see Fig. 2 in Noodleman *et al.*, 1995). This will destabilize the reduced state, and is manifested in a reduced reduction potential. And a further reduction (to $[Fe^{II}_2S_2]$ state) is still harder.

The $[Fe_4S_4]$ cluster in bacterial ferredoxins and other $[Fe_4S_4]$ clusters operates between $[Fe_4S_4]^{2+}$ and $[Fe_4S_4]^{1+}$; that is, $Fe^{III}_2(Fe^{II}_2)/Fe^{III}Fe^{II}(Fe^{II}_2)$ under usual conditions. Hence the reduction potential of this pair is low, similar to that of the plant type $[Fe_2S_2]$ mentioned earlier. Unlike $[Fe_2S_2]$, however, the valence electrons are delocalized on the entire cluster in the case of $[Fe_4S_4]^{2+}$, so that the oxidation state of each Fe may be expressed as "$+2.5$", or $[Fe^{2.5}_4S_4]^{2+}$ (Beinert *et al.*, 1997; Noodleman *et al.*, 1995). In the reduced state $[Fe_4S_4]^{1+}$, one pair is $(Fe^{2.5})_2$, and the other pair is $(Fe^{II})_2$.

The so-called HiPIP protein is a small soluble electron transport protein and has been isolated from photosynthetic and nonphotosynthetic bacteria such as *Chromatium*, *Rubrivivax gelatinosus*, *Rhodoferax fermentans*, and *Rhodocylus tenius* (Lieutaud *et al.*, 2003). It is also found in an acidophilic bacterium, *Acidithiobacillus ferroxidans*, which functions in a very low pH medium (Nouailler *et al.*, 2006). They have unusually high reduction potentials (around $+0.35\,V$) and yet their $[Fe_4S_4]$ clusters are very similar to those of the bacterial ferredoxins. It turned out that this protein operates with $[Fe_4S_4]^{3+}$ and $[Fe_4S_4]^{2+}$; nominally $Fe^{III}_3Fe^{II}/Fe^{III}_2Fe^{II}_2$ instead of the ordinary couple $[Fe_4S_4]^{2+}/[Fe_4S_4]^{1+}$. In view of this fact, the stability of different oxidation states seems to be $[Fe_4S_4]^{3+} < [Fe_4S_4]^{2+} > [Fe_4S_4]^{1+}$. A first question would be, "Why this order?" The next question might be why HiPIP alone uses the pair of $[Fe_4S_4]^{3+}/[Fe_4S_4]^{2+}$ rather than the usual pair of $[Fe_4S_4]^{2+}/[Fe_4S_4]^{1+}$.

Let us make a very simple molecular orbital argument, assuming that four atomic orbitals (one each from each Fe) interact equally for simplicity sake. This might be justified as all four Fes have been found to be equivalent in $[Fe_4S_4]^{2+}$ (Beinert *et al.*, 1997). A more sophisticated treatment is found in a review by Noodleman *et al.* (1995). A simple molecular orbital calculation indicates that four molecular orbitals arise from such an interaction resulting in a bonding orbital (with energy $\alpha + 3\beta$) and three equivalent antibonding orbitals (with energy $\alpha - \beta$), where the α represents the coulomb energy of the atomic orbital and β the exchange energy, and both are of negative values.

Assuming that each Fe(II) contributes a single electron to these molecular orbitals, the energy of each oxidation state (in addition to the energy due to the other interactions common to all the different oxidation states) would be as follows: $(\alpha + 3\beta)$ for $Fe^{III}_3Fe^{II}$ $[Fe_4S_4]^{3+}$, $(2\alpha + 6\beta)$ for $Fe^{III}_2Fe^{II}_2$ $[Fe_4S_4]^{2+}$, $(3\alpha + 5\beta)$ for $Fe^{III}Fe^{II}_3$ $[Fe_4S_4]^{1+}$, and $(4\alpha + 4\beta)$ for Fe^{II}_4 $[Fe_4S_4]^0$. Therefore, the energy change from $Fe^{III}_3Fe^{II}$ to $Fe^{III}_2Fe^{II}_2$ is $(\alpha + 3\beta)$, whereas that from $Fe^{III}_2Fe^{II}_2$ to $Fe^{III}Fe^{II}_3$ would be $(\alpha - \beta)$. The former energy value governs the reduction potential of HiPIP and the latter that of bacterial ferredoxins. The former energy value is negative, resulting in a positive reduction potential, whereas the latter value is much higher (by $4|\beta|$) than the former (and could be positive) and would result in a much lower reduction potential than that of the former pair (perhaps negative reduction potential). It is also to be noted that $[Fe_4S_4]^{2+}$ state is the lowest in energy if $|\beta| > |\alpha|$. This assumption is consistent with a recent experimental result (Dey *et al.*, 2004). This is an answer to the first question. The reality is much more complicated with a nonregular cubic structure, but the basic difference between bacterial ferredoxin and HiPIP is consistent with this argument.

The second question raises the issue of accessibility of the different oxidation states and is related to the specific functions. It has been shown that HiPIP in a purple bacterium *Rubrivivax gelatinosus* serves as electron donor to the photosynthetic reaction center (Lieutaud *et al.*, 2003). It is a replacement for cytochrome c_2 in the species that lacks cyt c_2. Cyt c_2 has a reduction potential of about $+0.34\,V$, which is comparable to that of HiPIP. These entities, cyt c_2 and HiPIP, are not good reducing agents as their reduction potentials are high, and yet function as such. This is because the photo-excited electron-deficient chlorophyll center in the photoreaction center has a higher reduction potential $(+0.4 \sim +1.0\,V)$. It is feasible, therefore, to function as a reductant only if the HiPIP can bind specifically to the photoreaction center. This indeed has been shown to be the case (Lieutaud *et al.*, 2003).

5.1.1.3. *Copper Proteins*

Several different types of copper centers are present in copper enzymes and proteins. One is typically responsible for the characteristic blue color; proteins containing such a center are called blue copper proteins. Many copper electron transfer agents are of this type. The second type is an ordinary coordination complex, mostly coordinated by histidine nitrogens, and does not show the characteristic blue—called nonblue copper. The third is a dinuclear one, which binds an O_2 molecule. Hemocyanin is a typical example that contains a dinuclear center. The second and the third type seem to combine to form a trinuclear center as well in some proteins. Blue copper oxidases contain all the three types of copper center. All copper centers operate as a Cu^{II}/Cu^{I} redox pair, and this process usually is sufficiently facile, hence copper enzymes do not require other electron transfer agents such as iron-sulfur clusters. The copper electron transfer agents are blue copper, which usually takes structures close to tetrahedron rather than square planar. This structure stabilizes the Cu(I) state, resulting in an increase of reduction potential. This also has an implication for the facileness of the oxidation and reduction of such a center (entatic state). Copper proteins and enzymes emerged more recently in the evolution of life (see Chapter 1), and as such operate typically in a relatively high reduction potential range as seen in Figure 5.1.

5.1.1.4. *Molybdenum and Tungsten Proteins*

The catalytic centers of these enzymes are either Mo or W. Mo or W in these enzymes typically operate in the redox pair Mo^{VI}/Mo^{IV} (or W^{VI}/W^{IV}). The reduction potential of this pair seems to be compatible with the reduction potentials of the substrates. For example, xanthine (or aldehyde) is the reductant and O_2 is the oxidant in the reaction catalyzed by xanthine oxidase (or aldehyde oxidase). The critical redox center, Mo, thus has a redox potential between those of the reductant and the oxidant (see Fig. 5.1). In the case of DMSO reductase and (dissimilatory) nitrate reductase, the oxidants are either DMSO and nitrate, respectively, and the reductant is the respiratory hydroquinone. Hence the reduction potential of the Mo comes between the reduction potentials of these compounds as seen in Figure 5.1.

The reduction potential of W^{VI}/W^{IV} is lower than the corresponding Mo pair, as seen in Figure 5.1. Indeed the reduction potential of W-enzymes seems to cover a range lower than that of Mo.

The rate of the change of oxidation states between VI and IV in these entities is fast enough to perform the catalytic function, mostly transfer of an O atom, which involves a two-electron process. Such a two-electron process may not easily be accommodated by the regular one-by-one electron transfer mechanism. Hence molybdenum enzymes usually contain iron-sulfur cluster(s) (and other agents such as FAD) to help transfer electrons from one substrate to another (in one-by-one fashion).

5.1.2. **Kinetic Factors—Electron Transfer between and in Protein(s)**

The electron transfer between chemical entities involve essentially two processes. One is the change of the structure/conformation in the chemical species involved, so that an electron transfer can take place at reasonable speed at the origin and the destination. The second is the movement of electrons through a medium between the origin and the destination.

The first factor is, for example, the energy required to change the original Cu(I) state to prepare for the electron transfer, in which it is turned into Cu(II). The stable structure of Cu(I) is typically tetrahedral whereas that of Cu(II) is square planar. Therefore, if the Cu(I) center is tetrahedral, the resulting Cu(II) will not be very stable without structural change. Because of this factor, the Cu(I) center of tetrahedral structure will be reluctant to release the electron, slowing the electron transfer process. A solution is to make the structure of such a center compatible with both oxidation states (I) and (II). Indeed an intermediate structure between tetrahedron and square plane is chosen in most of electron-transfer blue-copper proteins. This kind of structure was coined *entatic state* (Vallee and Williams, 1968). Even then some degree of reorganization needs to take place for the electron transfer; this energy is called reorganization energy λ.

As seen before, electron-transferring cytochromes usually take six-coordinate, low-spin structures. With this structure, the difference in the overall energy of two oxidation states (II) and (III) is relatively small, because the reorganization energy is minimal. As a result their electron transfer processes would be fast. The same comment can be applied to the situation of most of iron-sulfur proteins. In other words, the structural change in iron-sulfur clusters is relatively small when the oxidation states of Fe's change, though the interatomic distances increase slightly with reduction. This in turn guarantees fast electron transfer, aside from the long distance transport in or between proteins.

Taking account of the second factor (long range electron movement), the overall electron transfer rate constant k can be, to a first order approximation, given by the following relationship (Page *et al.*, 2003):

$$\log k = 13 - 0.6(R - 3.6) - 3.1(\Delta G + \lambda)^2/\lambda$$

where k is in units of s^{-1}, R is the edge-to-edge distance in Å, ΔG is the driving force in eV (the free energy difference between the two centers). The electron is supposed to be transferred through a tunneling mechanism between the edge of a donor site (e.g., the periphery of a cytochrome) to the edge of the acceptor site. The second term reflects the fact that the tunneling rate depends on the spread of wave function that decays exponentially with distance. The third term represents an activation free energy and arises from simple assumptions that the potential curves of the reactant system and the product system are simple parabolas, and also that the transition is a kind of Franck-Condon type (Marcus, 1964). Theoretically, the maximum rate will be obtained when $\Delta G = -\lambda$; that is, the driving force matches the reorganization energy. The edge-to-edge distance in many proteins surveyed range mostly from van der Waals contact distance of $\sim 3\,\text{Å}$ to about $15\,\text{Å}$, though distances of as large as $28\,\text{Å}$ are known in some (Page *et al.*, 2003). Bioevolution can control the electron transfer rate by modifying R, ΔG, and λ values so as to optimize the electron transfer process. However, natural selection appears to have brought about inherently robust frameworks, so that the system had become tolerant to sequential changes (Page *et al.*, 2003).

5.2. **IRON ENZYMES AND PROTEINS**

Many iron-containing proteins and enzymes deal with O_2. They either carry O_2 as in the case of myoglobin, hemoglobin, and hemerythrin, or incorporate one or two of the O_2 in the product, as in the case of cytochrome P-450 dependent monooxygenases, pyrocatechase, or methane monooxygenase. These proteins and enzymes are dealt with in the next chapter.

The only iron-containing enzyme that uses O_2 as the electron acceptor is cytochrome c oxidase, which contains cytochrome a and a_3 (see earlier). Catalase, hydroperoxidase, and cytochrome c peroxidase use HOOH or ROOH as the oxidizing agent. Several types of nitrite reductase are known; among them are multiheme c enzymes, and heme cd_1 enzyme. The Fe of heme in these nitrite reductases binds and reduces NO_2^-. The enzymes of this last category are to be discussed here. A nitric oxide (NO) reductase also is a heme-enzyme.

The other widespread use of iron is to transfer electron(s); they function as such an agent either in the form of cytochromes or iron-sulfur proteins. They were discussed in terms of reduction potential earlier. A few other types of iron-sulfur are known, particularly in association with hydrogenase and nitrogenase. Nitrogenase will be discussed in Chapter 8.

5.2.1. **Cytochromes and Iron-Sulfur Electron Transfer Proteins**

The major function of these proteins is one-electron transfer; particularly cytochromes cs and cytochrome bs. Issues relevant to the functions of these proteins are reduction potential and kinetic factors, which were discussed earlier.

As electron transfer agents, they constitute parts of many enzymes, as will be seen in the subsequent sections. Fumarate reductase is used in microbial anaerobic growth, and succinate dehydrogenase (which catalyzes the reverse of fumarate reductase reaction) is used in aerobic growth. Both contain $[Fe_2S_2]$ as well as $[Fe_4S_4]$ units and FAD-proteins, and heme bs are also used as electron transfer agents. The reducing agent is usually NADH. As seen in Figure 5.1, these entities (fumarate/succinate, $FAD/FADH_2$, $NAD^+/NADH$) and the electron transfer agents all have commensurate reduction potentials. FAD is believed to be the catalytic site.

5.2.2. **Nitrite Reductase and Nitric Oxide Reductase**

One kind of nitrite reductases (assimilatory), that from *Desulfovibrio desulfricans* and others, consists of two identical subunits. Each subunit contains five heme groups (c-type), spaced closely together (Moura and Moura, 2001; Averill, 1996). This enzyme carries out the reaction:

$$NO_2^- + 7H^+ + 6e \rightarrow NH_3 + 2H_2O$$

The reductant is NAD(P)H. The Fe in one of the hemes is in a high-spin state, suggesting that it is five-coordinate, and provides the binding site of the substrate, NO_2^-.

Another nitrite reductase (dissimilatory or respiratory)—for example, from *Pseudomonas aeruginosa* or *Paracoccus denitrificans*—contains heme c and heme d_1. The structure of heme d_1 is significantly different from other hemes a, b, and c-type, and is shown in Figure 5.3. The Fe(II) of heme d_1 binds first NO_2^-, and reduces it to NO, as in

$$(\text{Heme } d_1)Fe^{II}-NO_2^- + 2H^+ - H_2O \rightarrow (\text{Heme } d_1)Fe^{II}-|N{=}O^+| \leftrightarrow$$
$$(\text{Heme } d_1)Fe^{III}-\cdot NO \rightarrow (\text{Heme } d_1)Fe^{III} + \cdot NO$$

An electron is then transferred through a heme c from an electron source (for example, ascorbate) restoring Fe(II) state of heme d_1.

The reduction of NO to N_2O (dissimilatory process) is carried out by a heme-enzyme nitric oxide reductase. The reaction involves a sort of dimerization:

$$2NO + 2H^+ + 2e \rightarrow N_2O + H_2O$$

The catalytic site is a single Fe-heme, which is of P-450 type, and the enzyme structure containing the NO bound to P-450 has been determined (PDB 1CL6, Shimizu *et al.*, 2000). It shows that NO binds to the Fe in the form of Fe^{III}—NO^- (bent Fe—N—O). It has been hypothesized that Fe—NO as in Fe—OO of P-450 (see Chapter 6) turns to Fe—N (equivalent of ferryl) (see Hille, 1996). Then another NO reacts with this Fe—N entity to form N_2O.

5.2.3. Horseradish Peroxidase (HRP), Catalase, and Cytochrome C Peroxidase

HRP catalyzes reactions of the following type:

$$ROOH \text{ (hydroperoxide)} + AH_2 \rightarrow ROH + A + H_2O$$

R can be H, that is hydrogen peroxide. A variety of substrates can be oxidized by this enzyme; such substrates include pyrogallol, *p*-cresol, aniline, *p*-toluidine, ascorbic acid, NADH, and dihydroxyfumarate. Catalase catalyzes similar reactions, acting on similar substrates, but its main function is decomposition of H_2O_2. In this reaction both ROOH and AH_2 are HOOH. The reaction sequence for HRP and catalase has been fairly well understood. As seen in Figure 5.5, ROOH first oxidizes Fe^{III}-Porphyrin to compound-I (CpI), which looks green. That is:

$$ROOH + Fe^{III}—P \rightarrow ROO^-—Fe^{III}—P + H^+ \rightarrow RO\cdot + O{=}Fe^{IV}—P + H^+$$

$$RO\cdot + O{=}Fe^{IV}—P + H^+ \rightarrow ROH + O{=}Fe^{IV}—P\cdot^+ (CpI)$$

$P\cdot^+$ is a porphyrin π radical in the case of HRP and catalase. The H of ROOH is first removed by an H^+-accepting amino acid residue, and then ROO^- binds to Fe^{III}. One electron oxidation of Fe^{III} by ROO^-

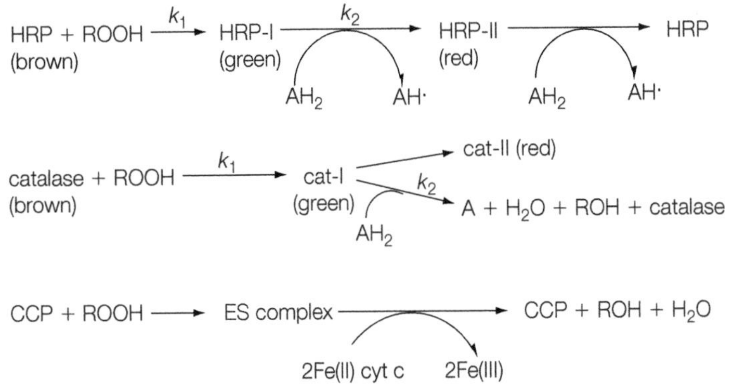

■ Figure 5.5. The reaction mechanisms of HRP (horseradish peroxidase), catalase, and CCP (cytochrome c peroxidase).

leads to the formation of the ferryl ion, nominally $Fe^{IV}=O^{-II}$ and of $RO\cdot$. The latter then abstracts a hydrogen atom (or a single electron) from the porphyrin entity, forming $P^{\cdot+}$, HRP-I, or cat-I (CpI in general), then reacts with an AH_2, abstracting a hydrogen atom, thereby turning to HRP-II or cat-II. This state is believed to be $O=F^{IV}-P$. This intermediate further reacts with another AH_2 or $AH^{\cdot}$, returning the enzyme back to the initial state $F^{III}-P$ and H_2O. This suggests that the O in $O=Fe^{IV}-P$ should be described as $\cdot O^{-}(-Fe^{III})$. This seems to be consistent with a quantum chemical result by Solomon and his coworkers (Decker and Solomon, 2005).

The electronic state of $P^{\cdot+}$ is slightly different between cat-I and HRP-I (Dolphin and Felton, 1974). This difference is reflected in their absorption spectra and also in the slightly different behaviors as seen in Figure 5.5. Besides, HRP cannot decompose hydrogen peroxide; this implies that $O=Fe^{IV}-P^{\cdot+}$ (HRP-I) cannot react with H_2O_2. On the contrary, the reaction of $O=Fe^{IV}-P^{\cdot+}$ with H_2O_2 is the major event in the case of catalase. This could be due to the difference in the electron distribution in $P^{\cdot+}$ ring between HRP and catalase.

In the case of cytochrome c peroxidase, the ES complex that corresponds to HRP-I or cat-I does not show an absorption spectrum similar to those of HRP or cat-I. The ES complex consists of $Fe^{IV}=O$ and an amino acid radical. In other words, the $RO^{\cdot}$ produced extracts a hydrogen atom from a tryptophan residue nearby rather than the porphyrin, forming an indoyl cation radical (Sivaraja et al., 1989).

A crucial issue in these enzymes is how and why a higher oxidation state Fe(IV) is involved. Under ordinary circumstances, it is rather difficult to oxidize Fe beyond the Fe(III) state. The fifth ligand for the Fe of these heme enzymes is all negatively charged; tyrosine's O^{-} in catalase and N^{-} of histidine in horseradish peroxidase and cytochrome c peroxidase. These negatively charged ligands must sufficiently stabilize the Fe(IV) oxidation state.

The O^{-} of tyrosine and the N^{-} of histidine are "hard" bases, whereas the S^{-} of cysteine is a soft base. A cysteinyl thiolyl negative ion (S^{-}) is the fifth ligand in the case of cytochrome P-450; and here the ferryl O behaves differently. Chloride peroxidase and nitric oxide reductase (see earlier) also have a S^{-} of cysteine at their fifth coordination site. These enzymes will be discussed in the next chapter.

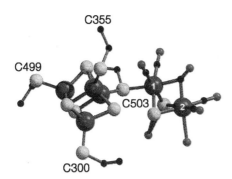

C355

C499

C503

C300

1

2

■ **Figure 5.6.** The active site of the Fe-only hydrogenase from *Desulfovibrio desulfuricans*. (PDB 1HFE: Nicolet, Y., Piras, C., Legrand, P., Hatchikian, C.E., Fontecilla-Camps, J.C. 1999. *Desulfovibrio desulfuricans* iron hydrogenase: The structure shows unusual coordination to an active site Fe binuclear center. *Structure Fold. Des.* **7**, 13–23.)

5.2.4. **Hydrogenase**

The structure of the hydrogenase from *Clostridium pasteurianum* was determined by x-ray crystallography (Peters *et al.*, 1998). The active site consists of two iron clusters combined (known as H-cluster) (see Fig. 5.6). A $[Fe_4S_4]$ cluster is bound to the special dinuclear iron cluster through a cysteinyl thiol. The unusual dinuclear cluster has two S^{2-}s and one CO bridging the Fe atoms. In addition, the Fe that is connected to the $[Fe_4S_4]$ cluster has one each of CO and CN^- and the other Fe has CO, CN^-, and one H_2O as the ligands.

This enzyme usually reduces $2H^+$ to H_2. It has been proposed (Peters *et al.*, 1998) that the proton is provided by a nearby free cysteine residue. It is more likely that the water molecule that is bound to the Fe will obtain a proton from a nearby proton source and the proton now on H_3O^+ is reduced by the Fe. In this process the water molecule will come off. This is to occur when the Fe is in a low oxidation state, perhaps Fe(I) or Fe(0). This low oxidation state is made accessible because of the presence of π-acid ligands such as CO and CN^-. The reaction is an example of oxidative addition; e.g., Fe(0) + $H^+\longrightarrow$ $Fe^{II}(H^-)$. Oxidative addition is discussed in Chapter 2. The hydride captures another proton to form H_2. The oxidized Fe (Fe(II) in Fig. 5.6) can then be reduced through the attached iron-sulfur cluster. The mechanism of the reverse reaction is discussed in conjunction with the Ni-Fe enzyme. A further discussion of the mechanism of these intriguing enzymes is found in Armstrong (2004).

5.3. **COPPER ENZYMES AND PROTEINS**

Representative copper proteins and enzymes are given in Table 5.1. A variety of different types of copper sites have been found. Some prominent ones will be discussed here.

Table 5.1. Representative Copper Enzymes and Proteins

Category	Enzyme/Protein	Function
Blue copper protein	plastocyanin	electron transfer
	stellacyanin	electron transfer
	azurin	electron transfer
Blue copper enzyme	ascorbate oxidase	oxidation of ascorbate
	laccase	oxidation of catechols
	ceruloplasmin	oxidation of Fe(II)
	nitrite reductase	$NO_2^- \rightarrow NO$
Nonblue copper enzyme	cytochrome c oxidase	
	amine oxidase	oxidation of amine
	superoxide dismutase	$2H_2O_2 \rightarrow 2H_2O + O_2$
	nitrous oxide reductase	$N_2O \rightarrow N_2$
	catechol oxidase (tyrosinase)	
	D-galactose oxidase	

■ **Figure 5.7.** The coordination structure of Cu in stellacyanin. (PDB 1JER: Hart, P.J., Nersissian, A.M., Herrmann, R.G., Nalbandyan, R.M., Valentine, J.S., Eisenberg, D. 1996. A missing link in cupredoxins: Crystal structure of cucumber stellacyanin at 1.6 Å resolution. *Protein Sci.* **5**, 2175–2183.)

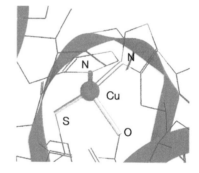

5.3.1. **Blue Copper Proteins**

Blue copper proteins are electron transfer agents characterized by deep blue color. Plastocyanin and stellacyanin both have four-coordinate Cu as the electron transfer center; two histidine Ns, cysteine S, and methionine S are the ligands in the case of plasto-cyanin, and those for stellacyanin are two histidine Ns, cysteine S, and serine O (see Fig. 5.7). This type of Cu is designated as Type 1. The coordinating structures are not quite tetrahedral, but not square planar either. This structure is favorable for electron transfer, that is, change in the oxidation state (Cu^{II}/Cu^{I}), as discussed earlier. The

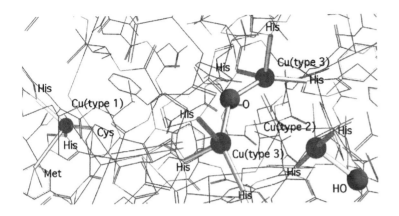

■ **Figure 5.8.** The catalytic site of ascorbate oxidase. The enzyme consists of two subunits, each of which contains a blue Cu(Cu(type 1)), a dinuclear site (type 3)and a single nonblue site (type 2), but these latter three (2Cu(3) and Cu(2)) are close together and likely work together; the two subunits are bound together by a single Cu ion that binds a histidine from each subunit. (PDB 1AOZ: Messerschmidt, A., Ladenstein, R., Huber, R., Bolognesi, M., Avigliano, L., Petruzzelli, R., Rossi, A., Finazzi-Agro, A. 1992. Refined crystal structure of ascorbate oxidase at 1.9 Å resolution. *J. Mol. Biol.* **224**, 179–205.)

blue color seems to be due to a charge transfer absorption between Cu^{II} and cysteinyl thiol.

5.3.2. **Blue Copper Oxidases**

The reactions catalyzed by ascorbate oxidase and laccase are:

Blue copper oxidases typically contain a blue Cu unit (type 1) in addition to one nonblue Cu (type 2) and a dinuclear Cu unit (type 3), as illustrated for ascorbate oxidase (see Fig. 5.8). The latter three Cus are close to each other, almost forming a trinuclear unit. The dinuclear Cu is believed to be the binding site for O_2. Laccase contains almost the same set of four Cu atoms. The details of biophysical studies on these enzymes and other multicopper enzymes are found in the article by Solomon *et al.* (1996).

In both cases, a single electron at a time is removed from the substrate to the blue copper center, forming an intermediate free radical species. The blue copper center swiftly transfers away the electron turning back to Cu(II) state. Another electron then is removed from the intermediate free radical to the same site resulting in the dehydrogenated product. An alternative pathway is that the intermediate disproportionates into ascorbate and dehydroascorbate. Whichever the case may be, the result is the same. This reaction is repeated, and altogether four electrons are deposited onto the four Cu atoms in the system. The dinuclear $[Cu^I_2]$ site then binds O_2, and two more electrons stored on the other two Cu locations are transferred to the O_2 moiety, reducing O_2 to $2H_2O$. Another blue oxidase, ceruloplasmin (also known as ferroxidase), contains the same sets of copper centers.

5.3.3. **Cytochrome C Oxidase**

Cytochrome c oxidase is the last and most oxidizing enzyme in the mitochondrial electron transfer cascade; four cytochrome cs are oxidized by O_2, and O_2 is reduced to $2H_2O$. This enzyme contains two hemes (a and a_3), and one single Cu_b and a dinuclear Cus center (Cu_a). Cu_b is associated with the Fe of heme a_3, and an O_2 molecule is to bind between them forming a peroxide bridge; this is the function of the dinuclear unit in the blue copper enzymes mentioned earlier. The dinuclear Cu_a center consists of two Cus bridged by two cysteinyl thiols (see Fig. 5.9). This is reminiscent of the dinuclear iron-sulfur cluster, though the bridging groups are S^{2-} in ferredoxin. The Cu_a site appears to accept electron(s) from cytochrome c. The electrons then are quickly transferred to the oxygen binding site: cytochrome a_3 and Cu_b through heme a.

5.3.4. **Nitrite Reductase and Nitrous Oxide Reductase**

Several different types of nitrite reductase are known: multiheme c containing enzyme, heme cd_1 protein, and a Cu-containing protein (Moura and Moura, 2001). The multiheme c enzyme from *sulfurospirillum delenyanum* or *Desulfovibrio desulfuricans* contains five heme cs. This assimilatory enzyme catalyzes the reduction of NO_2^- to NH_3 (see earlier).

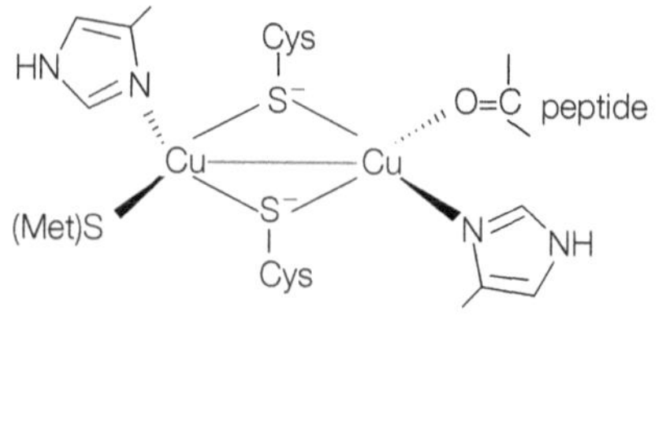

■ **Figure 5.9.** A schematic representation of a Cu-S cluster found in cytochrome c-oxidase. (PDB IOCC: Tsukihara, T., Aoyama, H., Yamashita, E., Tomizaki, T., Yamaguchi, H., Shinzawa-Itoh, K., Nakashima, R., Yaono, R., Yoshikawa, S. 1996. The whole structure of the 13-subunit oxidized cytochrome c oxidase at 2.8 Å. *Science* **272**, 1136–1144.)

The other enzymes function in a dissimilatory fashion, and reduce NO_2^- to NO. One is a heme enzyme (heme cd_1) (see earlier), and another is a copper enzyme. The enzyme from *Alcaligenes xylosoxidans* contains two types of mononuclear Cu sites (type 1 and 2) per sub-unit. The Cu of type 1 is bound to two histidine Ns, one methionine S, and one cysteine S^-, and is responsible for the blue color. The Cu of type 2 is bound with three histidine Ns, and the fourth ligand seems to be a water. Hence, NO_2^- appears to bind the type 2 Cu, and be reduced by Cu(I). The Cu(II) of type 2 site then is reduced by an electron transferred from cytochrome c through the type 1 site. The formation of a side-on NO-Cu (type 2) intermediate has been observed in an x-ray crystallographic study (Tocheva *et al.*, 2004).

No copper enzyme is known to carry out the reduction of NO to N_2O. All the NO-reducing enzymes are heme-enzymes. This is suggested by the affinity of NO to the heme iron; this enzyme was discussed in an earlier section.

The reduction of nitrous oxide is the last step of denitrification. The reaction is:

$$N_2O + 2H^+ + 2e \longrightarrow N_2 + H_2O$$

The enzyme, nitrous oxide reductase from *Alcaligenes xylosoxidans*, contains unusual Cu clusters. The one cluster is a dimer and is the same as the dimeric unit found in cytochrome c oxidase discussed earlier (see Fig. 5.9). The other is a $[Cu_4S]$ cluster (see Fig. 5.10), where four Cus are bridged by a single S^{2-} (Moura and Moura, 2001).

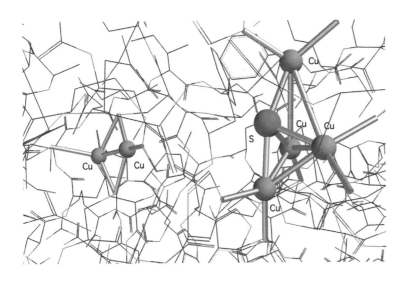

■ **Figure 5.10.** A Cu_4S cluster in nitrous oxide reductase. (PDB 1FWX: Brown, K., Djinovic-Carugo, K., Haltia, T., Cabrito, I., Saraste, M., Moura, J.J., Moura, I., Tegoni, M., Cambillau, C. 2000. Revisiting the catalytic CuZ cluster of nitrous oxide (N_2O) reductase. Evidence of a bridging inorganic sulfur. *J. Biol. Chem.* **275**, 41133–41136.)

The terminal O in N_2O may bind to one of the four Cus of the cluster. If the cluster provides two electrons to the bound N_2O and two protons are provided by nearby acidic entities, N_2O will be split between N and O, resulting in the formation of N_2 and H_2O. The reason why such an unusual Cu cluster is used may be that the reaction requires simultaneous donation of two electrons from the cluster to N_2O.

5.3.5. Amine Oxidases

Amine oxidase is an example of nonblue copper enzymes, and catalyzes the oxidation of an amine to form aldehyde and H_2O_2. That is,

$$RCH_2\!-\!NH_2 + O_2 + H_2O \longrightarrow RCH\!\!=\!\!O + NH_3 + H_2O_2$$

Interestingly, the same reaction is catalyzed by noncopper, flavin-dependent amine oxidases. This suggests that a compound similar to flavin may also be required for the copper enzyme. The copper amine oxidase does require a cofactor, which has been identified as TPQ (topaquinone = paraquinone form of trihydroxyphenylalanine or 6-hydroxydopa) (Janes *et al.*, 1990). The structure of the enzyme with this cofactor bound was determined by x-ray crystallography (Wilce *et al.*, 1997). The ligands on the single Cu are three histidines and one water.

The cofactor will react with a substrate amine (e.g., 2-phenyl ethylamine) and forms a Schiff base. But then it further reacts with H_2O resulting in the formation of the aldehyde product and the amine derivative of TPQ (reduced) under anaerobic conditions. This has been demonstrated by a freezing technique of an intermediate state and determining its structure by x-ray crystallography (Wilmot *et al.*, 1999). When this was exposed to O_2, hydrogen peroxide formed; this was also confirmed by x-ray crystallography. During the entire reaction cycle, the copper seems to remain as Cu(II); at least no significant formation of Cu(I) has been observed by regular techniques. ESR would be a powerful tool to examine the oxidation state in enzymatic reactions.

An interpretation of ESR data regarding this kind of issue (i.e., changes in oxidation states during the enzymatic process), however, needs to take into consideration a general basic problem. The problem stems from the fact that ESR measurements typically are done at very low temperatures such as $20\,K$, whereas the enzymatic reactions are carried out around room temperatures, say, $300\,K$.

There is a huge temperature gap between them. If the chemical entity of interest shows a strong temperature dependence, the situation at a low temperature (e.g., 20 K) may not necessarily reflect that at 300 K. One ought to be wary of this fact in interpreting and extrapolating the data at low temperature to room temperature (Ochiai, 1987).

It turned out that the ESR signal does change when the ESR measurement is done at room temperature (Dooley *et al.*, 1991). That is, the Cu(II) signal diminishes and a free radical signal appears when a substrate amine is added to the enzyme under an anaerobic condition (Dooley *et al.*, 1991). This radical signal is likely to be the semiquinone form of TPQ.

A possible reaction mechanism is shown in Figure 5.11 (Ochiai). The first half (a) is the anaerobic reaction, where TPQ is reduced. The products at this stage have been captured in the x-ray study

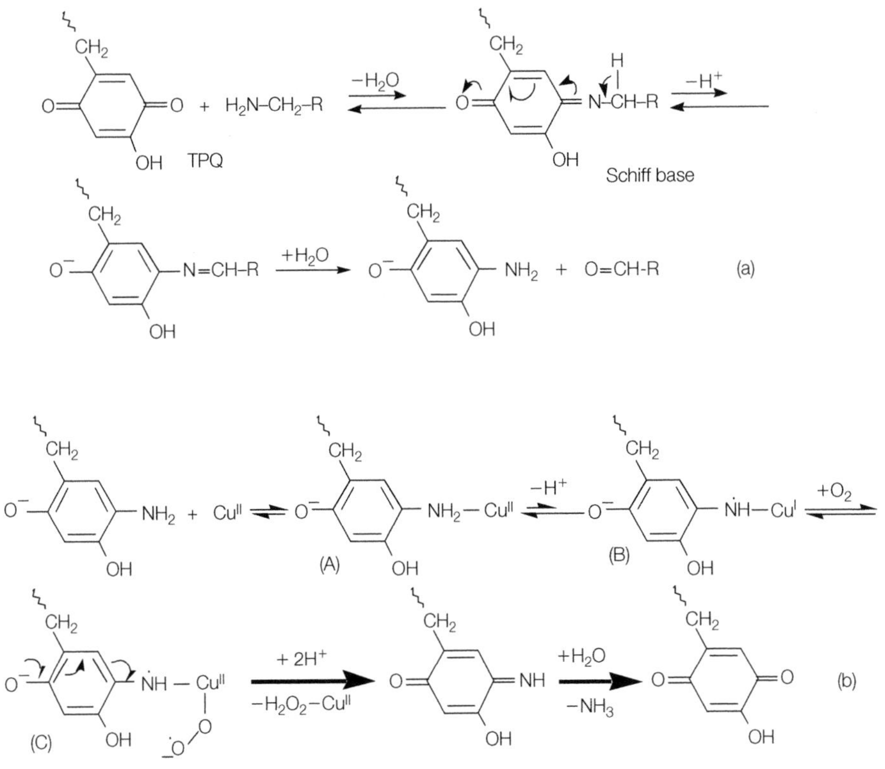

■ **Figure 5.11.** A possible mechanism of amine oxidase reaction (Ochiai).

(Wilmot *et al.*, 1999). The second half is more difficult. It is assumed here that the product amine in reaction (a) binds to Cu(II) (A in Fig. 5.11), and further that the amine reduces Cu(II) to Cu(I) (B). The extent of this reaction is small in thermodynamic sense, and besides, Cu(I) formed would swiftly bind O_2, turning to Cu(II)—O_2^- (C). Hence the Cu(I) state, if any, would be insignificant in both thermo-dynamic and kinetic sense in the presence of O_2. Cu(II)—O_2^- will abstract the hydrogen atom as shown, and the result is the formation of an imine and HO_2^-. The imine will be hydrolyzed spontaneously or catalyzed by the Lewis acid Cu(II). More discussion on the mech-anism is found in Knowles and Dooley (1994).

Methylamine dehydrogenase, whose cofactor had been presumed to be PQ (2,7,9-tricarboxypyrroloquinoline quinone), has been dem-onstrated recently to contain instead another novel cofactor, tryp-tophan tryptophylquinone (McIntire *et al.*, 1991). Two tryptophan moieties are covalently bound, and one of them is converted to an *ortho*-quinone. This quinone is believed to participate in the redox reaction as in amine oxidase mentioned earlier.

5.3.6. **Superoxide Dismutase (SOD)**

SOD catalyzes the disproportionate reaction of superoxide $\cdot O_2^-$

$$2\cdot O_2^- + 2H^+ \longrightarrow O_2 + H_2O_2.$$

There are three types of SOD; one is either Fe or Mn-enzyme, which seems to be the same protein; the second is a Cu/Zn-protein; and the third is a newly discovered Ni-enzyme (see Maroney, 1999). The Cu/Zn protein has no sequential similarity to the Fe or Mn-one. Fe or Mn-SOD is found in the prokaryotes or the mitochondria in the eukaryotic cells, whereas the Cu/Zn-SOD is found in the cyto-plasm of eukaryotic cells. The Cu is the catalytic site and the Zn plays a structural role. The basic mechanism, however, appears to be the same. That is,

$$\cdot O_2^- + Cu^{II} \longrightarrow O_2 + Cu^I; \ Cu^I + \cdot O_2^- + 2H^+ \longrightarrow Cu^{II} + H_2O_2$$

The reduction potentials (at pH7) of $O_2 + H^+/HO_2$ and $\cdot O_2^- + 2H^+/H_2O_2$ are $-0.45\,V$ and $+0.98\,V$, respectively. Therefore, any entity that has a reduction potential between $+0.98$ and $-0.45\,V$ would be able to perform this catalysis. Indeed any of Fe, Mn, or Cu fits the bill. It appears that Fe and Mn-SOD are ancient, originating in prokaryotes and being incorporated in mitochondrion through symbiosis, and that the Cu/Zn enzyme was more recently created when copper became relatively abundant, perhaps after about 1.8 billion years ago (Section 1.4; Ochiai, 1978).

5.4. MOLYBDENUM ENZYMES AND TUNGSTEN ENZYMES

According to Hille (1996), mononuclear molybdenum enzymes can be classified into three groups depending on the types of the Mo-binding pterin entity and the composition. Representative Mo-enzymes are listed in Table 5.2. A more complete list is found in Hille's article (1996). Nitrogenases will be discussed in Chapter 8.

5.4.1. Xanthine Oxidase and Aldehyde Oxidase

Enzymes of this class catalyze reactions of the following type:

$$RH + H_2O^* + O_2 \longrightarrow RO^*H + H_2O_2$$

RH is xanthine and ROH in uric acid in the case of xanthine oxidase; RH = aldehyde and ROH = carboxylic acid in the case of aldehyde oxidase. The reaction suggests that the hydroxylating O comes from a water molecule, not O_2, as has been demonstrated (Hille, 1996). Under a single turnover condition, the O comes from a Mo-bound O.

In xanthine oxidase from *Desulfovibrio gigas* molybdenum is coordinated by the two thiols of pterin and two Os and one S ligand,

Table 5.2. Mononuclear Molybdenum Enzymes

Class	Examples	Active site*	Composition
xanthine oxidase	xanthine oxidase	PMoOS	Mo/2[Fe$_2$S$_2$]/FAD
	xanthine dehydrogenase	"	"
	aldehyde oxidase	"	"
	aldehyde dehydrogenase	"	"
	CO-dehydrogenase	PCMoOS	"
	quinoline oxidoreductase	"	"
sulfite oxidase	sulfite oxidase	PMoO$_2$	Mo/cyt b-type
	nitrate reductase (assimilatory)	"	Mo/cyt b/FAD/NAD(P)H
dmso reductase	dmso reductase	(PG)$_2$MoX	Mo/4[Fe$_4$S$_4$]/cyt b-type
	formate dehydrogenase	"	"
	nitrate reductase (dissimilatory)	"	"

*P = pterin, PC = pterin cytosine dinucleotide, PG = pterin guanine dinucleotide.

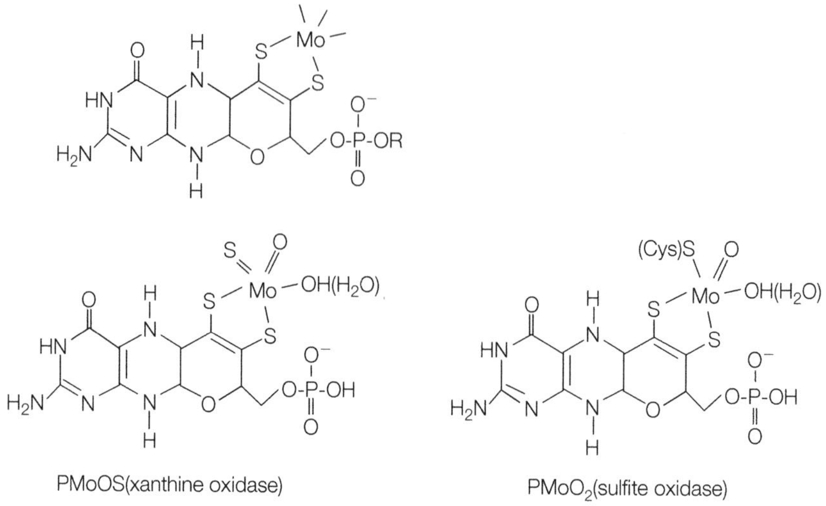

■ **Figure 5.12.** Pterin cofactor and its molybdenum complexes.

PMoOS(xanthine oxidase)

PMoO₂(sulfite oxidase)

as shown in Figure 5.12. The closest Fe_2S_2 cluster is located near the end of the pterin system, which is actually hydrogen-bonded to a cysteine that coordinates to the Fe of the cluster. The pterin appears to play the role of electron conductor between the iron-sulfur cluster and the Mo atom.

As suggested in Figure 5.13, one of the Mo—O bonds is described as Mo=O, and the oxidation state of Mo in the oxidized state is assigned as +VI. This implies that Mo=O (this O is often called "oxo") must be Mo^{VI}—O^{-II}, where Mo—O bond represents a coordination bond (to Mo(VI)) of one of the lone pairs on O^{-II}. However, the short Mo—O distance suggests some double bond character. Therefore, it can be described as a resonance between Mo^{V}—O^{-II} and Mo^{IV}=$\underline{O}|^{0}$. In the latter state, the O-entity may function as an electrophile (electron pair acceptor). It seems that a simple MoO_2 species reacts with a nucleophile through this kind of Mo=O. For example,

$$MoO_2L_2 + P(Ph)_3 \longrightarrow MoOL_2 + OP(Ph)_3$$

The enzyme's Mo=O might work similarly. Several possible reaction mechanisms for xanthine oxidase are discussed by Hille (1996), but none of the mechanisms mentioned incorporates this notion directly. The mechanism shown in Figure 5.13 is based on this notion (Mo^{IV}=$\underline{O}|$ component of Mo=O is the active species), and is not quite the same as any of the mechanisms proposed, but similar to scheme 4 of the Hille's article (1996). This mechanism (see

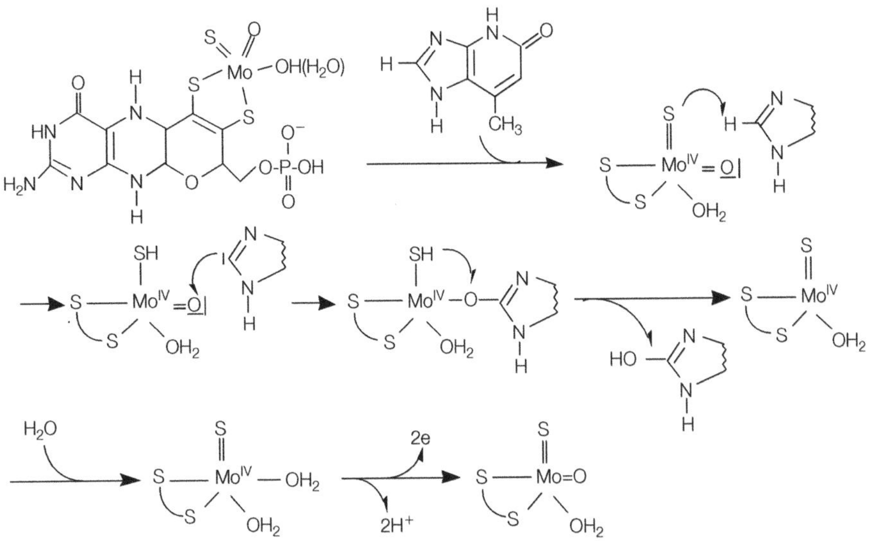

■ **Figure 5.13.** A mechanism of xanthine oxidase reaction (Ochiai; this is similar to one of the mechanisms discussed in Hille, 1996).

$$(2e + 2H^+ + FAD \rightarrow FADH_2,\ FADH_2 + O_2 \rightarrow FAD + H_2O_2)$$

Fig. 5.13) has the same essential feature as that for sulfite oxidase and nitrate reductase discussed later.

In these and all the other Mo-enzymes, the first part of the catalytic cycle seems to be the change of Mo^{VI}/Mo^{IV} accompanying the O (oxo)-transfer. The second part is the reverse process to either oxidize Mo(IV) to Mo(VI) or reduce Mo(VI) to Mo(IV) through the other substrate. This part appears to proceed step by step, so that the intermediate state Mo(V) becomes often observable through a number of spectroscopic techniques, particularly ESR (electron spin resonance).

5.4.2. **Sulfite Oxidase and Nitrate Reductase (Assimilatory)**

The catalytic center of these enzymes is essentially the same as that of xanthine oxidase, except for a cysteine S^- replacing S^{2-} (see Fig. 5.14). Perhaps this is because there is no need for abstraction of a proton from the substrate in these enzymes. It appears that the essential process in these enzymes is transfer of O-atom between the substrates and the catalytic Mo-center. Tentative mechanisms are shown in Figure 5.14, described in Ochiai (1987), but are essentially

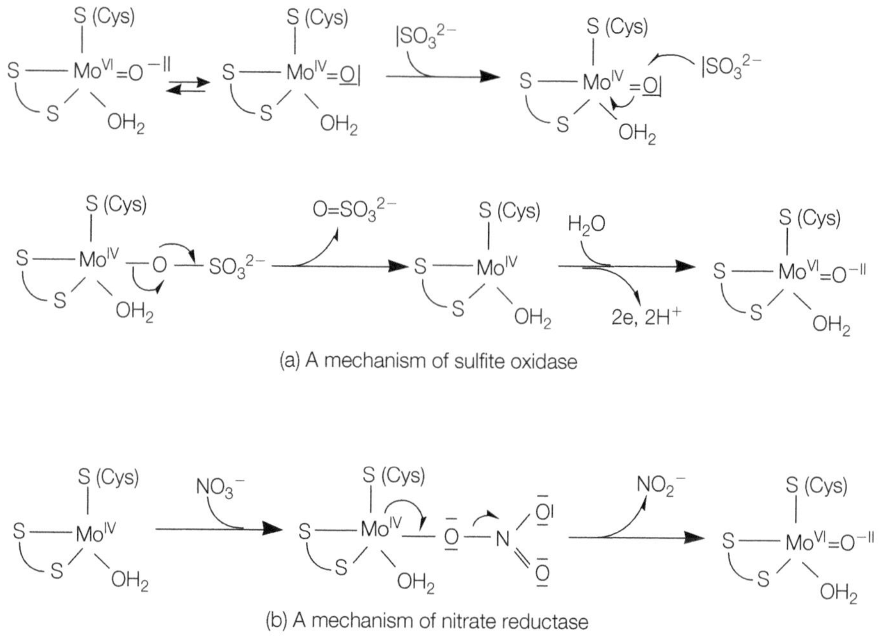

■ **Figure 5.14.** A mechanism of sulfite oxidase and that of nitrate reductase (Ochiai, 1987).

(a) A mechanism of sulfite oxidase

(b) A mechanism of nitrate reductase

the same as some mechanisms discussed in Hille (1996). This is basically consistent with the mechanism of xanthine oxidase or aldehyde oxidase discussed earlier (see Fig. 5.13). All the mechanisms presented here are based on the assumption that $Mo^{VI}{=}O^{-II}$ is in resonance with $Mo^{IV}{=}O^{0}$. In other words essentially a single mechanism is operating in these enzymes.

As seen in Figure 5.14, the sulfite oxidase is operating in the direction of $Mo(VI) \rightarrow Mo(IV)$, whereas the nitrate reductase is operating in the opposite direction $Mo(IV) \rightarrow Mo(VI)$. The reduction potentials of the Mo-center in these enzymes have not been reported, but they can be predicted to be somewhere in the range of 0 to 0.3 V. This potential is much higher than that in xanthine oxidase. The part of this difference seems to be due to the difference in the coordination sphere of Mo in these two different groups of enzymes.

5.4.3. DMSO Reductase and Nitrate Reductase (Respiratory or Dissimilatory)

DMSO reductase physiologically catalyzes the following reaction.

$$(CH_3)_2S{=}O + 2H^+ + 2e \rightarrow (CH_3)_2S + H_2O$$

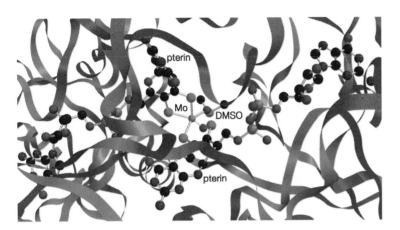

■ **Figure 5.15.** The Mo-site of DMSO reductase. (PDB 4DMR: McAlpine, A.S., McEwan, A.G., Bailey, S. 1998. The high resolution crystal structure of DMSO reductase in complex with DMSO. *J. Mol. Biol.* **275**, 613–623.)

This is essentially the removal of an O (oxo), as in nitrate reductase (assimilatory) discussed earlier. The 2e and $2H^+$ are provided by a compound of hydroquinone type in both of these enzymes. Since the reduction potential of the Mo-center in a respiratory nitrate reductase, $+0.34\,V$, is between those of two substrates, NO_3^- and hydroquinone (see Fig. 5.1), the enzyme is thermodynamically fit.

DMSO reductase has been shown to effect the following chemical reaction as well (Hille, 1996).

DMSOReductase(oxidized) (contains Mo$=$O) + PR$_3$ $\rightarrow$
    DMSOR(reduced) + OPR$_3$
DMSOR(reduced, deoxo) + R$_2$SO $\rightarrow$ DMSOR(oxidized, Mo$=$O) + R$_2$S

These reactions suggest a mechanism similar to that shown in Figure 5.14. The x-ray crystal structural data are available for the oxidized and the reduced states of this enzyme (PDB 1EU1, 4DMR). The catalytic sites are shown in Figure 5.15. The O of DMSO is bound to the Mo in the reduced state. Two pterins are bound with Mo in the reduced state. Only one of them is bound in the oxidized state, though the second pterin remains in the vicinity. Formate dehydrogenase has also a similar Mo(pterin)$_2$ active site (PDB 1QP8).

An interesting question would be why the ligation of two pterin entities is required in this enzyme in the reduced state while only one pterin is sufficient for the other enzymes mentioned earlier. Since the Mo-center is coordinated by a single pterin entity in the oxidized state, the reduction potential would be similar to those in the previous section.

5.4.4. **Tungsten Enzymes**

Several tungsten analogues of molybdenum enzymes are known (Johnson *et al.*, 1996). Some of the Mo-enzymes are produced in certain species, whereas the equivalent enzymes in other species are W-enzymes. For example, formate dehydrogenase, which catalyzes the reaction HCOOH $\rightarrow$ CO_2 + $2H^+$ + 2e, is a Mo-enzyme in *Pseudomonas oxalaticus* and *E. coli* (the one from *E. coli* contains also selenocysteine; see Chapter 9), whereas formate dehydrogenase in *Clostridium pasteurianum*, *C. thermoaceticum*, and *C. formicoaceticum* is a W-specific enzyme. The organisms that use the Mo-enzyme use it to dehydrogenate the formic acid (the forward reaction), and the organisms that produce the W-enzyme use it to hydrogenate CO_2 (the reverse reaction). The reduction potential CO_2/HCOOH is –0.42 V; that of Mo^{VI}/Mo^{IV} is around –0.35 V and that of W^{VI}/W^{IV} is around -0.45 V. Therefore, the choice of the metal seems to match the reaction effected (Ochiai, 1987).

Aldehyde ferredoxin reductase, formate ferredoxin reductase, and glyceraldehyde-3-phosphate ferredoxin reductase from *Pyrococcus furiosus*, and formate ferredoxin reductase from *Thermococcus litoralis* are W-specific. These organisms live at temperatures as high as 100°C in the case of *P. furiosus*. Another W-specific enzyme is carboxylic acid reductase. This reaction is the reverse of aldehyde oxidation, which is catalyzed by aldehyde oxidase or dehydrogenase, Mo-enzymes. This enzyme is also found in thermophiles such as *C. thermoaceticum*. It is possible that Mo-entities may not be stable at high temperatures and that W-entities may function properly only at elevated temperatures (Johnson *et al.*, 1996). The same comment in the previous paragraph also applies here; Mo-enzymes tend to oxidize (a few exceptions) and W-enzymes tend to reduce because of their reduction potential ranges.

An x-ray crystallographic study revealed the active site structure of aldehyde ferredoxin reductase from *P. furiosus* to be the same as that found in DSMO reductase, a Mo-enzyme (see Fig. 5.15). That is, two pterins bind to W through the pterin's thiol groups.

5.5. **MANGANESE OXIDOREDUCTASES**

Manganese can also readily change its oxidation states; it takes +II, +III, +IV, and +VII oxidation states under ordinary conditions. Hence Mn can play catalytic functions in oxidation–reduction reactions, as well as Lewis acid catalysis discussed in Chapter 4. A list of Mn-enzymes is found in an article by Dismukes (1996).

5.5.1. **Manganese Catalase**

The major type of catalase is a heme enzyme as discussed earlier. Thermophilic and heme-deficient microorganisms such as *Thermus thermophilus* and *Lactobacillus plantarum* rely on Mn-catalase. The enzyme's catalytic site is a dimeric Mn-complex; its structure is shown schematically in Figure 5.16.

The cluster can take several different oxidation states, but only two—(II, II) and (III, III)—are catalytically active (Dismukes, 1996). In the (II, II) state, the Mn-Mn distance is about 3.7 Å. The mechanism of disproportionation of HOOH is not understood. But it is possible that ^-OOH binds to one of the Mn's and is converted to two molecules of H_2O by withdrawing two electrons from the Mns: (II, II) + $3H^+$ + $^-OOH \rightarrow$ (III, III) + $2H_2O$. Another ^-OOH then binds to one of the Mn(III)s, and is oxidized; that is, ^-OOH + (III, III) $\rightarrow O_2$ + (II, II) + H^+. This presupposes a rapid electron transfer through the bridging O^{2-}s between the two Mns.

An alternative mechanism is that HOOH binds perpendicularly between the two Mns. When the Mns are in (II) state, HOOH will be reduced. The two Os bridging the Mn(III)s as seen in Figure 5.16 could be remnants of this process, and need to be removed by adding protons. Then another HOOH will bind likewise and be oxidized by (III, III).

5.5.2. **Water Oxidase**

This is the pivotal reaction in the water decomposing photosynthetic process, and is being intensely studied. Electrons will be removed from a cluster of chlorophylls in photosystem II by light radiation. This electron is transferred first to plastoquinone, and eventually to photosystem I. This electron(s) will then be replenished by a water molecule; the electrons from the reaction: $2H_2O \rightarrow O_2 + 4H^+ + 4e$ are used to do so. The enzyme that performs this reaction under light radiation is a Mn-enzyme. This enzyme is known as oxygen-evolving center (enzyme) or water-plasquinone photo-oxidoreductase, but here we call it simply water oxidase.

A simplest Mn-compound that oscillates between oxidations II and IV is $MnO_2/Mn(II)$. The reduction potential of this system is +1.23 V at pH 0 and +0.40 V at pH 7. The oxidation potential of H_2O

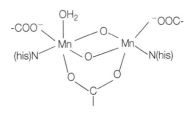

■ **Figure 5.16.** The catalytic site of Mn-catalase. (PDB 1JKU: Barynin, V.V., Whittaker, M.M., Antonyuk, S.V., Lamzin, V.S., Harrison, P.M., Artymiuk, P.J., Whittaker, J. W. 2001. Crystal structure of manganese catalase from *Lactobacillus plantarum. Structure* **9**, 725–738.)

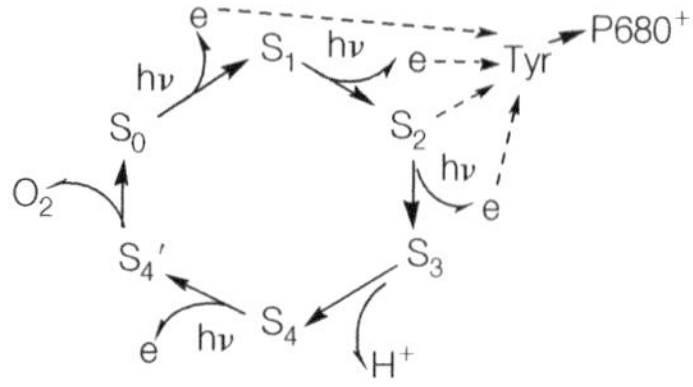

Figure 5.17. The cycle of O_2-evolving system (water oxidase).

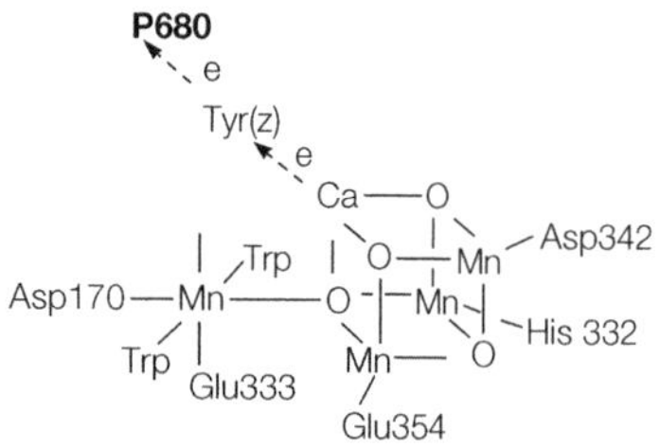

Figure 5.18. The Mn_4Ca cluster of oxygen evolving system (water oxidase) (Ferreira et al., 2004).

is $-1.23\,V$ at pH 0, and $-0.82\,V$ at pH 7. Hence the ΔG for the redox system—$2MnO_2 + 4H^+ \rightarrow 2Mn^{II} + 2H_2O + O_2$—is relatively small ($\Delta G = 0$ at pH 0 or $+162\,kJ/mol$ at pH 7). The reverse involves the process $Mn^{II} + 2H_2O \rightarrow MnO_2 + 4H^+ + 2e$ that has $E = -02.V$, $\Delta G = +77\,KJ/mol$, and can readily be affected by irradiation. If these two processes are combined, H_2O is oxidized to O_2 (i.e., $2H_2O \rightarrow O_2 + 4H^+ + 4e$). These considerations suggest that the water oxidase system may be based on this kind of redox system: Mn(IV)/Mn(II). A simplest possible mechanism may be that a dinuclear Mn-complex would bind two water molecules between the Mns. When Mns are oxidized to Mn(IV) state, each Mn(IV) oxidizes one water to $O + 2H^+ + 2e$, being itself reduced to Mn(II) (Ochiai, 1987). Recently Sauer and Yachandra (2002) suggested that MnO_2 might have been the origin of this enzyme.

It has been well established that the water oxidation proceeds in four steps of one-electron oxidation, which is affected by a single photon each. It has recently been proposed (Rutherford and Boussac, 2004) that there are six intermediates, S_0, S_1, S_2, S_3, S_4, S_4' as shown in Figure 5.17. O_2 is evolved at the fourth flash starting with S_0 in this scheme.

Two groups (Ferreira et al., 2004; Biesiadka et al., 2004) reported the x-ray crystallographic determination of the crystal structure of photosystem II of *Thermosynechococcus elongatus* in 2004. The structure of the Mn cluster that is believed to carry out the oxidation of water is shown schematically in Figure 5.18 (after Ferreira et al., 2004). The cluster is essentially cubic with three Mns and one Ca, and four O^{2-} occupying the alternate points, and the fourth Mn is located outside of this cube and is bonded to one of the bridging O^{2-} as shown. A more recent structural study (Yano et al., 2006) suggests that Ca is not a part of the cube, but located a little apart from the Mn_4 cluster, and that the extra Mn is doubly bridged by two O^{2-}.

Each of the S_n states represents the Mn_3Ca—Mn cluster $+2O$ entity in a certain oxidation state. It is not known which of the four Mn's is involved in the water oxidation. Ferreira et al. (2004) are of the opinion that the extra Mn is the catalytic site. The mechanism of water oxidation reaction is still very obscure (see Ferreira et al., 2004; Siegbahn, 2003; Iwata and Barber, 2004). Figure 5.19 is thrown in here simply to offer a possible mechanism based on these findings. Here are assumed that (1) Ca(II) acts as water molecule carrier, and (2) the extra Mn outside of the cluster is oxidized by light (through

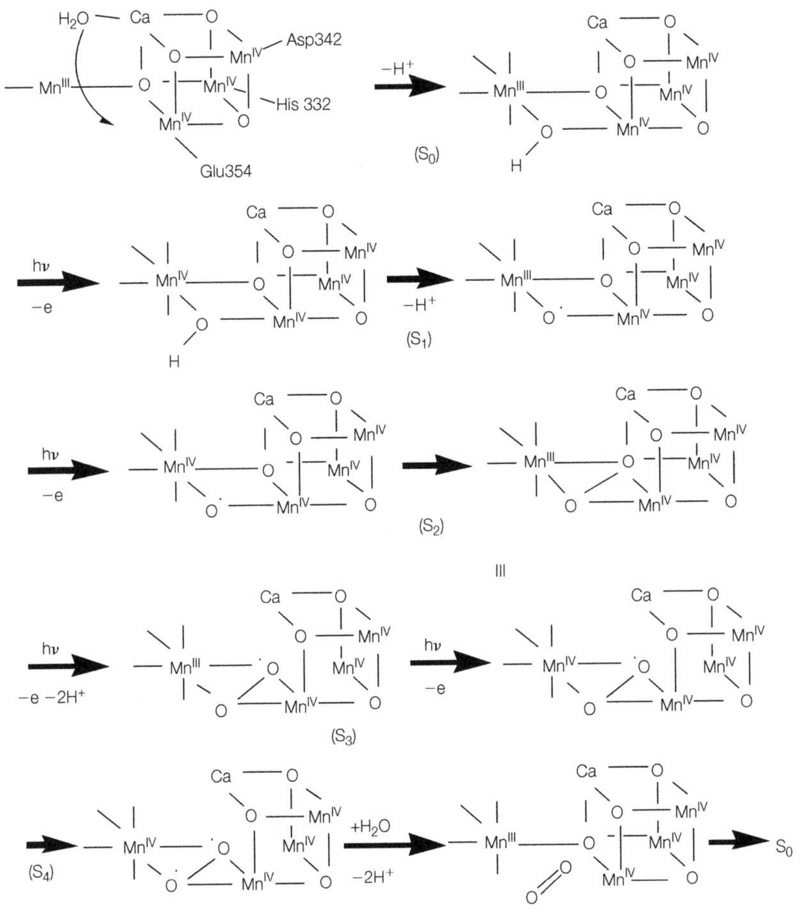

■ **Figure 5.19.** A mechanism of oxygen evolving system (water oxidase) (Ochiai). More recent reviews on oxygen evolving center are found in a special issue of Phil. Trans. of the Royal Soc., B: Biological Science, 363 (2008), 1129–1303; and a special issue of Coord. Chem. Rev., **252** (2008), 244-455.

P680 and a tyrosine) and then Mn(IV) removes electrons one by one from the two (originally) water molecules. Alternatively, the two Mn's involved may operate their action through $2Mn(IV) + 2H_2O \rightarrow 2Mn(II) + O_2 + 4H^+$ and $2Mn(II) \rightarrow 2Mn(IV) + 4e$ by four photons.

5.6. **Ni-CONTAINING REDOX ENZYMES**

Nickel metal is well known for its catalytic effect for hydrogenation. In other words, it can activate H_2. In view of this fact it is not surprising to find that a type of hydrogenase depends on Ni. Superoxide dismutases (SOD) long have been known to be of several different types: Cu/Zn, Mn or Fe-dependent. Yet another type has been discovered recently that contains Ni as its active site (see Maroney, 1999). Another interesting Ni-dependent oxidoreductase is carbon monoxide dehydrogenase (Ragsdale and Kumar, 1996). Acetyl Co-A synthase and methyl-CoM reductase may involve a nickel-to-carbon organometallic bond formation (see Maroney, 1999; Ermler *et al.*, 1998). Ni enzymes are not necessarily widely distributed, but their chemistry is very interesting, unlike the ordinary Ni-chemistry, which is virtually confined to Ni(II) chemistry. Lewis acid catalyses of Ni-containing enzymes are discussed in Chapter 4, and indeed are restricted to Ni(II) chemistry.

Nickel can take any oxidation states between Ni(0) (even lower as Ni(-I)) to perhaps Ni(III). Its congeners, Pd and Pt, can readily take the oxidation states up to (IV), but it is not so with Ni. Ni-SOD is believed to operate with Ni(III)/Ni(II) couple and its mechanism seems to be similar to those of Fe or Mn-SOD and Cu/Zn-SOD.

5.6.1. **Ni-Fe (Se) Hydrogenase**

Three types of hydrogenase are known; the first type includes metal-free enzymes, the second type the Fe-enzyme (see earlier), and the third type includes the Ni-Fe enzyme. The Fe-only enzyme usually works in the reduction of protons, whereas the Ni-Fe enzyme typically oxidizes H_2 ($\rightarrow 2H^+ + 2e$) or splits H_2 heterolytically (i.e., $\rightarrow H^- + H^+$). The catalytic site of the Ni-Fe hydrogenase is a Ni-Fe dinuclear complex (see Ermler *et al.*, 1998). The Fe (in the enzyme from *Desulfovibrio gigas*) is coordinated by two CN^- and one CO ligands, the Ni is coordinated by two cysteine S^-s, and the Ni and Fe are bridged by two cysteine sulfurs. A third bridging group will be discussed later. This dimeric structure is similar to the active dinuclear Fe-cluster (attached to a $[Fe_4S_4]$ cluster) in the Fe-only hydrogenase (see earlier, and Fig. 5.6). In the Ni-Fe hydrogenase the iron-sulfur clusters are separated from the Ni-Fe dinuclear center. It is interesting to note that the enzyme from *Desulfomicrobium baculatum* has one Se of selenocysteine and one S(Cys) instead of two S(Cys)s coordinating to the Ni (Garcin *et al.*, 1999).

The Ni-Fe enzyme takes various oxidation states during the catalytic cycle (see Stein and Lubitz, 2002). A state designated as Ni-B is supposed to be the active one. The oxidation state of Ni-B is believed to be Ni(III)/Fe(II) (Stein and Lubitz, 2002). They also identified an intermediate (designated as Ni-C) as the dimer bridged by a H^- entity. In a recent report Volbed *et al.* (2005) suggested that one of the groups that bridge the Ni and Fe could be a hydroperoxide in the active oxidized form (Ni-B) of the enzyme from *Desulfovibrio fructosovorans*. Stein and Lubitz (2002) assume that the bridging group is a hydroxide. The mechanism is still obscure despite an intensive investigation.

5.6.2. Carbon Monoxide Dehydrogenase (CODH)

The carbon monoxide dehydrogenase is a misnomer; carbon monoxide itself cannot be dehydrogenated. CO along with H_2O is oxidized to form CO_2 ($+2H^+ + 2e$). The reaction $CO + H_2O \rightarrow CO_2 + H_2$ is known as a water gas shift reaction in synthetic chemistry and is catalyzed by Fe-metal. The reverse is the reduction of CO_2 to CO. One type of CODH is a Mo-enzyme, which is found in aerobic organisms, and has the regular Mo-pterin active site (see earlier). The Ni-enzyme (NiCODH) is found in anaerobically growing organisms. NiCODH is present as an independent enzyme or in conjunction with Ni-containing acetyl Co-A synthase (ACS, see later).

It turned out that NiCODH from *Rhodospirillum rubrum* contains an unusual cluster—a [NiFe$_3$S$_4$] cubic cluster to which the fourth Fe is attached through a cysteine S and a S^{2-} of the cubic cluster (Drennan *et al.*, 2001). The mechanism of the reaction is obscure. A hypothetical mechanism is offered here based on the mechanism of the water gas shift reaction (see Fig. 5.20). Scheme (a) is the oxidation of CO, and (b) is the reduction of CO_2. The reduced or oxidized Ni center will be restored to the initial oxidation state by the electron transfer system: the cluster and the other associated Fe-S clusters. Step (1) is a kind of *cis*-migration, step (2) a reductive elimination, and step (3) is an example of oxidative addition reaction; all of these reactions are known to occur with organometallic compounds (see

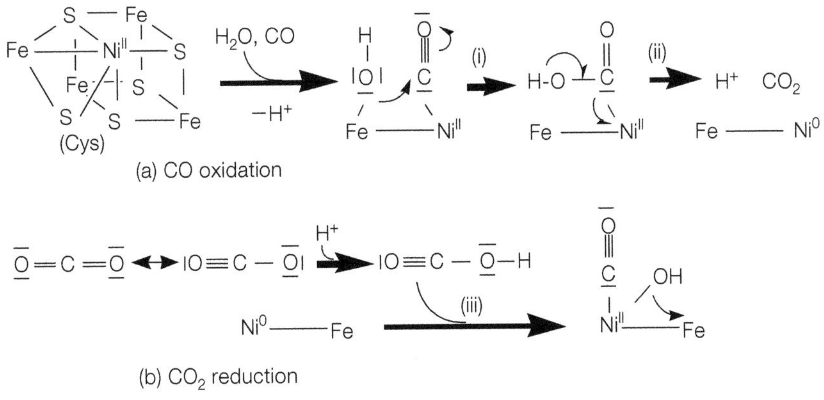

(a) CO oxidation

(b) CO$_2$ reduction

Figure 5.20. A mechanism of carbon monoxide dehydrogenase (Ochiai).

Chapter 2). A similar mechanism has been proposed recently based on x-ray crystallographic data (Jeoung and Dobbek, 2007).

5.6.3. Acetyl CoA Synthase (ACS)

An enzyme from *Moorella thermoacetica* consists of $\alpha_2\beta_2$. The subunit β converts CO_2 to CO (i.e., CODH activity), and subunit α catalyzes the reaction to produce acetyl CoA from CO, CoA and a methyl group source; that is, acetyl CoA synthase function. The methyl group source is a corrinoid iron-sulfur protein (CFeSP), and the overall reaction is:

$$CO + HSCoA + CH_3 - Co^{III} - CFeSP \rightarrow CH_3(CO) - SCoA$$
$$+ Co^{I} - CFeSP + H^+$$

The organometallic chemistry of corrinoid derivatives will be discussed in Chapter 7. The α subunit (ACS) contains an unusual cluster called A-cluster, which consists of Fe, Ni, and Cu (Doukov *et al.*, 2002). One of the Fes of the $[Fe_4S_4]$ cluster is connected via a cysteinyl S^- to a Cu, which constitutes a dinuclear cluster where the other metal is Ni (Doukov *et al.*, 2002). In other words, structurally this cluster is very similar to that found in Fe-only hydrogenase (see Fig. 5.6). The reaction mechanism proposed by Doukov *et al.* (2002) is summarized as follows. CO binds to the Cu site, and then a methyl group is transferred to the Ni site from the corrinoid entity; then a *cis*-migration takes place, in which the CH_3 group migrate to the CO group, resulting in the formation of an acetyl ($-(C=O)CH_3$) group attached to the Cu; the final step is an attack by S^- of HSCoA on the carbonyl carbon on the Cu.

5.6.4. Methyl-Coenzyme M Reductase

Methane-forming archaea bacteria extract energy from the reduction of CO_2 to CH_4. The last step of methane formation is described by:

$$\underset{\text{(methyl-coenzyme M)}}{CH_3S(CH_2)_2SO_3^-} + \underset{\text{(coenzyme B)}}{HS(CH_2)_6\overset{O}{\overset{||}{C}}\text{-}\overset{H}{\overset{|}{N}}\text{-}\overset{COO^-}{\overset{|}{CH}}CHCH_3} \longrightarrow CH_4 + CoM\text{-}S\text{-}S\text{-}CoB$$
$$\underset{}{OPO_3^{2-}}$$

This reaction is catalyzed by a nickel-containing methyl-coenzyme M reductase. The nickel entity is a porphinoid complex, F430 (see Fig. 5.21). X-ray crystallographic structures of various states of the enzyme have been determined (Grabarse *et al.*, 2001). The nickel seems to be in Ni(I) state under the resting conditions. A mechanism proposed by Grabarse *et al.* (2001) is as follows.

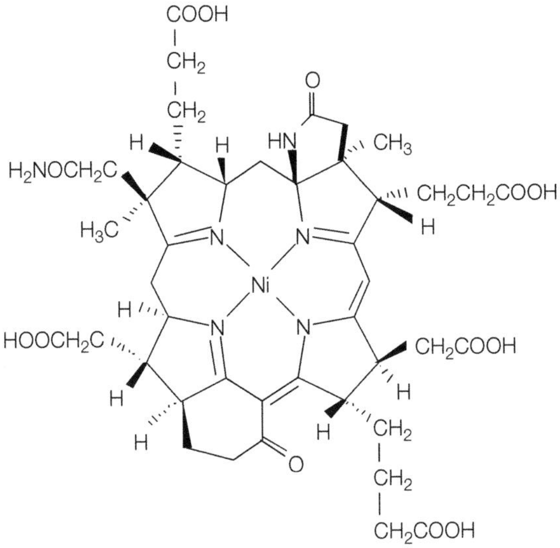

■ **Figure 5.21.** F 430.

(1) Ni(I) is assumed to attack the S—CH$_3$ of methyl CoM, resulting in CH$_3$—Ni(III) and HSCoM.

(2) HSCoM then reacts with CH$_3$—Ni(III) resulting in the formation of CH$_3$—Ni(II) and ˙SCoM.

(3) Proton addition to CH$_3$—Ni(II) forms CH$_4$ (the product) and Ni(II).

(4) ˙SCoM reacts with HSCoB, forming a disulfide anion radical: (CoMS—SCoB)˙⁻.

(5) The disulfide anion radical, being a strong reducing agent, reduces Ni(II) to Ni(I).

In this mechanism, Ni(I) in the first reaction is assumed to act like Co(I) in cobalamin reactions; that is, a nucleophilic attack on the C of CH$_3$—SCoM. This is rather unlikely on two accounts. The one is that Ni(I), being d^9, would not behave like Co(I) d^8, and that the resulting CH$_3$—Ni(III) is one electron-short of the 18 electron rule (see Chapter 2). The second is that the C—S bond is much more covalent than C—X (X = halogen or the like) on which Co(I) is supposed to attack in the cobalamin reactions, and hence that it is not likely that the C—S is cleaved heterolytically in the manner of the cobalamin reactions. It may be more likely that the first and second reactions assumed is one step reaction; in other words, the reaction between Ni(I) and CH_3—SCoM is one electron transfer, which splits the bond into CH_3—Ni(II) and ˙SCoM.

REVIEW QUESTIONS

1. As suggested, coordination of a negatively charged ligand such as ¯NHR lowers the reduction potential of a metal ion as compared to the corresponding neutral ligand such as NH_2R. Why?

2. Would a hydrophobic environment about a cationic entity increase or decrease (in general sense) its reduction potential as compared to a hydrophilic environment?

3. It has been found that the structure about Cu in the blue copper protein/enzymes is four-coordinate, and not quite tetrahedral nor square planar. Discuss the reason for this.

4. Iron-sulfur clusters are used widely in living organisms. Perhaps, an iron-sulfur protein, ferredoxin of plants is one of the most abundant proteins in biosphere. When it was first identified as such, it was quite a surprise, as it has an unusual composition and structure. What does it (composition and structure) suggest about the origin and evolution of the protein?

5. Explore literature for the experimental evidence in the reaction of xanthine oxidase (a Mo-enzyme) that the O incorporated into the product does come from H_2O rather than O_2.

6. The reaction catalyzed by catalase is $2HOOH \rightarrow 2HOH + O_2$. Is this an oxidation–reduction reaction? If so, what oxidizes what, and what reduces what?

7. Two different catalases have been found: one is Fe-enzyme and the other Mn-enzyme. In fact, many other entities (including Cu(II)/Cu(I), I^-, etc.) can function as a catalyst for the catalase reaction, though no enzyme has been found to contain other than Fe or Mn. This suggests that the reaction itself is quite facile. Indeed the Gibbs free energy change for the reaction is quite negative. Estimate it using the reduction potential data given in Table 6.1.

8. The metallic cations used for oxidation–reduction enzymatic reactions include V, Mn, Fe, Ni, Cu, Mo, and W, as discussed

in this chapter and in Chapter 9. Interestingly no enzyme for oxidation–reduction reactions has been found to employ Cr (which lies between V and Mn), nor Co (between Fe and Ni). Discuss possible reasons. (Note: Co in adenosyl cobalamin does change its oxidation state during the catalytic process; see Chapter 7.)

9. It is interesting to note that heme-enzymes, copper enzymes, and Mo-enzymes are involved in reduction of N-oxides (NO_3^-, NO_2^-, NO, N_2O). List all the reactions and associate them with specific enzymes. Refer to Chapter 5 for Mo.

PROBLEMS TO EXPLORE

1. Hemes (Fe-porphyrin) have found a wide-ranging use in biological systems. Any thought on why?

2. The blue copper sites in blue copper proteins and enzymes shows a strong absorption around $\nu_{max} = 16 \times 10^3$ cm^{-1} (with molar extinction coefficient ($M^{-1}cm^{-1}$) of about 3500). Explore the basis of this light absorption.

3. Hydrogenase from *Clostridium pasteurianum* catalyzes a seemingly simple reaction: $2H^+ + 2e \rightarrow H_2$. Yet the structure of the active site of this enzyme is one of the most complicated, involving COs and CN$^-$s bound to Fe(s). Discuss why it might require such a complicated structure.

4. Cu-containing oxidoreductases are believed to function typically with the Cu(II)/Cu(I) pair. However, Cu(III) has been suggested to be involved in certain enzymatic systems. What evidence has been obtained to indicate it? Search literature and discuss.

5. An analogy is suggested in the text regarding the intermediate between (dissimiratory) nitric oxide (NO) reductase and cytochrome P-450 monooxygenation reactions. Explore and discuss.

6. An involvement of the resonance $[Mo^{VI}O^{-II} \Leftrightarrow Mo^{IV}{=}O^0]$ is suggested as a basis for all the Mo-enzymes. Discuss pros and cons for this mechanism.

7. Explore possible mechanisms for nitric oxide reductase that catalyzes the reaction $2NO + 2H^+ + 2e \rightarrow N_2O + H_2O$.

Oxygen Carrying Processes and Oxygenation Reactions

Reactions involving oxygen, dioxygen, and related entities are diverse. Particularly the modes of reactions of O_2 are so many that no clear-cut systematic understanding of all those reactions has emerged. This chapter tries to give such a systematic understanding, but it must be admitted that such an attempt cannot be complete yet. I tried to provide the same understanding in my earlier books (Ochiai, 1977, 1987).

6.1. THE CHEMISTRY OF OXYGEN, DIOXYGEN, AND RELATED ENTITIES

6.1.1. Electronic Structures

The O-atom is in the triplet ground state ^3P ($\cdot\underline{\text{O}}\cdot$); that is, it is a biradical. As such, it combines readily (i.e., in a kinetic sense) with any entity having a lone electron and forms $\cdot O^-$ or $\cdot OR$. The formation of these entities is also thermodynamically favored. However, the reaction of the ground state $\cdot O\cdot$ (^3P) with a molecule in the singlet state is spin-forbidden. The lowest excited state of O is a singlet state, ^1D, in which one of the p-orbitals ($|\overline{\text{O}}$) (or one of sp^3 hybrid orbitals) is unoccupied. This empty orbital can act as an electron pair acceptor. That is, the singlet oxygen (^1D) can act as a Lewis acid or electrophile.

The ground state of dioxygen is a triplet state: $^3\Sigma_g$. This is also a biradical. Some of the excited states of O_2 molecule (in order of energy) are $^1\Delta_g$ (92 kJ above the ground state) and $^1\Sigma_g$ (155 kJ). In terms of the molecular orbitals, the wave functions for $^3\Sigma_g$ are:

$$(1\sqrt{2})\{\pi_x{}^*(1)\pi_y{}^*(2) - \pi_y{}^*(1)\pi_x{}^*(2)\} * \alpha(1)\alpha(2),$$

the same orbital function$*(\alpha(1)\beta(2) + \beta(1)\alpha(2))/\sqrt{2}$

the same orbital function $*\beta\beta$

that for $^1\Delta_g$ is:

$$(1\sqrt{2})\{\pi_x{}^*(1)\pi_x{}^*(2) + \pi_y{}^*(1)\pi_y{}^*(2)\} * (\alpha\beta - \beta\alpha)/\sqrt{2},$$

and that for $^1\Sigma_g$ is:

$$(1\sqrt{2})\{\pi_x{}^*(1)\pi_y{}^*(2) + \pi_y{}^*(1)\pi_x{}^*(2)\}(\alpha\beta - \beta\alpha)/\sqrt{2},$$

where α and β represent spin functions and the O—O bond is defined to be along the z-axis.

Like the ground state O-atom $O(^3P)$, the ground state O_2 ($^3\Sigma_g$) reacts readily (in a kinetic sense) with an entity having a single electron available (e or $\cdot$R). On the other hand, its reaction with an organic molecule in a singlet ground state is spin-forbidden. O_2 in the singlet-excited state(s), on the other hand, reacts with an entity in a singlet state.

6.1.2. **Basic Reactions of O and O_2**

Since we are interested in reactions in aqueous media, the thermodynamic reactivity in oxidation–reduction reactions can be represented by reduction potential(s). They are given in Table 6.1. The reduction of O, O_2, and their derivatives is mostly thermodynamically favorable. The first reduction of O_2 yields the superoxide $\cdot O_2{}^-$ (OOH upon addition of H^+). The $\cdot O_2{}^-$ species is very unstable in the thermodynamic sense and reactive in the kinetic sense, because of its electronic configuration (two electrons in one of the antibonding π^*-orbitals and a single electron in another). This is reflected in the low reduction potential of the addition of a first electron to O_2. The second reduction to produce HOOH is very favorable, as the electron configuration of HOOH is stable. The addition of another electron to HOOH ($O_2{}^{2-}$) inevitably splits the O—O bond, because the extra electron has to enter the σ-anti-bonding orbital.

The thermal decomposition of HOOH takes place in two modes (a) and (b):

$$\text{HOOH} \rightarrow \text{HO} \cdot + \cdot \text{OH} \quad \Delta H = +146\,\text{kJ/mol (in gas phase)} \quad \text{(a)}$$

$$\text{HOOH} \rightarrow \text{HOH} + \text{O} \quad \Delta H = +101\,\text{kJ/mol (in gas phase)};$$
$$+150\,\text{kJ/mol (pH 7)} \quad \text{(b)}$$

The large difference in reduction potential between the fourth and fifth reaction in Table 6.1 reflects this enthalpy for bond cleavage (b).

Table 6.1. Reduction Potential (E_o' at pH 7) of Oxygen and Its Derivatives (*vs* NHE)

Reaction (reduction)	E_o' (v)
O (ground state) $+ e + H^+ \rightarrow$ OH	$+2.0*$
O $+ 2e + 2H^+ \rightarrow$ HOH	$+2.13$
O_2 (ground state) $+ e + H^+ \rightarrow$ OOH	-0.45
OOH $+ e + H^+ \rightarrow$ HOOH	$+0.98$
OOH $+ e + H^+ \rightarrow$ HOH $+$ O	-0.67
HOOH $+ e + H^+ \rightarrow$ OH $+$ HOH	$+0.38$
HOOH $+ 2e + 2H^+ \rightarrow$ 2HOH	$+1.36$
$O_2 + 2e + 2H^+ \rightarrow$ HOOH	$+0.27$
$O_2 + 4e + 4H^+ \rightarrow$ 2HOH	$+0.82$

*Estimate.

6.1.3. **Reactions of Ground State O and O_2**

The ground states of O and O_2 both are spin-triplet. Their reactions with organic compounds in singlet state are (spin-) forbidden. Strictly speaking, a combustible organic compound (RH) would not spontaneously burn in the air for this reason. Any substance that provides a free radical ($\cdot$X) can induce a free radical chain reaction, and thus combustion will ensue. That is:

$\cdot O \cdot + \cdot X \rightarrow \cdot OX, \cdot OX + RH \rightarrow HOX + \cdot R, \cdot R + \cdot O \cdot \rightarrow \cdot OR$, etc.
$\cdot O_2 \cdot + \cdot X \rightarrow \cdot OOX, \cdot OOX + RH \rightarrow HOOH + \cdot R, \cdot R + O_2 \rightarrow \cdot OOR$, etc.

The second step is a hydrogen abstraction that can readily take place with most C—H entities (see Chapter 7).

Entities of another type that contain available unpaired electron(s) are metallic elements, particularly of transition metals. The reaction of O_2 with metals of low reduction potential is often vigorous, as an enormous number of free electrons are available and the reactions are highly exothermic, as in the case of metallic sodium. Some metals may form an impenetrable thin film of oxides on the surface, and spontaneous oxidation may not continue any further. Spontaneous oxidation with O_2 would not take place, of course, with metals of sufficiently high reduction potential.

■ **Figure 6.1.** The modes of O_2-binding to (transition) metals.

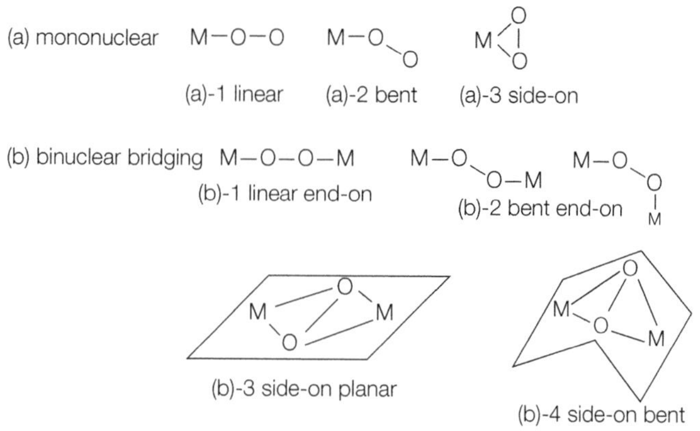

(a) mononuclear   M—O—O   M—O / O   M(O/O)

                    (a)-1 linear   (a)-2 bent   (a)-3 side-on

(b) binuclear bridging M—O—O—M   M—O / O—M   M—O / O

              (b)-1 linear end-on   (b)-2 bent end-on   M

(b)-3 side-on planar

(b)-4 side-on bent

6.1.4. Interactions of Ground State O_2 with Compounds of Transition Metals

The ground state O_2 can react with compounds of transition metals in various ways. First, it may bind with transition metals in various manners as shown in Figure 6.1. The mode (a)-1 and mode (b)-1, both linear, are not known. The nature of the bonding may be understood in terms of molecular orbital interactions between the molecular orbitals of O_2 and atomic orbitals on the metal; important interactions are depicted schematically in Figure 6.2.

The main contribution in the bonding mode (B) is an electron transfer from the atomic orbital of the metal to the half-empty π^*-orbital of O_2; hence it can be described as $M^+—O_2^-$. The O-atoms on O_2^- now have a lone pair in an sp^2 hybrid orbital; this would result in a $120° < M—O—O$ angle. This is typically the case when O_2 binds to an M in the end-on fashion.

The bonding of a side-on type ((a)-2) consists of molecular orbital interactions of (A) and (C): a partial donation of the pair of electrons in π-orbital of O_2 to the empty atomic orbitals on the metal (A) and a partial (back) donation of a pair of electrons in (e.g., d_{zx} orbital to the now empty π^*-orbital on O_2 (C)). Here, the two degenerate π^*-orbitals on O_2 are presumed to split in energy upon binding to the metal. If indeed this is the case, the O_2 now would behave as a singlet state. The alternative scheme involves sharing electrons (one each from d_{zx} and π^*) in the molecular orbital formed between d_{zx} and π^*. This will leave an unpaired electron in π_y^*. This latter situation has not been observed.

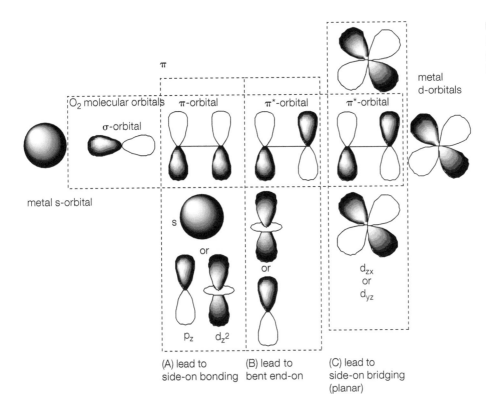

■ **Figure 6.2.** The (simplified) MO description of the binding between O_2 and a metal.

Scheme (C) is the planar side-on bridged binding ((b)-3); there must be an interaction of (A) kind as well on both sides of O_2. On each interaction, three molecular orbitals (i.e., the set of $[d_{zx}(1) + \pi^* + d_{zx}(2)]$, $[d_{zx}(1) - d_{zx}(2)]$, and $[d_{zx}(1) - \pi^* - d_{zx}(2)]$ or set of $[s(1) + \pi + s(2)]$, $[s(1) - s(2)]$, and $[s(1) - \pi - s(2)]$) would form, and the lowest orbitals would accommodate two electrons each. If the interactions $[d_{zx}(1) + \pi_x^*]$ and $[d_{yz}(2) + \pi_y^*]$ predominate, a side-on bridged bent structure ((b)-4) will result.

The further reactions of these O_2-bound entities depend on many factors. If the bound O_2 remains intact and the O_2 in the system is removed (by pumping or otherwise), the O_2-complex would give up the O_2. Hence the reaction is reversed. The binding reactions are either of:

$$M + O_2 \rightleftharpoons MO_2 \qquad\qquad (a)$$

$$2M + O_2 \rightleftharpoons M_2O_2 \qquad\qquad (b)$$

ΔG for these reactions should be in the range of roughly -8 to -50 kJ/mol for reactions (a) and (b) to be reversible under practical conditions (Ochiai, 1973). If $\Delta G > -8\,kJ/mol$, O_2 would hardly bind, whereas if $\Delta G < -50\,kJ/mol$, O_2 binds too strongly to be removed by reducing the O_2 pressure.

If O_2 binds in a mononuclear end-on fashion and the resulting complex can be described formerly as, say, $Fe^{III}-O_2^-$, this entity can undergo further reactions such as:

$$Fe^{III}-O_2^- + H_2O \text{ (solvent)} \rightarrow Fe^{III}-H_2O + O_2^-$$

or

$$Fe^{III}-O_2^- + Fe^{II} \rightarrow Fe^{III}-O_2^{2-}-Fe^{III}, \quad Fe^{III}-O_2^{2-}-Fe^{III}$$
$$+ 2H_2O \rightarrow 2Fe^{III}-H_2O + O_2^{2-}$$

or

$$Fe^{III}-O_2^{2-}-Fe^{III} + 2Fe^{II} \rightarrow 4Fe^{III}-O^{2-}-Fe^{III}, \quad Fe^{III}-O^{2-}-Fe^{III}$$
$$+ 2H_2O \rightarrow 2Fe^{III}-H_2O + O^{2-}$$

That is, the result is the oxidation of Fe(II) to Fe(III). From the standpoint of oxygen, O_2 is reduced to superoxide (one electron reduction), peroxide (two), and/or water (four electrons). To prevent these reactions (i.e., the oxidation), approach of solvent molecules and another entity containing Fe(II) needs to be made difficult. However, if the ligand substitution reaction is intrinsically slow as in Cr(III) complexes or low-spin Co(III) complexes, this condition may not be necessary and the process would be reversible, provided that it is thermodynamically feasible.

In the mode of a side-on binding, the metallic ion would formally be oxidized by $+2$, and O_2 would become formally O_2^{2-}. The mononuclear O_2-complex of side-on type is known, particularly with the second and third series of d^8 or d^{10} metal (Ir(I), Pt(0), etc.). The reversibility is governed mainly by the thermodynamic factors.

The dinuclear end-on bridged MO_2M is well known with $M=Co$. Because the Co is in low-spin Co(III) state, the $Co^{III}O_2^{2-}Co^{III}$ is substitution inert and is usually reversible if thermodynamically feasible.

MO_2M (side-on bridged) was first observed with a model compound for hemocyanin (Kitajima et al., 1989). Later it was determined that O_2-hemocyanin has the same binding mode. That is:

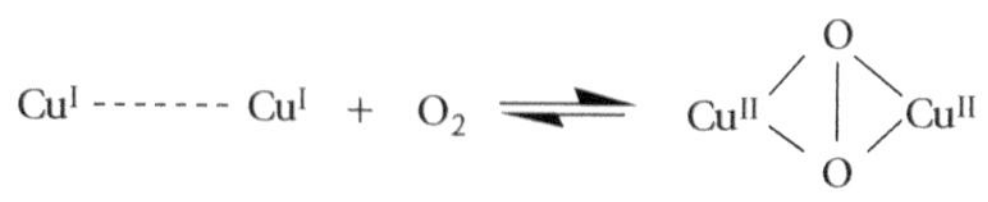

The molecular orbital description of this binding is shown in Figure 6.2, as previously discussed. Solomon and his coworkers (1996) studied this kind of bonding quantum-mechanically in details. It has been suggested by these authors that an interaction between the σ^* orbital of O—O and d_π makes a significant contribution in addition to the two molecular orbital interactions (A) and (C) of Figure 6.2. By the way, both types of binding—(b)-2 and (b)-3 of Figure 6.2—have been found with different Cu-dinuclear complexes (e.g., see Klein Gebbink *et al.*, 1999).

6.1.5. **Reactions of Oxygen Derivatives**

The O-atom in the singlet excited state ^1D behaves as a Lewis acid, but the reaction product with such an O-atom is regarded as an oxidized product. Examples are:

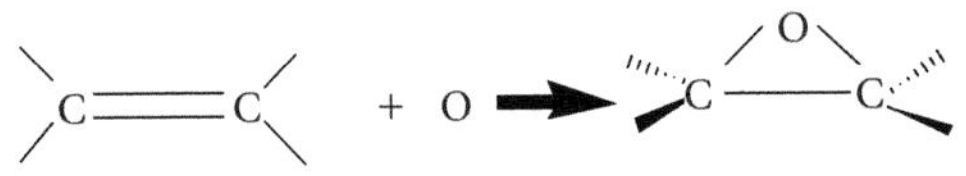

Singlet oxygen $O_2(^1\Delta_g)$ reacts quite differently from the ground state O_2 (triplet). A few examples of reactions of $O_2(^1\Delta_g)$ are:

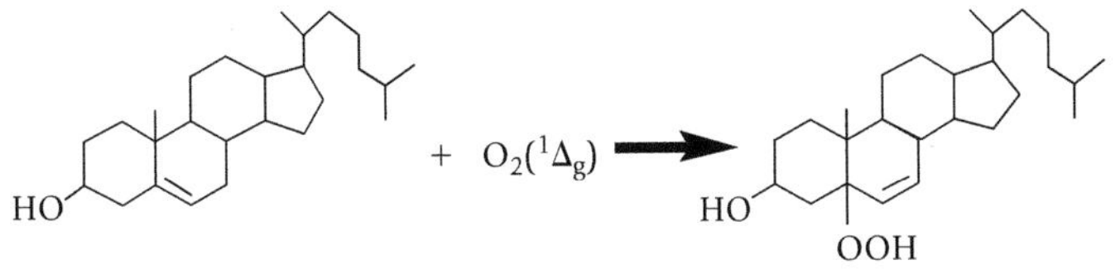

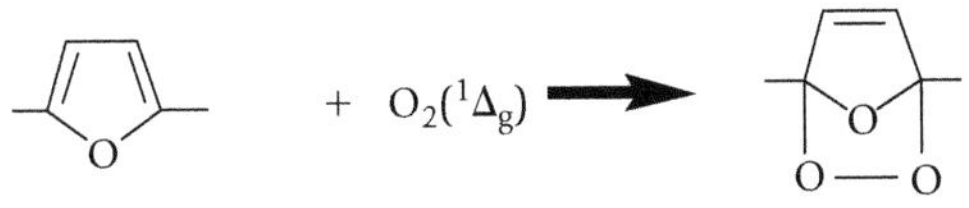

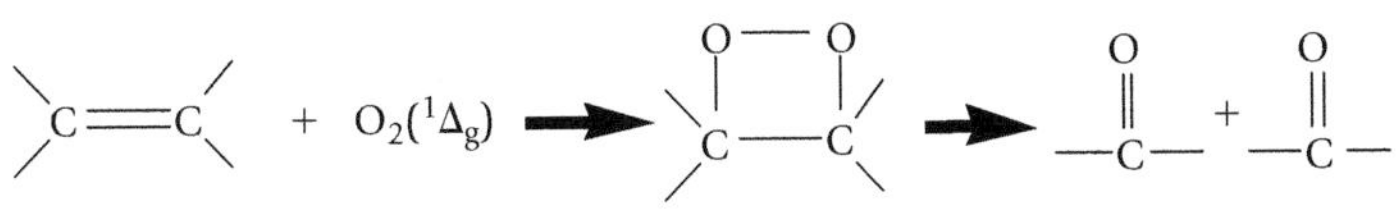

Important reactions of superoxide (O_2^-) include abstraction of H-atom from a hydrocarbon and others, disproportionation $(2O_2^- \rightarrow O_2 + O_2^{2-})$ and reduction to peroxide.

Peroxides can be simply O_2^{2-} or HOOH or ROOH. They have a single O—O σ-bond, and the bond can thermally be split homolytically (to $HO\cdot + \cdot OR$, etc.) or heterolytically (to $O^{2-} + O$ or $ROH + O$, etc.). An electron added to ROOH would go into the σ^*-antibonding orbital, and hence the O—O bond would split inevitably. That is:

$$ROOH + e \longrightarrow RO^- + \cdot OH \quad (\text{or } RO\cdot + OH^-)$$

Addition of two electrons simultaneously would produce RO^- and HO^-. A provider of the previous electron may be a transition element; then, for example

$$ROOH + Fe(II) \longrightarrow RO^- + \cdot OH + Fe(III)$$

If a further oxidation of Fe(III) is thermodynamically accessible,

$$ROOH + Fe(III) \longrightarrow ROO^--Fe(III) + H^+ \longrightarrow RO\cdot + (O^{-II}-Fe^{IV})^{2+} + H^+ \tag{a}$$

$$OO^{2-}-Fe(III) + 2H^+ \longrightarrow (FeO)^{3+} + H_2O \tag{b}$$

$$ROO^--Fe(III) + H^+ \longrightarrow (FeO)^{3+} + ROH \tag{b}$$

The first of these (mode (a)) involves (formal) one-electron oxidation of Fe(III), and the second/third (mode (b)) is nominally a two-electron oxidation of Fe(III). In a way, the first reaction can be viewed as "homolytic cleavage of O—O bond," and the second and third are "heterolytic cleavage" (Decker and Solomon, 2005).

The reaction of the first type is found in hydroperoxidase, catalase, and cytochrome c peroxidase as discussed in Chapter 5. In these enzyme systems, the free radical RO· immediately abstracts a hydrogen atom from the heme ring or another nearby amino acid residue, resulting in the formation of a free radical at the heme ring or the amino acid residue; that is, a species like $(O{=}Fe^{IV})(P\cdot)$, where $P\cdot$ is a porphyrin radical. This entity often is called compound I (CpI) and the entity $(Fe^{IV}{=}O)^{2+}$ is designated as ferryl ion.

The $(FeO)^{3+}$ is one oxidation step higher than $(OFe)^{2+}$, and it is nominally Fe^V-O^{-II}; this species often is expressed as $Fe^V{=}O$, indicating a double-bond character. This species can be designated as superferryl. The oxo species of this type can play the crucial role in the P-450 dependent monooxygenating enzymes.

6.2. REVERSIBLE O$_2$ BINDING: OXYGEN CARRIERS

The reversible binding of O_2 has been discussed earlier. Hemoglobin and myoglobin are heme-proteins and bind O_2 in an end-on fashion:

$$(\text{heme})Fe^{II} + O_2 \rightleftharpoons (\text{heme})Fe^{III}-O_2^-$$

Myoglobin is a single polypeptide, and hemoglobin consists of four subunits $\alpha_2\beta_2$. Cooperative effects among the four subunits increase the oxygen affinity of the heme groups in such a way that the saturation curve of oxygen uptake takes an S-shape. At lower oxygen pressure, the oxygen affinity of myoglobin is greater than that of hemoglobin, but hemoglobin's oxygen affinity exceeds that of myoglobin as the oxygen pressure increases. This situation suits the functions of both proteins. Hemoglobin accepts oxygen at a high pressure and yields it to myoglobin at a lower oxygen pressure. Hemoglobin is found in all vertebrates and some other animals and some plants.

Another iron-containing O_2-carrier is hemerythrin, which is found in some invertebrates such as annelids and arthropods. The O_2 binding site in hemerythrin is a dinuclear (nonheme) Fe^{II}—Fe^{II}. O_2 binds in an end-on manner to one of the Fe's, and the loose end of O_2 is believed to hydrogen-bond to the bridging OH group.

Hemocyanin is found in such invertebrates as gastropods (snail, etc.), cephalopods (octopus, etc.) and some crustaceans (malacostraca, etc.). It is a large protein consisting of multiple subunits. The molecular weight of gastropod's hemocyanin is up to 9×10^6 Da. The hemocyanin of vineyard snail (*Helix pomatia*) consists of 20 subunits. Hemocyanin binds O_2, as described earlier, in a side-on fashion across the two Cu's.

The further details of these oxygen carriers will be found in standard textbooks of biochemistry and bioinorganic chemistry. The reason for the different binding modes and what factors determine the modes are not very well understood.

6.3. **MONOOXYGENASES**

There are two kinds of oxygenase reaction. One is dioxygenase reaction, in which both Os of O_2 are incorporated into a single substrate. The other is monooxygenation reaction, in which only one O of O_2 is incorporated into the major substrate.

A variety of monooxygenases are known; they are dependent on either Fe (of both heme and nonheme) or Cu. They insert one O of the molecular O_2 into a substrate. One major type of monooxygenation reaction involves a reducing agent such as NADPH. Hydroxylation takes place at an aliphatic C—H. That is,

$$R\text{—}H + H^+ + NADPH + O_2 \rightarrow ROH + H_2O + NADP^+$$

■ **Figure 6.3.** Typical monooxygenation reactions catalyzed by P-450 enzymes.

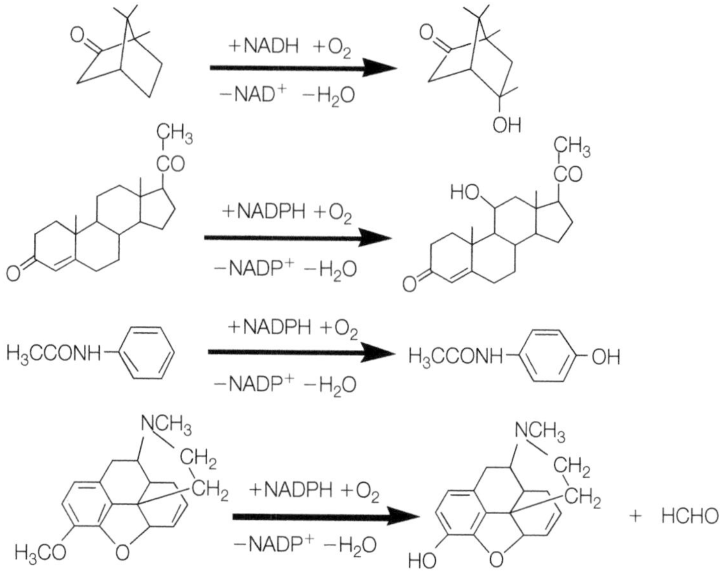

The other type of monooxygenation can be represented by:

$$R'{-}H + X \text{ (cofactor)} + O_2 \longrightarrow R'{-}OH + X'O$$

R'—H is mainly aromatic compounds (in which ring C—H is hydroxylated) and X' is derived from X.

6.3.1. **Monooxygenases Dependent on Cytochrome P-450**

The most versatile monooxygenases are those that are dependent on cytochrome P-450. A variety of substances are monooxygenated or dealkylated. A few examples are shown in Figure 6.3. The first three reactions are typical monooxygenation, where the oxygen is inserted into an aliphatic C—H bond. Hence reactions of this type often are characterized as hydroxylation. The fourth is a demethylation reaction, but it seems that it consists of monooxygenation of the methyl carbon and a subsequent spontaneous decomposition.

The catalytic site is a Fe-protoporphyrin-IX with a cysteinyl S⁻ negative ion as the fifth ligand. An outline of the mechanism is shown in Figure 6.4. This mechanism has been well established up to the crucial intermediate (vi) (see Groves, 2005 for a recent review). The controversy surrounds two issues: (1) the nature of intermediate (vi), and (2) how the oxo O is inserted into C—H bond.

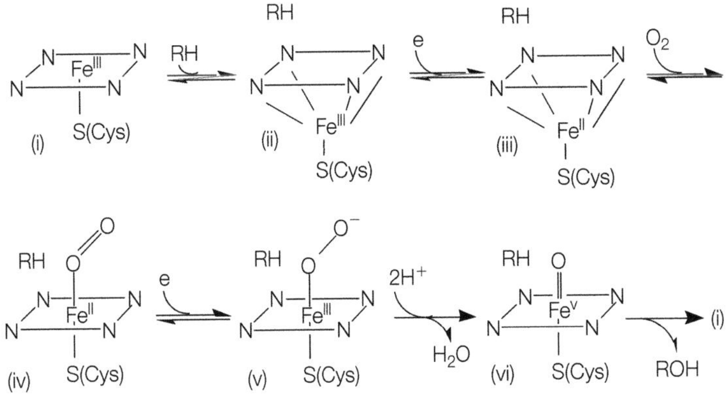

■ **Figure 6.4.** A mechanism of monoxygenation reaction by P-450 enzymes.

First, the nature of intermediate (vi): Is it similar to CpI found with horseradish peroxidase and catalase (see Chapter 5) or is it indeed $Fe^V{=}O$? Many researchers, including Mennier *et al.* (2004), opt for the first mode—intermediate (vi) is equivalent to $(Fe^{IV}{=}O)(P\cdot)$, where $P\cdot$ is the porphyrin ring from which a hydrogen atom is abstracted away. In other words, one of the two electrons to be given to O_2^{2-} is supposed to come from the Fe and the other from the porphyrin ring. The O atom of ferryl ion ($Fe^{IV}{=}O$) is then supposed to be inserted into a C—H bond. The ferryl ion associated with horseradish peroxidase or catalase has not been demonstrated to insert the oxo O into a C—H bond.

The alternative is to assume that it is nominally $Fe^V{=}O$ (superferryl). The formal charge v on Fe is based on the assumption that the O carries -2 charge (or oxidation state) as in any other situation. If so, the O has four lone pairs. One of them coordinates to the Fe as σ-donor. One of the remaining lone pairs can also be donated to the Fe through π-type interaction. The electronic configuration of this species can then be described as a resonance of three structures:

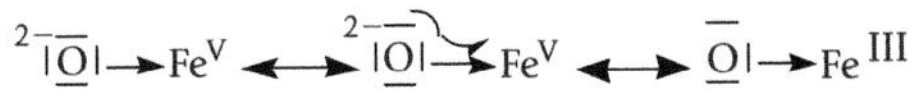

$$^{2-}\overline{|\underline{O}|}{\rightarrow}Fe^V \longleftrightarrow \, ^{2-}\overline{|\underline{O}|}{\rightsquigarrow}Fe^V \longleftrightarrow \underline{O}|{\rightarrow}Fe^{III}$$

$$(O^{2-}-Fe^V) \qquad (O{=}Fe^V) \qquad (O^0-Fe^{III})$$

The contribution of the last resonance structure may not be very large, but could be significant. This O is equivalent to the singlet O ($^1\Delta_g$), and would behave as such. Besides, it would be more electrophilic than a free O ($^1\Delta_g$), because it is coordinated to a strongly electron withdrawing Fe(III).

The next question is how the O-atom is inserted into a C—H bond. One mechanism proposed and believed by many is called oxygen rebound mechanism. It assumes the following. Here the process is expressed in terms of superferryl, but an identical process can be drawn for the case of ferryl ion:

$$\left(|\overline{\underline{O}}|{\rightarrow}Fe^{IV \text{ or } V}\right)^{2-} \longleftrightarrow \left(\cdot\overline{\underline{O}}|{\rightarrow}Fe^{IV}\right) \xrightarrow{\quad R{-}H \quad} {}^{R\,\cdot}$$

(ferryl or superferryl)   (assume superferryl)

$$\left(H{-}\overline{\underline{O}}|{\rightarrow}Fe^{\,IV} \longleftrightarrow H{-}\overline{\underline{O}}\cdot Fe^{III}\right) \xrightarrow{\quad R\,\cdot \quad} Fe^{III}$$
$$\searrow ROH$$

Experiments based on ultra-fast radical clock indicated that the apparent lifetime of the alkyl radical (70 fs) is far too short for the rebound mechanism (see Wong, 1998). A more recent article (Mennier *et al.*, 2004) discusses the mechanism (based on this model and CpI) in detail.

An alternative mechanism may be thrown in here. It is based on the super electrophilicity of the oxo O in the superferryl entity as mentioned earlier. If this supposition is reasonable, this oxo O may be assumed to be sufficiently electrophilic so that it attracts the pair of electrons of C—H σ-bond. That is:

$$(Fe^{V}{=}O\longleftrightarrow)\ Fe^{III}{\leftarrow}|\overline{\underline{O}} \xrightarrow{\quad RC{-}H \quad} Fe^{III}{\leftarrow}|\overline{\underline{O}}{\leftarrow}\overset{R}{\underset{H}{\overset{|}{\underset{|}{C}}}} \longrightarrow Fe^{III} + RCOH$$

This mechanism is essentially the same as the so-called concerted mechanism (see Bugg, 2001), but is not quite the same, for the insertion was not considered to be synchronous in the original mechanism (Newcomb *et al.*, 1995).

A final question is the difference between cytochrome P-450 and peroxidase (and catalase). As discussed in Chapter 5, the reaction mechanism involving peroxidase and catalase seems to be fairly well established; it is based on CpI, including the ferryl entity and a free radical on the porphyrin ring. However, the oxo O in this ferryl in peroxidase and catalase is simply reduced to O^{2-} state by receiving one more electron from a substrate. This implies that the oxo O in CpI type intermediate would not insert into C—H. The reason could be that the oxo O in the ferryl ion of CpI is not electrophilic enough due to the lower oxidation state of Fe, nominally Fe(II).

Then the question can be restated: Why does $Fe^{IV}=O$ form in the case of peroxidase (or O—O hemolytic splitting) while $Fe^{V}=O$ in the case of P-450 (O—O heterolytic splitting)? This difference is very likely attributable to the difference in the fifth ligand. Horseradish peroxidase and cytochrome c peroxidase have N^- of histidine at the fifth coordination site, and catalase has O^- of tyrosine, whereas the fifth ligand of cytochrome P-450 enzymes is S^- of cysteine. N^- and O^- are the so-called hard bases, and S^- is a soft base. S^- (cys) may be soft enough to allow the Fe to attain a higher oxidation state than N^- or O^- can. Delocalization of electron on S^- (cysteine) onto Fe in the high oxidation state (such as formally V) may contribute to the stabilization of the high oxidation state. It is interesting to note that a heme-chloroperoxidase has S^- (cys) at the fifth position as in P-450 heme, and oxidizes halide(s) with HOOH. This enzyme will be discussed in Chapter 9 in terms of halogen biochemistry.

6.3.2. **Nonheme Mononuclear Iron Monooxygenases**

Several different types of nonheme mononuclear iron mono-oxygenases are known. They are pterin dependent Fe(II)-enzyme, α-ketoglutarate (or 2-oxo glutarate) dependent Fe(II) enzyme, and others. The active site of many of these enzymes is said to have a motif of facial arrangement of two histidine and one carboxylate ligands around Fe atom (in an octahedral arrangement) (Lange and Que, 1998).

Let's discuss first pterin-dependent aromatic amino acid hydroxylase such as phenylalanine hydroxylase. The active site of this enzyme consists of a biopterin and Fe(II). The quinoid form of dihydrobiop-terin is reduced to tetrahydrobiopterin by NADH, and this reduced biopterin together with Fe(II) carries out the catalytic hydroxylation of the aromatic ring. The overall reaction is:

Phenylalanine $+ O_2 +$ NADH $+ H^+ \longrightarrow$ tyrosine $+ NAD^+ + H_2O$

But the immediate reaction is:

Tetrahydrobiopterin $(XH_2) + O_2 +$ phenylalanine $\longrightarrow$
X + tyrosine $+ H_2O$

Biopterins are shown in Figure 6.5. The structure is similar to that of flavin, and it can undergo two one-electron redox reactions like quinone and flavin. This suggests that it is capable of providing two electrons to the reaction system. This is similar to the situation of

■ **Figure 6.5.** Biopterins (tetrahydropterin and dihydropterin).

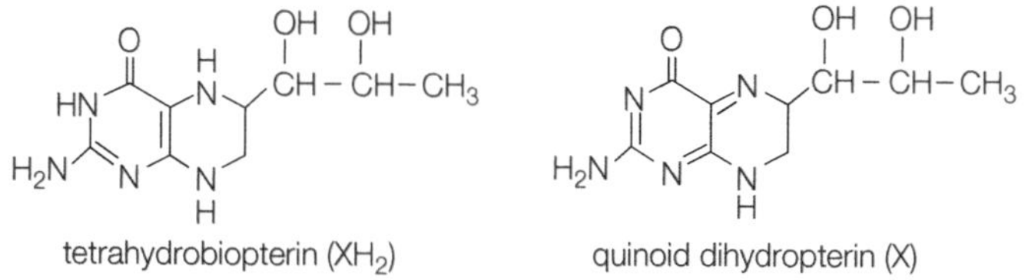

tetrahydrobiopterin (XH$_2$)   quinoid dihydropterin (X)

■ **Figure 6.6.** A mechanism of biopterin-dependent monooxygenation reaction (Ochiai).

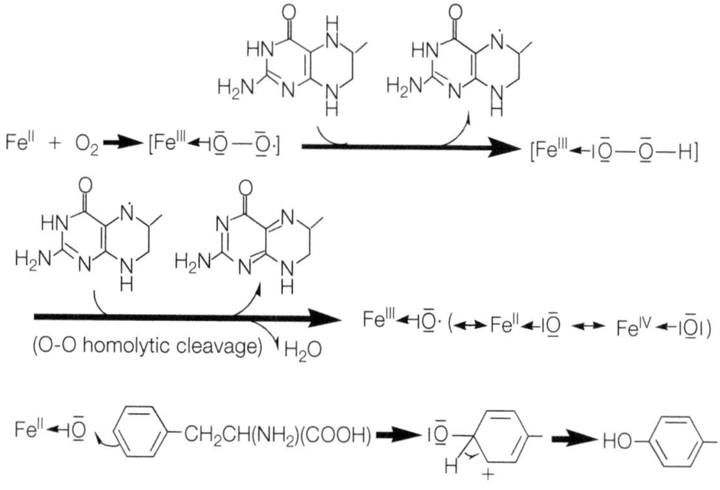

cytochrome P-450 enzyme reactions. A difference is that the iron is simply Fe(II) coordinated by two histidine and one carboxylate ligands unlike P-450, which is a heme with cyteine S⁻ at the fifth site. Besides, the iron in P-450 is in Fe(III) state initially. The ligands around Fe(II) in this enzyme suggest that a very high oxidation state (i.e., Fe(V)) may not be attainable here. A speculative mechanism is shown in Figure 6.6. Here the active oxygenating entity is assumed to be ferryl (FeO)$^{2+}$ (see Bugg, 2001, though the mechanism shown in this article does not make chemical sense). The oxo O of the ferryl may not be as electrophilic as that of superferryl (P-450), but may be sufficient in this case, because the substrate in this case is a stronger nucleophile (i.e., an aromatic ring). Alternatively, a rebound mechanism may be operative here. This suggestion is consistent with the mechanism to be discussed next for α-keto glutarate dependent Fe(II) enzymes, because the ligands structure about Fe(II) is approximately the same in both pterin-enzymes and α-keto glutarate dependent enzymes.

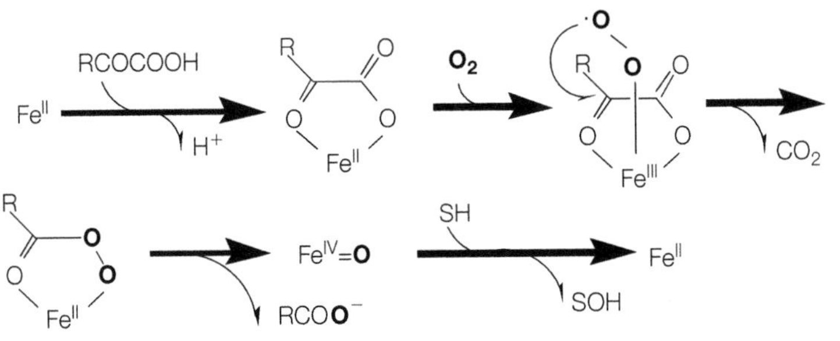

■ **Figure 6.7.** A mechanism of 2OG (2-oxo-glutarate) dependent monooxygenation (Que and Ho, 1996).

α-Keto glutarate (2-oxoglutarate, 2OG) along with Fe(II) is involved in a number of monooxygenation and related reactions (see Que and Ho, 1996; Schofield and Zhang, 1999). The monooxygenation reaction can be expressed by:

$$RH + HO_2CCH_2CH_2(C{=}O)CO_2H + O_2 \rightarrow ROH$$
$$+ HO_2CCH_2CH_2(C{=}O)OH + CO_2$$

Substrates for this type of enzyme reaction include proline, lysine, γ-butyrobetaine, and thymine. Hydroxylation reactions take place at an aliphatic C—H. The active site has the same motif of facial coordination of two histidine and a carboxylate ligands. A mechanism proposed by Que and his coworkers (Que and Ho, 1996) is shown in Figure 6.7. According to this mechanism and a similar one (by Schofield and Zhang, 1999), the oxygenating agent is the ferryl ion ($Fe^{IV}{=}O$). In the previous paragraph, the ferryl ion is assumed to be a strong enough electrophile for aromatic ring hydroxylation. However, it was assumed that a higher oxidation state $Fe^V{=}O$ (superferryl) was necessary for hydroxylation of an aliphatic C—H (in the case of P-450 enzymes).

If both of these assumptions are reasonable, then the ferryl ion as depicted in Figure 6.7 should not be able to hydroxylate aliphatic C—H bonds; the fact is that it does so. One possibility is that the electronic state of $Fe^{IV}{=}O$ may be significantly different in the composition of resonance structures, due to the different ligands' systems. Out of the three structures ($Fe^{IV}{-}O^{-II}$; $Fe^{III}{-}O\cdot^{-I}$, $Fe^{II}{-}O^0$), $Fe^{III}{-}O\cdot^{-I}$ may be predominant, and hence the reaction mechanism could be the rebound mechanism as described earlier for a P-450 mechanism. In accord with this suggestion, several other 2OG-dependent reactions such as those shown in Figure 6.8 involve a cyclization reaction that seems to require a free radical mechanism.

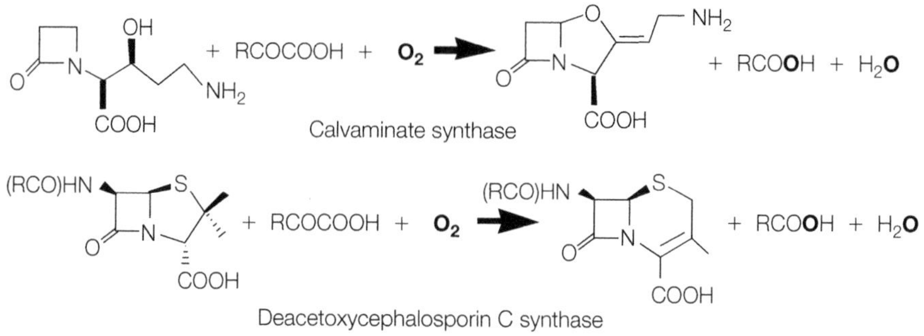

■ **Figure 6.8.** Two examples of 2OG dependent monooxygenation that involve cyclization (from Que and Ho, 1996).

Calvaminate synthase

Deacetoxycephalosporin C synthase

In conclusion, the catalytic entity is $Fe^{IV}=O$ (ferryl) in all these enzymes, and the facial (2-histidine and 1-carboxylate) structure is to stabilize mostly the $[Fe^{III}O\cdot^-]$ state of the three possible resonance structures.

6.3.3. Nonheme Dinuclear Iron Monooxygenases

Methane monooxygenase consists of three subunits; the catalytic unit is hydroxylase (MMOH) and the active site is a nonheme iron dinuclear complex. The structure is deceptively simple; each of Fe(II) in the reduced form is coordinated by one histidine and one glutamate, and they are bridged by the carboxylate of glutamate, and a μ-O that belongs to a glutamate. In addition, one of the Fes has one H_2O attached. In the oxidized form, the two Fes are bridged by the carboxylate of glutamate and two OH^-s. A mechanism proposed (see Lange and Que, 1998) is shown in Figure 6.9. The monooxygenating agent is believed to be intermediate Q. How Q hydroxylates methane has not been determined as in all the oxygenation reactions. One possibility (Lange and Que, 1998) is to assume that one of the bridging Os is to become a super ferryl (as defined earlier) as shown in Figure 6.9, and this superferryl is the hydroxylating agent. If this is reasonable, the final question is why the superferryl is possible. With the ligands available on the Fes, it is highly unlikely that Fe attains such a high oxidation state as +V (nominally though). It might be possible that the electron to be removed (to become V) is accommodated into the dinuclear system, nominally $Fe^{III}—Fe^V=O$. The electron actually is distributed over the dinuclear system. An alternative is to assume $Fe^{IV}=O$ and the rebound mechanism (see also Guallar *et al.*, 2002; Ryle and Hausinger, 2002).

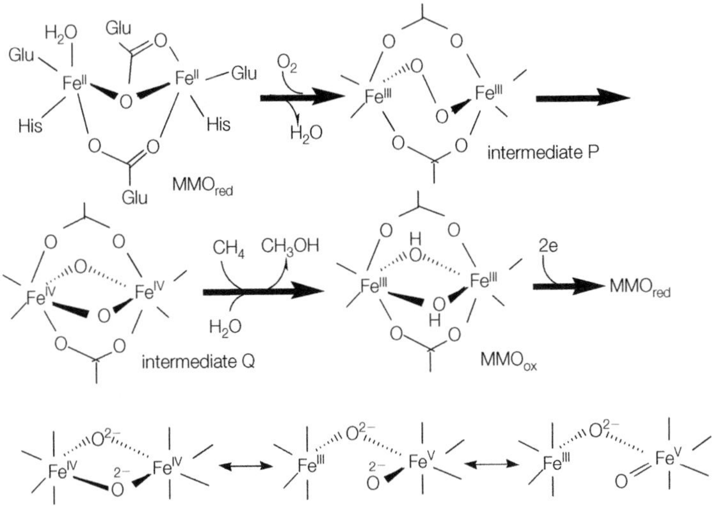

■ **Figure 6.9.** A mechanism of reaction catalyzed by methane monooxygenase (Ochiai).

6.3.4. **Copper Monooxygenases**

Tyrosinase has two functions: monooxygenase and oxidase. The reactions catalyzed by it are:

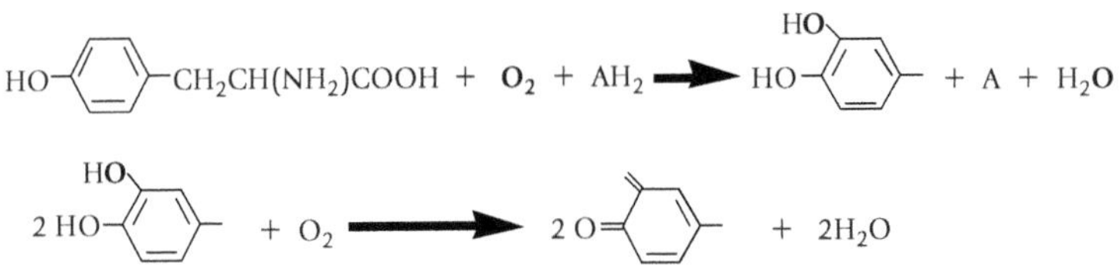

Tyrosinase has a dinuclear Cu catalytic site similar to the oxygen carrying hemocyanin. Each of the Cus is coordinated by three histidine residues, and two Cus are magnetically coupled. As a monooxygenase it hydroxylates an aromatic C—H. Dopamine β-monooxygenase (DβM) has also a dinuclear Cu site, but the two Cus are far apart and not coupled. This enzyme hydroxylates an aliphatic C—H.

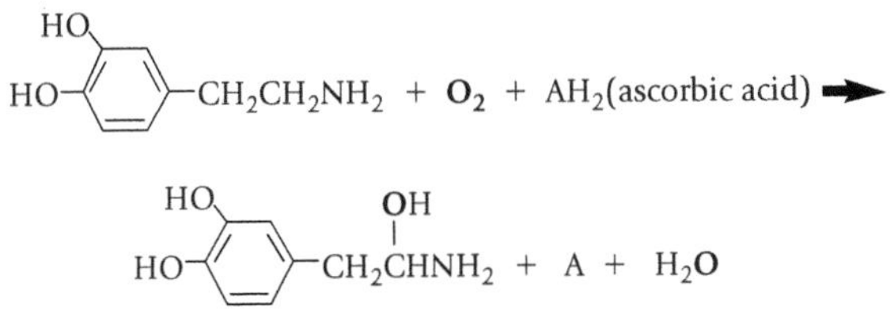

Decker and Solomon (2005) argue that (a) O_2 binds in a side-on bridging fashion and forms $Cu^{III}—O^{2-}$ as a result of the O—O hemolytic cleavage in the case of tyrosinase, but that (b) a single

Cu(I) binds O_2 forming Cu^{II}—O_2^- in the case of DβM. Their quantum chemical calculation (DFT theory) suggests that Cu^{III}—O^{2-} is indeed Cu^{II}—$O\cdot$, and this oxo O radical is said to abstract H from the aromatic ring (assuming a rebound mechanism). Alternatively, a Cu^{II}—O^{2-}—Cu^{IV}=O may form in a tyrosinase system, and the O of Cu^{IV}=O (which is in fact close to Cu^{II}—O^0) could function as an electrophile, and the reaction can then be described as an electrophilic aromatic ring substitution reaction. The electrophilic character of the reaction has been demonstrated (Mirica et al., 2005).

In the case of DβM, the superoxy radical bound to Cu^{II} is assumed to abstract an H-atom from the aliphatic C—H bond. A second electron is provided from the second Cu(I) nearby and is to split the O—O bond of HOO—Cu^{II} homolytically, forming a ·OH radical that combines with the free radical C of the β position (Decker and Solomon, 2005). In peptidylglycine-α-hydroxylating monooxygenase, O_2 has been found to bind to one of the two Cus (Cu_B) in an end-on manner (Prigge et al., 2004). It forms $Cu_B(II)$—$O_2^{\cdot-}$ after an electron transfer from $Cu_A(I)$, and that is supposed to abstract an H-atom from the substrate (Prigge et al., 2004).

6.4. DIOXYGENASES

There are two different kinds of catechol dioxygenases: Fe(II)-dependent extradiol dioxygenases and Fe(III)-dependent intradiol dioxygenases. Typical reactions are shown below: (a) is an extradiol dioxygenation catalyzed by protocatechuate 4,5-dioxygenase, and (b) an intradiol dioxygenation by protocatechuate 3,4-dioxygenase. Both reactions are summarized as —C=C + O_2 → —C=O + O=C—.

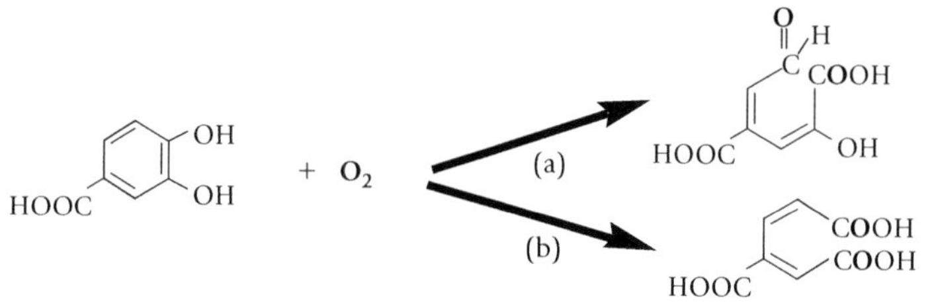

The ligation structure of the catalytic site is also different. The Fe(III) enzymes have two hisitidine and two tyrosine ligands about Fe(III), and the ligands around Fe(II) in the extradiol dioxygenases are two histidine and one glutamate ligands arranged in a facial manner,

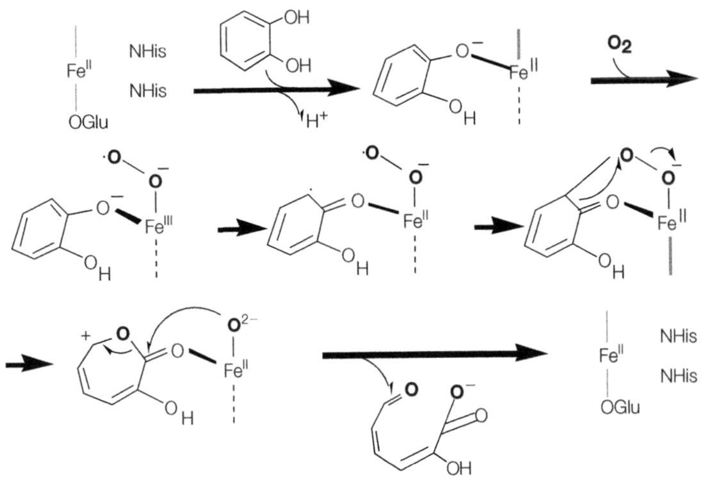

■ **Figure 6.10.** A mechanism of extra diol dioxygenation reaction (modified from Que and Ho, 1996; and Bugg, 2001).

the same motif of those mononuclear monooxygenases mentioned earlier.

Let's look at the extradiol dioxygenase, which has the same type of catalytic sites of nonheme Fe(II) monooxygenases. In monooxygenase systems, the basic reaction can be described as $O_2 + 2e \rightarrow O^{2-} + O$. The two electrons are provided by the cofactor, either 2-oxo-glutarate or tetrahydrobiopterin. The O ends up in the ferryl ion $Fe^{IV}=O$, which is to hydroxylate the substrates. In the extradiol dioxygenases, no cofactor is present. The substrate, a catechol itself binds to the Fe(II) in an extradiol dioxygenase through the two OHs but the binding is uneven; one of them with the proton removed binds more strongly than the other. This is in contrast to an intra-diol case where the deprotonated catechol chelates the Fe(III). O_2 is assumed then to bind to the Fe(II), becoming Fe^{III}—$O_2^{\cdot -}$. Then the catechol reduces the Fe(III) back to Fe(II), forming a semiquinone itself. The mechanism from this stage on has not been well delineated. A mechanism (modified from Que and Ho, 1996; and Bugg, 2001) is presented in Figure 6.10. In this mechanism the OH group that does not strongly bind to the Fe(II) remains untouched.

In the intradiol case, both deprotonated OH groups of a catechol bind to the Fe(III). The catechol is then to transfer an electron to the iron, forming an ortho semiquinone and Fe(II). O_2 binds across the Fe(II) and the free radical on the catechol ring. Figure 6.11 is a mechanism proposed by Que (Que and Ho, 1996) and modified by Bugg (2001).

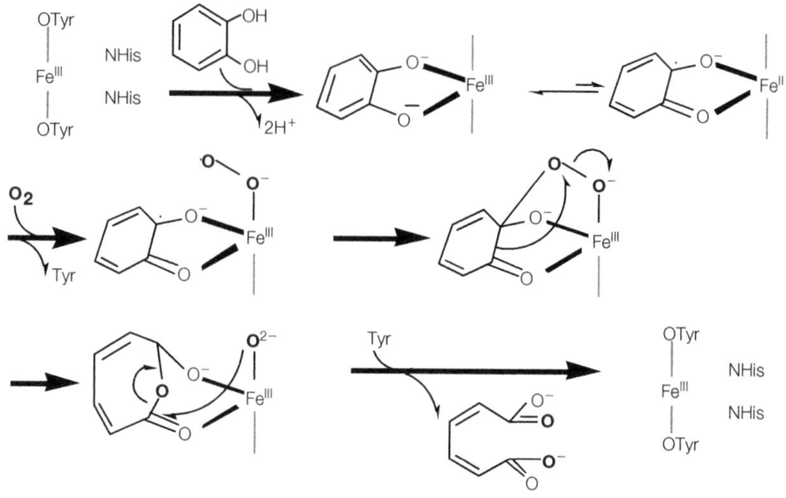

■ **Figure 6.11.** A mechanism of intra diol dioxygenation reaction (modified from Que and Ho, 1996; and Bugg, 2001).

Three intermediates in a Fe(II)-dioxygenase recently have been determined by x-ray crystallography (Kovaleva and Lipscomb, 2007). The first intermediate $[O_2Fe^{II}(orthodihydroxysubstrate)]$ seems to have the same structure as that shown in Figure 6.10. The second intermediate, the one in which the open end of O—O binds to the substrate, as interpreted by these authors, is not the same as that shown in Figure 6.10. The third intermediate is the open-ring end product bound with the Fe, which cannot shed light on the crucial step of ring-opening process. That is, the crystal structures of the intermediates are consistent with Figure 6.10, except for the crucial second intermediate. The difficulty lies in that the electron density map is the only information available to determine which atom binds to which atom.

Aside from this issue, there is another fundamental issue with Fe(II)/extradiol *vs* ital Fe(III)/intradiol dioxygenases. That is, the substrates for extradiol dioxygenases do not necessarily have two adjacent OH groups as the substrate in this x-ray study (Kovaleva and Lipscomb, 2007). Yet they are dioxygenated in extradiol manner. Hence a simultaneous coordination of two OHs (on adjacent positions) may not be necessary in the extradiol mechanism. This issue was discussed by this author (Ochiai, 1975), and is implied in the argument of Figure 6.12.

A crucial question is the selectivity; why one dioxygenates the intradiol position, and the other, the extradiol position. It has to do with

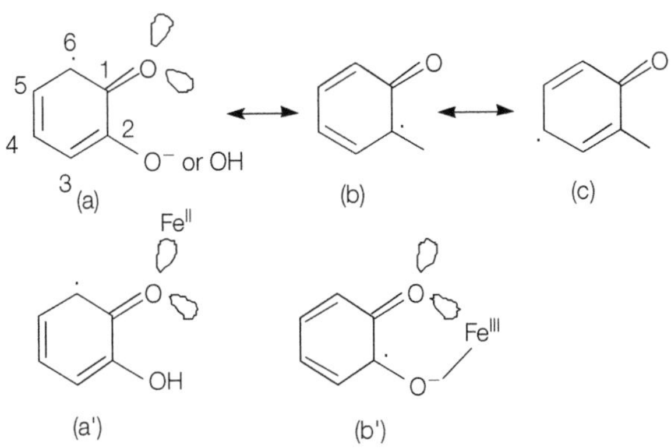

■ **Figure 6.12.** The difference between the two modes of dioxygenation (Ochiai).

the initial oxidation state Fe(II) or Fe(III) and the spin localization of the ortho semiquinone formed and its relationship to the free radical end of $O_2^{\cdot-}$ (bound to Fe). At the stage of semiquinone, two systems would arrange the components as shown in Figure 6.12. Fe(III) prefers to chelate both catechol Os, but Fe(II) is not strong enough to symmetrically chelate. The semiquinone has three resonance structures, and the carbonyl O is sp^2. Hence the iron can bind in two ways, as shown in Figure 6.12. In the intradiol case the Fe(III) chelates both Os and hence the odd electron at position 2 is nearest to the free radical end of $O_2^{\cdot-}$ (bound to Fe). In the extradiol case, the Fe(II) binds sufficiently to only one of the Os, which has become a carbonyl O, and could bind in the manner (a'). Why it would not bind in the manner (b') cannot be answered without further elucidation of the intimate structure of the active site. If this supposition is reasonable, then the free radical end of $O_2^{\cdot-}$ (bound to Fe) would bind to position 6 and lead to the extradiol type of dioxygenation. This is only a speculation and requires further elucidation.

6.5. **PROSTAGLANDIN ENDOPEROXIDE SYNTHASE**

Prostaglandin endoperoxide synthase (PGHS) catalyzes the first step in the conversion of arachidonic acid (AA) to prostaglandin derivatives. This step actually consists of two reactions: cyclooxygenase reaction (i) and peroxidase reaction (ii) (see the following reaction). Reaction (i) is a kind of dioxygenation, though it is quite different from those associated with nonheme dioxygenases mentioned earlier. Both reactions (i) and (ii) are believed to be catalyzed by the single enzyme (PGHS), which contains a heme group (Miyamoto *et al.*, 1974, 1976).

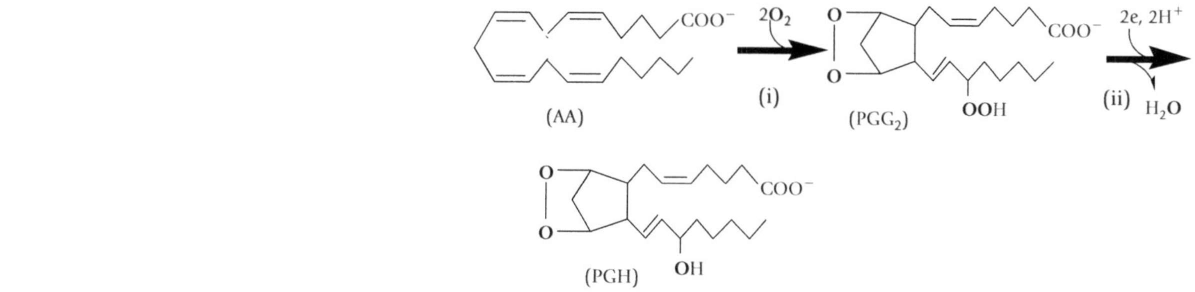

X-ray crystallographic studies (Picot et al., 1994; Malkowski et al., 2000) of the enzyme showed that the main active site is a protoporphyrin-IX heme in which the fifth ligand is N^- of a histidine. This is the same as that of peroxidases (see Chapter 5). The peroxidase activity (ii) is similar to that of cytochrome c peroxidase (see Chapter 5). It is believed to involve:

$$P—Fe^{III}—Por + ROOH \rightarrow P—(Fe^{IV}=O)—Por + ROH$$

where P = protein and Por = porphyrin. ROOH accepts an electron of Fe(III) and then splits in the RO—OH bond homolytically into RO· and $Fe^{IV}=O$. The former abstracts an H-atom from an amino acid residue. This residue has been identified to be tyrosine-385 (Karthein et al., 1988; Shimokawa et al., 1990). Dietz and coworkers (1988) postulated that the previous reaction produced Fe(V) (instead of RO· and Fe(IV)), and Fe(V) somehow removed an electron from tyrosine residue (by an intramolecular electron transfer according to the authors). This mechanism still seems to be endorsed by recent authors (Kulmacz, 2005). Mechanism of reactions of this type (peroxidase reaction) is discussed more fully in Chapter 5.

The mechanism of cyclooxygenase activity (i) is not well understood. It has been demonstrated that the ESR spectrum changed little when a substrate was added to the enzyme under an anaerobic condition, but that the ESR obtained upon addition of the substrate under an aerobic condition was very similar to that obtained when the product (PGH) was added instead. This ESR signal was reminiscent of a tyrosyl radical (Karthein et al., 1988). This tyrosyl radical was thought to be essential to the cyclooxygenase activity (Karthein et al., 1988).

A mechanism proposed a long time ago (Hamberg and Samuelson, 1967) is as follows. A free radical (now identified as tyrosyl-385) derived from a reaction such as the one earlier abstracts an H-atom from C-13 of AA and turns AA to a free radical ($\cdot$C—the odd elec- tron—is then transferred to C-11). O_2 can then add to C-11 and form the endoperoxide. This will result in the formation of a radical at C-15. O_2 adds to C-15, and the resulting ROO$\cdot$ would abstract the H-atom back from the tyrosine. This completes a reaction cycle. The same mechanism is found in recent reviews (Marnett, 2000; Kulmacz, 2005). The crucial issue here is how the first hydroperoxide may be produced. Once a small amount of hydroperoxide is formed, the reactions, both (i) and (ii), will proceed.

REVIEW QUESTIONS

1. The ground state of free O_2 is a triplet, a biradical. When it is bound to a metal ion in a "side-on" way (one-to-one) as in $[O_2Ir(PPh_3)_2Cl]$, in what state (triplet or singlet) would the bound O_2 be? Explain in terms of the molecular orbitals.

2. Fe(II) in hemoglobin is in the high-spin state (i.e., S = 2), and when it binds the ground state O_2 (S = 1), the resulting O_2—Fe (hemoglobin) is diamagnetic (no unpaired electron). Discuss.

3. How might one prove that the O incorporated in the product does indeed come from the O_2 molecule in monooxygenation reactions?

4. The ferryl ion often is expressed as $(Fe=O)^{2+}$. What does this description actually mean?

5. The variety of the ways in which O_2 reacts with biomolecules is rather large. List as many of them as possible.

6. A similar mechanism seems to be operating in both an iron enzyme, methane monooxygenase, and a copper enzyme, tyrosinase. Discuss.

PROBLEMS TO EXPLORE

1. One of the intriguing questions with regard to O_2-carriers (i.e., hemoglobin (myoglobin), hemerythrin, and hemocyanin) is how and why their oxygen-binding modes are different. Discuss this issue—what factors determine the different modes of O_2-binding?

2. O atom in the first excited singlet state reacts differently from the ground state O. Explore why. How may the singlet O be produced? In the processes of biochemical monooxygenation reactions, is there a possibility that a singlet O might be involved? How?

3. A few different mechanisms have been proposed for P-450 monooxygenases. Discuss pros and cons based on published data and theories available.

4. Several different types of nonheme mononuclear iron-dependent monooxygenases have been found. It has been shown that the active site has the same motif of facial coordination of a carboxylate and two histidine ligands. What significance does this feature have for the mechanism?

5. The difference between extradiol and intradiol dioxygenation reactions is quire subtle and interesting. The extradiol dioxygenases depend on Fe(II), and the intradiol dioxygenases use Fe(III). One explanation for the different reactions on Fe(II) or Fe(III) is presented in the text. Explore other likely mechanisms and compare.

Metal-Involving Free Radical Reactions

7.1. A SURVEY OF BIOLOGICALLY RELEVANT FREE RADICALS

A number of metalloenzymes that involve free radicals on their reaction mechanisms have been uncovered in recent years. In many of these enzymes an amino acid radical plays an essential role as the catalytic site. Examples of such enzymes include iron-dependent ribonucleotide reductase, galactose oxidase, and prostaglandin endoperoxide synthase. B_{12}-coenzyme dependent enzymes involve an adenosyl free radical that forms upon addition of a substrate.

The free radical species involved in biochemical systems can be classified into (a) the intermediates of quinones and flavins, (b) oxygen or dioxygen radicals ($O\cdot^-$ or $HO\cdot$; $O_2\cdot^-$ or $HO_2\cdot$) or its analogs ($RO\cdot$ or $RO_2\cdot$, which is derived from $ROOH$), (c) radicals on some amino acid residues such as tyrosyl (phenoxy radical), thiol radical (of cysteine) and alkyl (α-carbon of glycine in pyruvate formate-lyase), and (d) alkyl radicals $R\cdot$. The last group includes the free radical intermediate in enzymatic reactions that are formed through H-abstraction from a substrate.

Quinones and flavins are the important organic compounds that can relatively easily participate in one-electron transfer reactions, because semiquinone, the free radical intermediate in the reduction of quinone or oxidation of hydroquinone, is relatively stable. Flavins have the same type of chemistry as quinones (see Fig. 7.1.).

Quinones are employed as electron transfer agents in the mitochondrial or photosynthetic electron transfer system, and flavins function as an electron transfer intermediary in a number of enzymes that contain iron-sulfur units and molybdenum. The cofactor of mitochondrial amine oxidase recently has been determined to be a kind of quinone.

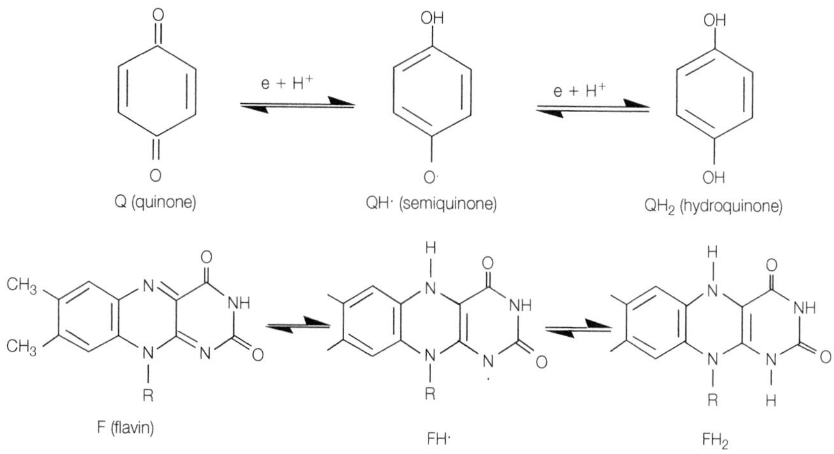

■ **Figure 7.1.** Quinone and flavin.

Oxygen and dioxygen radicals often are hypothesized to be involved in the monoxygenation and dioxygenation reactions. For example, cytochrome P-450 dependent monoxygenation involves the so-called ferryl ion $(FeO)^{3+}$, which is hypothesized to abstract H-atom from the substrate; thus

$$(FeO)^{3+} \Leftrightarrow (Fe^{IV}, O^{\cdot-}) + RCH_2{-}H \rightarrow$$
$$(Fe^{IV}, OH^-, RCH_2{\cdot}) \Leftrightarrow (Fe^{III}, HO{\cdot}, RCH_2{\cdot}) \rightarrow Fe^{III} + RCH_2OH$$

The oxygen radical postulated here needs to be associated with the heme iron throughout this reaction because an electron is assumed to be transferred between this O entity and the iron center as described. This mechanism is called *rebound* mechanism. Other examples of enzymes of this category include methane monooxygen-ase and numerous other iron- or copper-dependent oxygenases and peroxidases, and perhaps, isopenicillin N synthase. These reactions are dealt with in Chapters 5 and 6.

Oxygen and dioxygen radicals are also believed to participate in many physiological processes including the aging process of the cell mem-branes and the bacteriocidal effect of macrophage cells. Readers are referred to a review article by Valentine *et al.* (1998) as an example, and also to Chapter 11. Phenoxy radical (of tyrosine) and thiol radi-cal (of cysteine) have been identified as the essential radical entities in several metalloenzymes. Adenosylcobalamin (=vitamin B$_{12}$ coen-zyme) is believed to produce the adenosyl radical by splitting the

Co—C bond homolytically upon binding a substrate to the enzyme protein. There is some indication in certain enzyme systems such as ethanolamine ammonia lyase that the radical on the adenosyl residue actually is transferred to an amino acid residue, particularly a cysteine. The thiol radical thus created is considered to participate in the catalytic activity, abstracting H-atom from the substrate.

Pyruvate formate-lyase itself is a metal-free enzyme, but is activated by an iron-requiring enzyme containing adenosylmethionine and dihydroflavodoxin as cofactors. The activated enzyme has been demonstrated to have a free radical at α-carbon of a glycine residue (Gly-734), which is at the catalytic site (Wagner *et al.*, 1992).

A fair number of enzymes are now known to involve a free radical as a catalytic entity (a recent review: Frey, 2001). This implies that certain kinds of biochemical reactions would not take place without a participation of a free radical. We shall discuss briefly "why radicals?" and then describe briefly what is known about the reactivities of free radicals.

7.2. **WHY RADICALS?**

Free radical reactions usually involve a transfer of an entity containing an odd number of electrons. The most common entity of this kind is H-atom. Substrates such as hydrocarbons that have no polar bond are not very reactive and cannot easily be made to react in an acid/base manner. For example, a heterolytic cleavage of C—H bond in hydrocarbons to form a carbonium ion or carbanion is virtually unknown, unless the C—H is activated by an adjacent functional group. The only feasible means to render such a C—H bond reactive appears to be an abstraction of H-atom from it, leaving an unpaired electron on the carbon atom. This free radical site on the carbon atom is very reactive. It can abstract a H-atom from another moiety, recombine with another free radical species or undergo some other reactions. Indeed, many of the biologically important free radicals function as a hydrogen atom abstractor from a rather inert substrate such as a hydrocarbon or hydrocarbon-like compound.

Pyruvate formate-lyase has a free radical at α-carbon of a glycine residue (G-734) and catalyzes the following reaction:

$$CH_3COCOOH + CoA—SH \rightarrow CoA—S—(C{=}O)CH_3 + HCOOH$$

The regular acetyl-CoA formation from pyruvate by another enzyme, pyruvate dehydrogenase, requires NAD^+ and produces CO_2 as a product. This likely involves a hydride abstraction from COOH by NAD^+ forcing formation of $CH_3C^+{=}O$ (and release of CO_2). The following step is an attack of the $O{=}C^+$ by ^-S—CoA; a reaction of acid-base type.

In contrast, this reaction appears to be simply of an acid-base type. That is, the thiol of CoA—S^- attacks the carbonyl carbon of pyruvate and displaces $^-$COOH (and $^-$COOH + $H^+ \rightarrow$ HCOOH). However, the reaction does not seem to proceed this way, likely because the C—C bond (between C=O and COOH) is difficult to be cleaved heterolytically as needed. The reaction may require the involvement of free radicals. For example, a possible reaction scheme may be:

$$\cdot R \,(\text{glycine radical}) + H—S—CoA \rightarrow R—H + \cdot S—CoA$$
$$\cdot S—CoA + CH_3COCOOH \rightarrow CH_3CO—S — CoA + \cdot COOH$$
$$\cdot COOH + RH \rightarrow HCOOH + R\cdot$$

It seems necessary that the second and the third step are to take place simultaneously.

7.3. REACTIVITIES OF FREE RADICALS

Free radical reactions in general are not understood as well as reactions of polarized bonds (involving heterolytic bond cleavage and formation). Let us first examine the H-atom abstraction reaction by a free radical. A highly reactive radical species abstracts H-atoms rather indiscriminately, whereas a mildly reactive one does so more selectively. For example, Table 7.1 gives the relative reaction rates of H-abstraction from hydrocarbons by several free radicals (Isaacs, 1987). The variation in rate (except for H-abstraction from methane) is rather small in the cases of reactive free radicals such as $Cl\cdot$, $Me\cdot$, and t-BuO$\cdot$, whereas selectivity is evident in the cases of moderately reactive $Cl_3C\cdot$ and $Br\cdot$. The variations in the activation energy of H-atom abstraction by moderately reactive free radicals parallel somewhat the variation in the bond dissociation energy of C—H, as shown in Figure 7.2 (Isaacs, 1987). In this case the H-abstraction rate is determined to a large extent by the dissociation energy of the bond to be cleaved. That is, a linear relationship seems to hold between the activation energy and the enthalpy of reaction for these moderately reactive species.

Table 7.1. Relative Rate Constants k ($M^{-1}s^{-1}$) of Hydrogen Abstraction by Free Radicals

R—H	·CH$_3$	·O(t-Bu)	·Cl	·Br	·CCl$_3$
RCH$_3$	0.4	0.1	0.8	0.0001	
R$_2$CH$_2$	4.0	1.2	3	0.005	5
R$_3$CH	9.4	4.4	3.6	0.02	160
PhCH$_3$	(1)	(1)	(1)	(1)	(1)
PhCH$_2$R	13	6.9	5.5	37	260
PhCHR$_2$		4.7	2.0	10	50
Ph$_2$CH$_2$		9.6	7.2	18	160

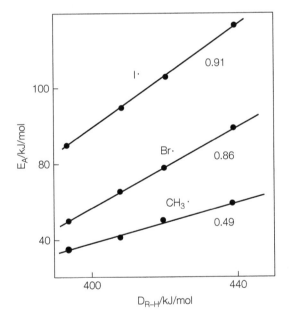

■ **Figure 7.2.** Reactivity-selectivity relationship between the activation energy E_A for hydrogen abstraction from R—H and the bond dissociation energy $D_{R—H}$. The numbers given (0.91, 0.86, 0.49) are the slopes, which are a measure of selectivity (from Ochiai, 1994a).

The activation energies for the H-abstraction from ethane by reactive radical species (reaction below) are given in Table 7.2 (March 1985). The activation energy is very low for F· and Cl·, and does not correlate well with the enthalpy of reaction (which is the difference in bond energy between C—H and X—H). The HO· (as well as O·$^-$) radical also behaves as a highly reactive indiscriminate radical (Neta, 1976). The ·CH$_3$ radical requires a substantial amount of activation energy in abstracting H-atom, but appears as reactive as Cl· or

Table 7.2. The Activation Energies for H-Abstraction from Ethane

X	F·	Cl·	Br·	$CH_3O·$	H·	·CH_3
Activation energy (kJ/mol)	1.3	4.2	55	31	38	49

Table 7.3. Bond (Dissociation) Energies (kJ/mol) of Biologically Relevant Compounds*

Bond	Energy	Bond	Energy	Bond	Energy
H—CH_3	435	H—CH_2CH_3	410	H—C_6H_5	431
H—$CH_2C_6H_5$	356	H—CH_2OH	389	H—CCl_3	402
H—(C=O)H	368	CH_3—CH_3	368	CH_3CO—$COCH_3$	347
H—OH	498	H—OCH_3	427	H—OC_6H_5	356
H—SH	377	H—SCH_3	390	H—NHR	444
H—F	569	H—Cl	431	(AdoCbl)Co—C	126

*Lowry and Richardson (1981), Ochiai (1994), and other standard textbooks.

t-BuO· (see Table 7.1). That is, a hydrocarbon free radical as represented by methyl radical is highly reactive in hydrogen abstraction
despite its relatively high activation energy.

The bond dissociation energy could provide some indication as to
which bond would be most susceptible to H-abstraction, or whether
or not a certain H-abstraction would be feasible. Table 7.3 lists the
bond dissociation energy values for biologically relevant bonds
(Lowry and Richardson, 1981). For example, a tyrosyl radical (a phenoxy radical) plays an important role in many biological systems.
Its bond dissociation energy is 356 kJ/mol, whereas the H—OH
bond energy is 498 kJ/mol. Therefore, the H-abstraction from the
OH group of tyrosine residue by an ·OH radical to form the tyrosyl
radical is thermodynamically favorable ($\Delta H = -142$ kJ/mol). The
formation of a thiol radical is also easy, because the S—H bond
energy (390 kJ/mol for CH_3SH) is relatively low. Now suppose
that the tyrosyl radical then abstracts H-atom from a substrate having C—H bond. This reaction is thermodynamically unfavorable,
because ΔH would be a positive value; for example, +ca 54 kJ (for

the H-abstraction from C—H in ethane by tyrosyl radical). In the case of a sulfhydryl radical, the corresponding ΔH value is +ca 20 kJ. Nonetheless, this type of reaction is postulated in many enzymatic reaction systems. Is the assumption reasonable?

Thermodynamic argument assumes that the reaction products remain in proximity immediately after a reaction and so a reverse reaction can take place. And if this reverse reaction is thermo-dynamically favored, no forward reaction appears to take place. If the reactants are supplied with enough energy to overcome the activation barrier and the products are immediately separated or the reverse reaction is somehow prevented, then the reaction could proceed in the forward direction even if it is not thermodynami-cally favored. The activation barriers are not exceedingly high in the cases of free radical reactions in general, and it is almost nil in the case of a reactive radical such as $F\cdot$, $Cl\cdot$, or $HO\cdot$, as seen in Table 7.2. However, should these conditions (i.e., the relatively low activation energy and prevention of the reverse reaction) not be met, the free radical H-abstraction reaction very likely would be controlled ther-modynamically as well as kinetically.

Another relevant issue is how the reactivity of a molecule would change when a H-atom is abstracted from it; that is, when it is converted to a free radical. For example, would the presence of an unpaired electron on a carbon atom affect the reactivity of the moi-ety bound to the same carbon atom or the one adjacent to it? Not much has been studied in this regard, unfortunately. One general effect that has been observed is that the acidity of alcoholic OH group increases when the alcohol becomes a free radical (Neta, 1976; Asmus *et al.*, 1966). The pK_a values of primary alcohols RCH_2OH is about 20 (i.e., $K_a = 10^{-20}$), whereas pK_a values of $RH(C\cdot)OH$ fall in the range of 11 to 12. That is, the K_a value of an alcohol increases by a factor of 10^8 to 10^9 when a H-atom is abstracted from the α carbon. The increase in pK_a values of alcohols is of the order of 10^4 to 10^5 when the unpaired electron is on the β-carbon (March 1985). An alcohol RCH_2OH does not bind strongly to a metal cation such as Cu(II), but it may bind like $RH(C\cdot)O^-$—Cu(II) when it becomes a free radical. Besides, in this state, an intra-molecular electron transfer may become facile; for example, $RH(C\cdot)O^-$—Cu(II) $\rightarrow$ RCH=O + Cu(I). This reaction will be pro-moted because of the presence of an unpaired electron on the car-bon atom and the double bond formation (more details provided in a later section).

7.4. **B$_{12}$-COENZYME (ADENOSYLCOBALAMIN)- DEPENDENT ENZYMES**

7.4.1. **Mutases, Diol Dehydratase, and Ethanolamine Ammonia Lyase**

Enzymes that are dependent on B$_{12}$-coenzyme include diol dehydratase, glutamate mutase, L-β-lysine mutase, methylmalonyl Co—A mutase, ethanolamine ammonia lyase, and ribonucelotide reductase. The first three enzymes consist of two subunits. Subunit E binds B$_{12}$-coenzyme, and component S contains sulfhydryl groups. The last three enzymes are, however, of a single polypeptide, which contains several cysteine sulfhydryl groups. Both sulfhydryl groups and B$_{12}$ coenzyme are believed to be essential for the enzymatic activity. B$_{12}$ coenzyme (or adenosyl cobalamin (AdoCbl)) contains a cobalt-to-carbon (C-5′ of adenosyl group) bond, as shown in Figure 7.3.

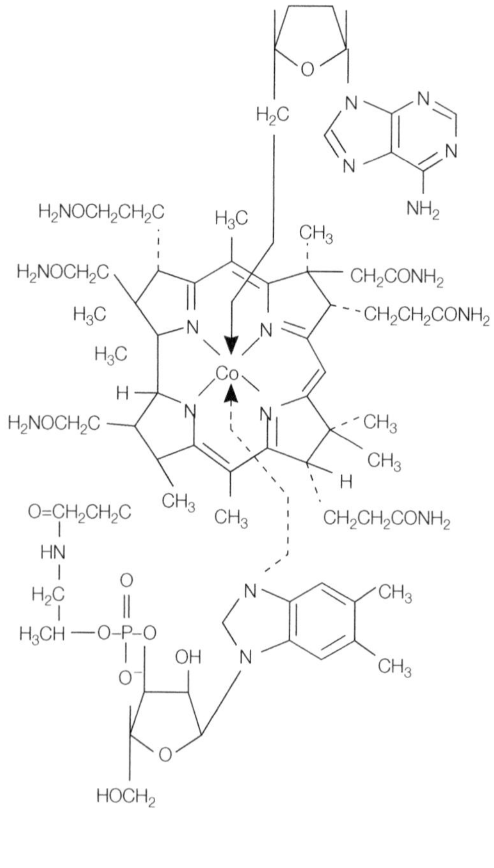

■ **Figure 7.3.** Vitamin B$_{12}$ coenzyme: adenosyl cobalamin (AdoCbl).

Some examples of reactions catalyzed by these enzymes are:

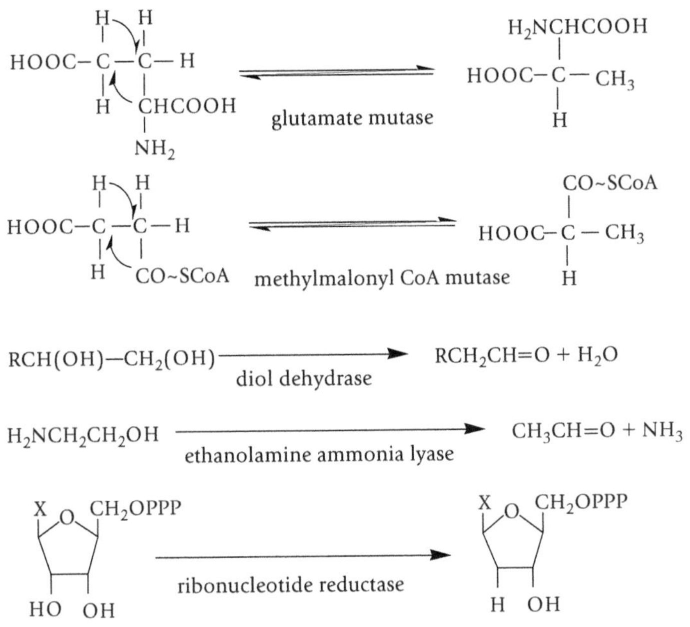

$$\text{HOOC}-\overset{\overset{\displaystyle H}{|}}{\underset{\underset{\displaystyle CHCOOH}{|}}{C}}-\overset{\overset{\displaystyle H}{|}}{\underset{\underset{\displaystyle NH_2}{|}}{C}}-H \rightleftharpoons \text{HOOC}-\overset{\overset{\displaystyle H_2NCHCOOH}{|}}{\underset{\underset{\displaystyle H}{|}}{C}}-CH_3$$

glutamate mutase

$$\text{HOOC}-\overset{\overset{\displaystyle H}{|}}{\underset{\underset{\displaystyle CO\sim SCoA}{|}}{C}}-\overset{\overset{\displaystyle H}{|}}{\underset{H}{C}}-H \rightleftharpoons \text{HOOC}-\overset{\overset{\displaystyle CO\sim SCoA}{|}}{\underset{\underset{\displaystyle H}{|}}{C}}-CH_3$$

methylmalonyl CoA mutase

$$RCH(OH)-CH_2(OH) \xrightarrow{\text{diol dehydrase}} RCH_2CH=O + H_2O$$

$$H_2NCH_2CH_2OH \xrightarrow{\text{ethanolamine ammonia lyase}} CH_3CH=O + NH_3$$

ribonucleotide reductase

The first two reactions are typical 1,2-shift reactions. The next two reactions also may proceed through 1,2-shift reaction, but they may not. The last reaction is a deoxygenation reaction, but could also proceed through a 1,2-shift reaction accompanied by reduction of a carbonyl group.

Essentially two different issues await elucidation regarding the reaction mechanisms of these enzymes. They are (1) the mechanism by which the cobalt-to-carbon bond splits homolytically upon binding a substrate, and (2) the mechanism of the subsequent rearrangement reactions. A recent review on the subject is found in Marsh and Drennan (2001).

7.4.1.1. *Homolytic Cleavage of the Cobalt-to-Carbon Bond upon Binding a Substrate*

Researchers agree on the first step of the enzymatic reactions. That is the homolytic splitting of the cobalt-carbon bond when the enzyme binds a proper substrate. However, the mechanism of this homolytic cleavage is far from well understood.

A most widely held mechanism postulates that upon binding a substrate, the corrin ring (of AdoCbl) changes its conformation in such a way that the steric interaction between the ring and the carbon atom bound to the cobalt atom becomes severe. Thus the steric interaction will force the cobalt-to-carbon bond to split (mechanochemical mechanism; Halpern, 1985). The cobalt-to-carbon bond dissociation energy of AdoCbl has been estimated to be approximately 126 kJ/mol (30 kcal/mol) (Hay and Finke, 1987). The Co—C bond dissociation energies of the so-called B_{12} coenzyme model compounds (of simpler structures) have been determined to be 84–126 kJ/mol (Ochiai, 1994). None of these values is exceedingly large for the bond dissociation energy. Therefore, the change in the hypothesized steric interaction has been thought by some to be sufficient to break such a weak bond. An attempt was made to estimate the possible magnitude of the steric interaction increase (Ochiai, 1994). Under a most favorable condition, it may amount to about one-third of the 126 kJ/mol. It is not nearly enough even under a most favorable circumstance likely to occur in the real system. This means that the steric effect may not be the major cause for the homolytic bond cleavage, though it may somewhat affect the bond splitting. The same conclusion was arrived at in the recent review by Marsh and Drennan (2001).

A suggestion has been made that the major portion of the binding energy of AdoCbl to the enzyme protein is utilized to weaken (lengthen) the Co—C bond (Holloway et al., 1978; Gaudemar et al., 1981). This weakened bond then would be readily split upon addition of an extra energy due to the binding of a substrate, as follows (Ochiai, 1994).

From a kinetic result of the binding of AdoCbl and a substrate to an enzyme protein (diol dehydratase), ΔG for the binding of AdoCbl was estimated to be -36 kJ/mol; ΔH then can be estimated to be about -86 kJ/mol (based on a rough estimate of $\Delta S = -40$ eu for protein-substrate binding; Jencks, 1975). ΔG for the binding of a substrate to the holoenzyme was -23 kJ/mol. Since the binding of AdoCbl to the protein is done mainly through hydrogen bonding (see Mancia et al., 1996), the overall hydrogen binding enthalpy can be estimated conservatively to be about -190 kJ/mol. Hence there is a discrepancy of about 100 kJ/mol (i.e., $-190 - (-86)$); a major portion of this energy may be used to modify the protein conformation in such a way that the Co—C bond is weakened. These are not accurate values in any way, and they represent ballpark figures. The binding of a substrate would supply additional energy, enough to split the bond by pulling away the adenine through another protein conformational change.

7.4.1.2. *Hydrogen Abstraction from Substrates*

The reactions that take place after the Co—C bond cleavage are believed to be H-abstraction from the substrate and then 1,2-shift rearrangement. That is:

$$(Cbl)Co-C(Ado) \rightarrow (Cbl)Co^{II} + \cdot C(Ado),$$
$$\cdot C(Ado) + RCH_2-CH_2X \rightarrow HC(Ado) + R(\cdot C)H-(C)H_2X$$
$$R(\cdot C)H-(C)H_2X \rightarrow R(C)HX-(\cdot C)H_2(+HC(Ado))$$
$$\rightarrow RCHX-CH_3(+\cdot C(Ado))$$

$\cdot C(Ado)$ is the adenosyl radical (at C-5′). In this mechanism, $\cdot C(Ado)$ itself is assumed to participate directly in the catalytic process.

An alternative mechanism suggested for ethanolamine ammonia lyase (O'Brien *et al.*, 1985) and ribonucleotide reductase postulates that the radical entity is first transferred to an amino acid residue, probably, cysteine. That is:

$$\cdot C(Ado) + HS(Cys) \rightarrow HC(Ado) + \cdot S(Cys)$$

This cysteine radical abstracts an H-atom from the substrate and the product radical then picks a H-atom from either H—S(Cys) or H—C(Ado). A problem here is whether the thiol radical $\cdot S(Cys)$ can abstract a H-atom from the H—C bond of a substrate such as diol or glutamate. The H-abstraction of H—S(Cys) by $\cdot C(Ado)$ would be exothermic (ca -20 kJ). However, the following reaction, that is, the H-abstraction from the substrate by $\cdot S(Cys)$, would be endothermic ($+$ca 20 kJ), though the ΔH is not exceedingly high. Thus, the idea that the thiol radical acts as the catalytic radical may be feasible from the energy point of view. This is the issue discussed in the previous section.

7.4.1.3. *1,2-Shift or Other Reactions of Substrate Free Radicals*

The major problem of these enzymatic reactions is the final step; that is, how the substrate that has lost a H-atom might transform into the product. Three questions arise:

(1) Are all the reactions (except that of ribonucleotide reductase) of 1,2-shift type?

(2) The mechanism of the 1,2-shift reaction itself.

(3) Is the metal (Co) involved in the 1,2-shift reaction?

Glutamate mutase, methylmalonyl CoA mutase, and other mutase reactions are definitely of 1,2-shift type. Diol dehyratase and ethanolamine ammonia lyase are also believed to involve 1,2-shift reaction. That is:

$$CH_2(OH)-CH_2(OH) \rightarrow \cdot CH(OH)-CH_2(OH)$$
$$\rightarrow CH(OH)_2-CH_2\cdot$$
$$CH(OH)_2-CH_2\cdot + X-H \rightarrow CH(OH)_2-CH_3(+X\cdot)$$
$$\rightarrow O{=}CH-CH_3\ (+H_2O)$$

or

$$CH_2(OH)-CH_2(NH_2) \rightarrow \cdot CH(OH)-CH_2(NH_2)$$
$$\rightarrow CH(OH)(NH_2)-CH_2\cdot$$
$$\rightarrow CH(OH)(NH_2)-CH_3$$
$$\rightarrow O{=}CH-CH_3(+NH_3)$$

However, the dehydration of the free radical of ethylene glycol has been shown to be spontaneous and very fast in aqueous media (Bansal *et al.*, 1973). This dehydration does not necessarily require 1,2-shift as described by the preceding equation. For example, the mechanism could be (Bansal *et al.*, 1973):

$$\cdot CH(OH)-CH_2(OH) + H^+ \rightarrow \ \cdot CH(OH)-CH_2(OH_2)^+$$
$$\rightarrow \cdot CH(OH)-CH_2{}^+(+H_2O) \rightarrow O{=}CH-CH_2\cdot(+H^+)$$

Alternatively,

$$\cdot CH(OH)-CH_2(OH) \rightarrow \cdot CH(O^-)-CH_2(OH)(+H^+)$$
$$\rightarrow O{=}CH-CH_2\cdot(+H_2O)$$

The driving force here would be the stability of $O{=}CH-CH_2\cdot$, which has an electronic structure equivalent to allyl radical.

A similar mechanism can be envisioned for ethanolamine ammonia lyase reaction. That is:

$$\cdot CH(OH)-CH_2(NH_2) + H^+ \rightarrow \cdot CH(OH)-CH_2(NH_3)^+$$
$$\rightarrow \cdot CH(OH)-CH_2{}^+(+NH_3) \rightarrow O{=}CH-CH_2\cdot$$

Thus, it is possible that the enzymatic mechanisms of diol dehydratase and ethanolamine ammonia lyase reaction are indeed different from those of mutases.

1,2-shift (rearrangement) reactions are known to be facile in compounds with a carbonium ion, but are rather rare and slow in the case of free radicals. Any organic chemistry textbook gives a plausible reason for this on the basis of a bridged intermediate mechanism. Therefore, 1,2-shift as postulated for glutamate mutase and other AdoCbl-dependent mutases is rather unique. How the 1,2-shift reaction might proceed in these enzymatic reactions is virtually unknown, though a number of model reaction studies have been attempted to gain insight into the mechanism (Finke *et al.*, 1984). These studies have focused on reactions of diol dehydratase and ethanolamine ammonia lyase type (Finke *et al.*, 1984). Studies by Dowd and coworkers (Choi *et al.*, 1990) do indicate that the mechanisms of glutamate mutase and methylmalocyl CoA mutase are indeed different from those of dioldehydratase in that they seem to involve the cobalt entity, whereas diol dehydratase reaction occurs independent of the cobalt, once the radical species is induced. A recent review article gives more recent empirical data for these enzyme systems (Reed, 2004), and recent versions of some proposed mechanisms are found in Marsh and Drennan (2001).

A mechanism I proposed in 1977, shown in Figure 7.4, may still be relevant (Ochiai, 1994). The crucial step is the formation of a bridged transition state. As is well known, 1,2-shift reaction through a bridged intermediate is facile in the case of carbonium ion, but it is difficult in a free radical case. The triangular bridged species would have, in simplistic terms, one bonding molecular orbital and two antibonding orbitals, and therefore, two electrons occupy the bonding orbital and the system (intermediate) would be relatively stable in the carbonium ion case. When three electrons have to be accommodated (i.e., in the case of free radicals), one electron has to go in the antibonding orbital. This is the basic reason for the difficulty to form a bridged intermediate in the case of free radicals (see Fig. 7.4a, c). It is hypothesized in the proposed mechanism that the one odd electron on Co(II) participates in the bridge formation as described in Figure 7.4(b). A very simplistic molecular orbital description for this state is given in Figure 7.4(c). The four electrons (three on the substrate free radical and one on Co) would occupy the bonding orbital and the nonbonding orbital. Hence, the energy of this bridged intermediate may not be as high as the noncatalytic free radical case.

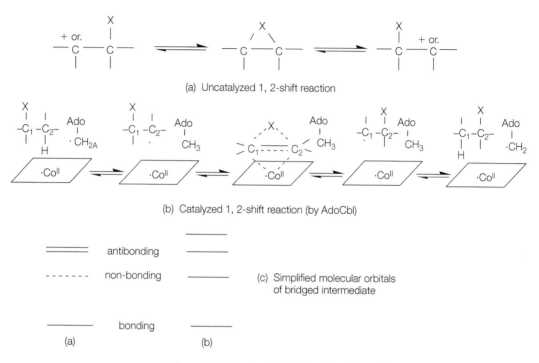

(a) Uncatalyzed 1, 2-shift reaction

(b) Catalyzed 1, 2-shift reaction (by AdoCbl)

	antibonding	
	non-bonding	
		(c) Simplified molecular orbitals
		of bridged intermediate
	bonding	
(a)		(b)

■ **Figure 7.4.** Mechanisms of 1,2-shift reactions: (a) uncatalyzed and (b) catalyzed by adenosyl cobalamin (Ochiai); (c) simplified molecular orbitals in mechanisms (a) and (b).

7.4.2. **Ribonucleotide Reductases (Cobalamin-Dependent)**

The reaction of ribonucleotide reductase is represented by:

$$-\mathrm{HC}^{3'}(\mathrm{OH})-\mathrm{C}^{2'}(\mathrm{OH})- \text{ (ribose of nucleotide)} + 2\mathrm{HSR}$$
$$\rightarrow -\mathrm{HC}(\mathrm{OH})-\mathrm{CH}- \text{ (deoxyribose)} + \mathrm{RS}-\mathrm{SR} + \mathrm{H}_2\mathrm{O}$$

There are three different types of ribonucleotide reductase (four if we include that from *E. coli* grown anaerobically): B_{12}-coenzyme (AdoCbl) dependent ones found in some fungi and anaerobic bacteria (Thelander and Reichard, 1979), Mn-dependent ones found in some bacteria (Auling and Follmann, 1994) and iron-dependent ones (Thelander and Reichard, 1979). The enzymes of the last category are found in a variety of organisms from virus and bacteria to mammals (Mann *et al.*, 1991).

The enzyme from *Lactobacillus leichmannii* is AdoCbl-dependent and consists of a single polypeptide of a molecular weight of 76,000. It contains a number of cysteine residues. The first step of the enzymatic reaction is believed to be the homolytic cleavage of the Co—C bond of AdoCbl, as in the cases of the other AdoCbl dependent enzymes discussed in the previous section. Isotopic studies with [3-^{3}H]UTP ([3'-^{3}H]ATP) as the substrate has established that the adenosyl radical formed in the first step does not directly abstract H-atom from the substrate (Stubbe, 1989, 1990). It has been demonstrated that the radical character is transferred from the adenosyl group to a thiol group, which then acts as the true catalytic site (Stubbe, 1989, 1990).

7.5. S-ADENOSYL METHIONINE (SAM)-DEPENDENT ENZYMES

Lysine 2,3-amino mutase does not depend on AdoCbl unlike L-β-lysine mutase, and instead requires Co(II), Fe(II)-sulfur protein, S-adenosylmethionine (SAM) and pyridoxal phosphate. The enzyme to activate pyruvate formate-lyase discussed earlier is dependent on SAM as well. SAM (see below) often is said to be a "poor man's adenosylcobalamin" and functions similarly (see Frey, 2001); that is, it will form an adenosyl radical upon a homolytic cleavage of the CH_2—S^+ bond.

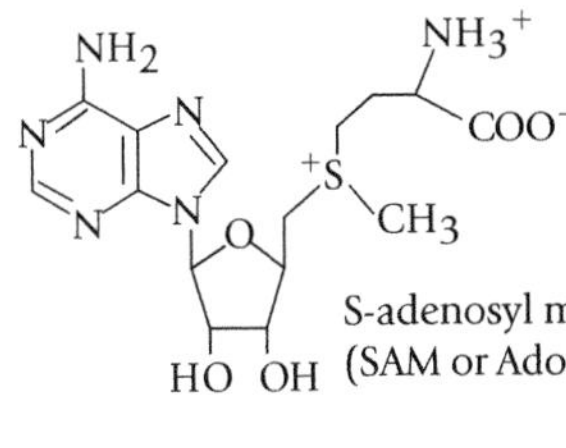

S-adenosyl methionine
(SAM or AdoMet)

Biotin synthase is another SAM-dependent enzyme (Berkovitch *et al.*, 2004). It catalyzes a remarkable reaction to insert S to dethiobiotin:

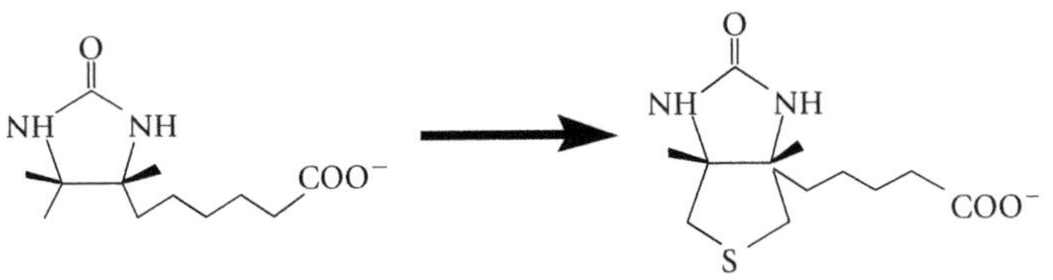

X-ray crystallography determined the structure of the enzyme with the substrate and SAM, and showed that there are two rather unusual iron sulfur clusters in it (Berkovitch *et al.*, 2004). One is a $[Fe_2S_2]$ cluster in which one cysteine and one arginine coordinate to one of the Fes and two cysteines to the other Fe. The three Fes of the $[Fe_4S_4]$ cluster are coordinated by three regular cysteines but the fourth Fes ligand is the amine N and carboxylate O^- of SAM. The authors suggested that the S^{2-}, which bridges the dinuclear cluster, is to be incorporated into the product biotin. The reaction mechanism is yet to be determined.

7.6. IRON-DEPENDENT RIBONUCLEOTIDE REDUCTASES

Iron-dependent ribonucleotide reductases consist of $\alpha_2\beta_2$. The larger subunit α_2 (R1) contains the substrate binding sites as well as allosteric sites. Three Cys's 225, 462, and 439 in *E. coli* R1 are suggested to be located near the interface between R1 and R2 and be directly involved in the reduction of ribose. It has been discovered (Stubbe, 1989, 1990) that R1 from *E. coli* and *Lactobacillus leichmannii*'s AdoCbl-dependent enzyme contain similar sequences containing two cysteine residues: -GC(754)ESGAC(759)KI- in *E. coli* R1 and -CEGGACPIK- in *L. leichmannii*. These sequences, it is believed, are recognized by thioredoxin (Stubbe, 1989, 1990).

Each of the smaller subunit β_2 (R2) contains one essential tyrosyl radical as well as a pair of antiferromagnetically coupled Fe(III)s. This tyrosyl radical was the first recognized free radical associated with metalloenzymes. B2 subunit (R2 subunit of *E. coli* enzyme) has been extensively studied and its structure has been determined by x-ray crystallography (Nordlund *et al.*, 1990). A more recent review (Stubbe, 2003) gives references to more recent x-ray data. The tyrosyl radical is located closely (about 50 pm) to the pair of Fe(III)s, and unusually stable.

We proposed a mechanism (mechanism (A) in Fig. 7.5) for the formation of tyrosyl radical based on a careful study of the stoichiometric relationships of the tyrosyl radical formation (Ochiai *et al.*, 1990). The major point of this mechanism is the assumption that an extraneous electron is required to split the $(O—O)^{2-}$ bond, and that this electron is supplied by a third Fe(II) ion. This stoichiometry (tyrosyl radical/Fe = 1/3) has been confirmed by a further

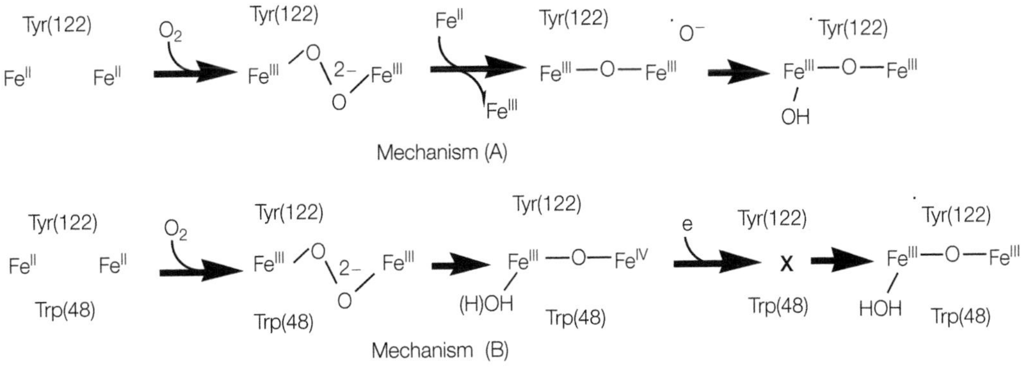

■ **Figure 7.5.** Mechanisms (A) (Ochiai) and (B) (Stubbe *et al.*) of the tyrosyl radical formation in the Fe-dependent ribonucleotide reductase.

study (Bollinger *et al.*, 1991). This is also consistent with the fact that only about 60% of a subunit has a tyrosyl radical (Stubbe, 2003) when two moles of Fe(II) are added to one mole of apo-protein.

Mechanism (B) (of Fig. 7.5) has been formulated based on many studies, particularly spectroscopic, of tyrosyl radical formation (Stubbe, 2003). A difference between mechanisms (A) and (B) is that the splitting of the O—O bond is assumed to take place after addition of another electron (from Fe(II)) in mechanism (A) while it does so before another electron is supplied in mechanism (B). A similar mechanism has been proposed for methane monooxygenase, which has very similar dinuclear Fe's at the active site (see Fig. 6.9). The nature of intermediate X as proposed is obscure.

The tyrosyl radical in *E. coli* R2 and the adenosyl radical in *L. leichmannii* enzyme, however, are not believed to be directly involved in the enzymatic reaction. As mentioned earlier, the radical entity is transferred to a cysteine residue. This cysteine residue is located on R1 subunit and the tyrosyl radical is on R2. Therefore, a free radical nature has to be transferred over quite a distance (35 Å) from R1 to R2 (Stubbe, 2003). Evidence is strong that this thiol radical first abstracts a hydrogen atom from C(3′) of the ribonucleotide:

$$—(HO)C^{3'}D—C^{2'}H(OH)— +(Cys)S· \rightarrow$$
$$—(HO)C^{3'}· —C^{2'}H(OH)— + (Cys)SD$$

The mechanism of the subsequent reaction has not been well elucidated despite an intensive study. One that has been proposed (Stubbe, 1989, 1990) assumes:

$$—(HO)C^{3'}\!·\!—C^{2'}H(OH)— \Leftrightarrow —(HO)C·—C^+H— (+H_2O)$$
$$\rightarrow —(O{=}C)—(C·)H \Leftrightarrow \rightarrow —(HO)(C·)—CH_2—(+(Cys)SD)$$
$$\rightarrow —(HO)CD—CH_2—(+(Cys)S·)$$

The last step in this mechanism is a hydrogen atom transfer from the catalytic thiol group back to the substrate (now in the form of product). This is to conform to an experimental finding that there was no hydrogen isotopic exchange at C(3′) position after a completion of one reaction cycle. This reaction sequence suggests the presence in the enzyme of a scheme of elaborate conformational changes that guarantee the specific hydrogen transfers between the important players at specific stages of reaction.

A new type of ribonucleotide reductase was recently reported (Jian *et al.*, 2007). The ribonucleotide reductase from *Chlamydia trachomatis* was found to lack a tyrosyl radical. Besides the dinuclear metal entity is Mn-Fe, rather than Fe-Fe. The authors proposed that the Mn-Fe behaves similarly to the Fe-Fe entity, but the radical entity is created directly on a cysteine residue on R1.

7.7. GALACTOSE OXIDASE

Galactose oxidase catalyzes the oxidation of primary alcohols as well as galactose to produce the corresponding aldehydes and hydrogen peroxide:

$$RCH_2OH + O_2 \rightarrow RCH{=}O + H_2O_2$$

An x-ray crystallographic determination (Ito *et al.*, 1991) has revealed that the active site contains a copper ion coordinated by two histidine Ns, one tyrosyl O (Tyr-272), and a water molecule. This tyrosine is covalently bonded through sulfur to a cysteine residue on the protein. A native enzyme from *Dactylium dendroides* can be activated by $Fe(CN)_6^{3-}$, and the active form can be reduced back to an inactive form by $Fe(CN)_6^{2-}$. The active form, though containing Cu(II), is ESR-silent, and shows an absorption spectrum indicative of a tyrosyl radical (Whittaker and Whittaker, 1990; Whittaker, 1994). The inactive reduced state shows a Cu(II)-ESR signal. Addition of a substrate to the active enzyme under an anaerobic condition abolishes both the Cu(II)-ESR signal and the tyrosyl radical optical absorption signal (Whitaker, 1994). Based on these studies, the following mechanism has been proposed:

$$(Cu^{II}, PhO\cdot)(ESR\text{-silent}) + RCH_2OH$$
$$\rightarrow (Cu^I, PhOH) + RCH{=}O + H^+$$

PhO$\cdot$ would first abstract hydrogen from the substrate. The resulting RH(C$\cdot$)OH is a stronger acid than the original alcohol and can thus reduce Cu(II) (see earlier); that is:

$$PhO\cdot + RCH_2OH \rightarrow PhOH + RH(C\cdot)OH$$
$$RH(C\cdot)OH + Cu^{II} \rightarrow RCH{=}O + Cu^I + H^+$$

The resulting Cu(I) moiety can readily be oxidized by O_2 to Cu(II), and $O_2\cdot^-$ may abstract hydrogen from the tyrosine, recreating the active catalytic entity:

$$(Cu^I, PhOH) + O_2 \rightarrow (Cu^{II}, PhOH, O_2\cdot^-) \rightarrow (Cu^{II}, PhO\cdot) + HO_2^-$$

Slightly different mechanisms are discussed by Whittaker (1994).

7.8. OTHER EXAMPLES

Diol dehydratase obtained from *Clostridium glycocalycum* does not contain AdoCbl, unlike the enzyme from the other sources. It shows an ESR signal at $g = 2.02$, which is sensitive to the action of hydroxyurea. Reagents that act on thiol groups diminished both the enzymatic activity and the radical signal intensity at the same time. This observation suggested that the radical is on a cysteine residue (Hartmais and Stadtman, 1987).

Copper-dependent amine oxidase involves a semiquinone type free radical of a cofactor topaquinone as discussed in Chapter 5. Z^+ in photosystem II has been identified as a tyrosyl radical (Barry *et al.*, 1987, 1990) (see Fig. 5.17).

REVIEW QUESTIONS

1. Use of free radicals tends to be avoided in the realm of organic synthesis. Discuss various reasons for it.

2. Why do organisms then use free radicals? Try to explain using example(s).

3. A number of mechanisms of the homolytic cleavage of Co—C bond in AdoCbl upon addition of a substrate in Ado-Cbl-dependent enzymes have been proposed. List them and discuss pros and cons for each of the mechanisms.

4. It often is assumed that a hydrogen atom abstraction by a free radical takes place even if the reaction is thermodynamically unfavorable; for example, as in a thiol free radical (cysteine-S˙) abstracting a H-atom from an alkane type C—H. Discuss this issue; how might a thermodynamically unfavorable reaction take place?

5. Discuss mechanisms proposed for the formation of tyrosyl radical in iron-dependent ribonucleotide reductase found in *E. coli* and others.

PROBLEMS TO EXPLORE

1. As shown in Figure 7.2, there is a good correlation between the hydrogen abstraction reaction by some free radicals and the bonding energy of the bond to be cleaved. This is reasonable and expected. However, it seems that no such correlation exists with highly reactive free radicals. Explore how this phenomenon might be explained.

2. It has been found that the pK_a value of an alcoholic OH (RCH_2—OH) increases from about 20 to 11 or 12 when it becomes a free radical, RC˙H—OH. Possible reasons?

3. The crucial issue of reactions catalyzed by AdoCbl-dependent enzymes is the mechanism, likely a free radical mechanism, of 1,2-shift reaction. The mechanism of this step is far from well understood. Explore relevant literatures and compare the proposed mechanisms including the one in the text.

4. Why have organisms not invented a way to perform 1,2-shift reactions by a cation-mechanism, which is known to be usually facile, rather than the free radical mechanism catalyzed by AdoCbl-dependent enzymes?

5. Propose mechanisms for the transfer of a radical character from one site to another in a protein. One such example is found in Fe-dependent ribonucleotide reductase. In this enzyme, a free radical character is supposed to travel over 20 Å from a tyrosine residue in subunit R2 to a cysteine residue in R1.

6. Biotin synthase reaction is quite unusual (see text). It is discussed in terms of free radical reaction. Find a recent literature on the enzyme, and discuss the current ideas on its mechanism.

Chapter

8

Nitrogen Fixation

8.1. **NITROGEN METABOLISM**

Nitrogen's simpler compounds range from the oxidation state of $-III$ all the way up to $+V$. They are:

$$NO_3^- \rightarrow NO_2^- \rightarrow NO \rightarrow N_2O \rightarrow N{\equiv}N \rightarrow HN{\equiv}NH \rightarrow NH_2NH_2 \rightarrow NH_3$$

| Oxidation state | $+V$ | $+III$ | $+II$ | $+I$ | 0 | $-I$ | $-II$ | $-III$ |

E^0 (pH 7) V $\quad +0.42 \quad +0.37 \quad +1.18 \quad +1.77 \quad |< \quad -0.64 \quad >| \quad +0.86$

Living organisms as a whole are involved in converting one form of nitrogen to another throughout these various oxidation states. However, organisms can assimilate directly NH_3 only. Two pathways are present in the biological systems to produce NH_3; one is by reducing NO_3^- (and NO_2^-) to NH_3 and the other is by reducing N_2 to NH_3. The first system consists of nitrate reductase ($NO_3^- \rightarrow NO_2^-$) and (assimilatory) nitrite reductase ($NO_2^- \rightarrow NH_3$), and is widely distributed among all the organisms, plants as well as animals. The enzyme nitrogenase ($N_2 \rightarrow 2NH_3$) is distributed in a limited number of microorganisms.

The enzymes involved in the nitrogen metabolism are listed in Table 8.1, and all employ transition metals. A review on these enzymes is found in Ferguson (1998). Most of the enzymes for reducing reactions are discussed in Chapter 5. Enzymes involved in the oxidation processes are not well understood, except for hydroxylamine oxidoreductase. It contains a number of Fe-heme groups, and an x-ray crystallographic structure of an enzyme is found in PDB 1FGJ.

Table 8.1. Enzymes Involved in Nitrogen Metabolism

Oxidation state change	Chemicals	Enzymes	Metal
$+V \longrightarrow +III$	$NO_3^- \longrightarrow NO_2^-$	nitrate reductase	Mo
$+III \longrightarrow +II$	$NO_2^- \longrightarrow NO$	(dissimilatory) nitrite reductases	heme-Fe, Cu
$+III \longrightarrow -III$	$NO_2^- \longrightarrow NH_3$	(assimilatory) nitrite reductase	heme-Fe
$+II \longrightarrow +I$	$NO \longrightarrow N_2O$	nitric oxide reductase	heme-Fe
$+I \longrightarrow 0$	$N_2O \longrightarrow N_2$	nitrous oxide reductase	Cu
$0 \longrightarrow -III$	$N_2 \longrightarrow NH_3$	nitrogenases	Mo, V, Fe
$-I \longleftarrow -III$	$NH_2OH \longleftarrow NH_3$	ammonia monooxygenase	Cu, Fe
$+III \longleftarrow -I$	$NO_2^- \longleftarrow NH_2OH$	hydroxylamine oxidoreductase	heme-Fe's
$+V \longleftarrow +III$	$NO_3^- \longleftarrow NO_2^-$	reverse of nitrate reductase	Mo

8.2. CHEMISTRY OF N_2 REDUCTION

First, the thermodynamic values for the following processes (gas phase) are given as:

$$N_2 + H_2 \longrightarrow HN{=}NH: \qquad \Delta H = +213\,kJ/mol$$

$$HN{=}NH + H_2 \longrightarrow H_2N{-}NH_2: \quad \Delta H = -114\,kJ/mol$$

$$H_2N{-}NH_2 + H_2 \longrightarrow 2NH_3: \qquad \Delta H = -191\,kJ/mol$$

$$N_2 + 3H_2 \longrightarrow 2NH_3: \qquad \Delta H = -96\,kJ/2mol\ of\ NH_3$$

In biological systems, the condition should be closer to that in aqueous media. In an aqueous medium (at pH 7), the ΔG values are estimated from the reduction potential values given at the beginning:

$$N_2 + 2e + 2H^+ \longrightarrow N_2H_2: \quad \Delta G = ?$$

$$N_2 + 4e + 4H^+ \longrightarrow N_2H_4: \quad \Delta G = +247\,kJ/mol$$

$$N_2H_4 + 2e + 2H^+ \longrightarrow 2NH_3: \quad \Delta G = -166\,kJ/mol$$

$$N_2 + 6e + 6H^+ \longrightarrow 2NH_3: \quad \Delta G = +80.6\,kJ/mol\ of\ NH_3$$

Considering these values together with the values in gas phase reactions, it can be concluded that the first step ($N_2 + 2e + 2H^+ \longrightarrow HN{=}NH$ in

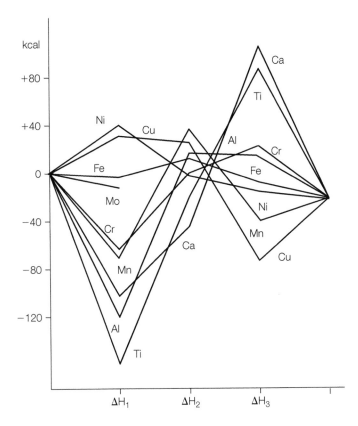

■ **Figure 8.1.** Enthalpy values for the
following reactions: ΔH_1 for 2K(catalyst) +
$N_2 \longrightarrow 2(KN)$; ΔH_2 for $2K + H_2 \longrightarrow 2(KH)$;
ΔH_3 for $(KN) + 3(KH) \longrightarrow 4K + NH_3$ (from
Ochiai, 1978b).

aqueous medium) is accompanied by a large positive $\Delta H/\Delta G$ value, and is the most difficult step of all the three steps. This implies a necessity of input of energy of a substantial magnitude in the nitrogen fixation; this is very likely to be used mostly for the first step ($N_2 \rightarrow N_2H_2$).

Another way of looking at the overall picture of the catalytic ammonia formation is as follows. In the case of solid state (industrial) catalytic process, the following three steps are hypothesized to be involved, where K represents a metallic catalyst.

$$2K + N_2 \longrightarrow 2KN \text{ (nitride):} \quad \Delta H_1$$
$$2K + H_2 \longrightarrow 2KH \text{ (hydride):} \quad \Delta H_2$$
$$(KN) + 3(KH) \longrightarrow 4K + NH_3: \quad \Delta H_3$$

The formation of ammonia would proceed smoothly if none of these ΔH_n values are exceedingly exothermic or endothermic (i.e., not very large negative or positive). ΔH_n values are collected in Figure 8.1

(Ochiai, 1978). It may be predicted that Fe provides a best catalyst, and that Ni is the second best, and that Cr and Cu may work. Indeed Fe is known to be the best catalyst for the industrial ammonia synthesis. Mo has a favorable ΔH for nitride formation, though other data are unavailable. This suggests that either Mo(0) or Fe(0) or both may have functioned as a primitive nitrogenase. MoO_4^{2-} was relatively abundant in ancient seawater, and it may be hypothesized that MoO_4^2 was reduced by FeS ore, and some Mo-Fe complexes resulted in the process. Indeed the most nitrogenases found today depend on Mo-Fe as the nitrogen reducing entity.

8.3. Mo-DEPENDENT NITROGENASE

The most extensively studied nitrogenase is the Mo-dependent one. A number of review articles are available: Rees and Howard (2000); Howard and Rees (1996); Burgess and Lowe (1996); Igarashi and Seefeldt (2003). As indicated earlier, it is essential that some sort of energy is supplied to the process. It is supplied by the hydrolysis of ATP, and hence a machinery to hydrolyze ATP is a component of the enzyme. The reaction stoichiometry has not been unequivocally established. A limiting relationship seems to be something like (Rees and Howard, 2000):

$$N_2 + 8H^+ + 8e + 16ATP \longrightarrow 2NH_3 + H_2 + 16ADP + 16P_i$$

That is, hydrolysis of two molecules of ATP is required for transferring an electron. The formation of H_2 is usually not stoichiometric, and may be a result of fortuitous reduction of H^+, as the nitrogenase has a capability of hydrogenase as well. Because of necessity of ATP hydrolysis and subsequent lengthy six-electron transfer, the nitrogenase reaction is rather slow, approximately one turnover per second.

The enzyme from *Azotobacter vinelandii* has been well studied and its x-ray crystallographic structure determined (see Rees and Howard, 2000 up to that date). The enzyme consists of two proteins: Fe-protein and Mo-Fe protein. The Fe-protein has a single $[Fe_4S_4]$ cluster for two identical polypeptides (γ_2), and each polypeptide binds a molecule of ATP. The Mo-Fe protein is a heterotetramer ($\alpha_2\beta_2$), and contains a special cluster (P-cluster) of 8Fes and a Mo-Fe cluster of 7Fe/1Mo-sulfur. The overall structure with ATP analogue (AlF_4–ADP) is shown in Figure 8.2 (Schindelin *et al.*, 1997) and the schematic pictures of the P-cluster and Mo-Fe cluster are given in Figure 8.3.

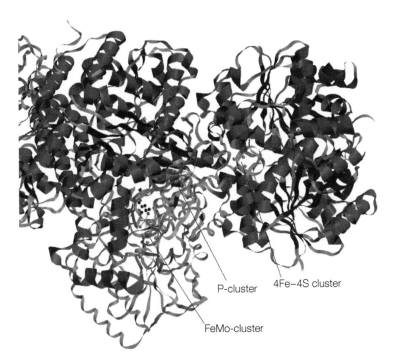

■ **Figure 8.2.** Half of the overall structure of a nitrogenase; there is another set of Fe and FeMo proteins symmetrically bound to the ones shown. (From PDB 1N2C: Schindelin, H., Kisker, C., Schlessman, J.L., Howard, J.B., Rees, D.C. 1997. Structure of ADP x AlF4(-)-stabilized nitrogenase complex and its implications for signal transduction. *Nature* **387**, 370–376.)

P-cluster

4Fe–4S cluster

FeMo-cluster

■ **Figure 8.3.** Mo-Fe cluster and P-cluster in Mo-nitrogenase (schematic; after Rees and Howard, 2000).

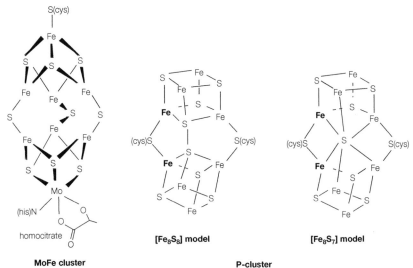

MoFe cluster

P-cluster

$[Fe_8S_8]$ model

$[Fe_8S_7]$ model

There are two models for resting P-cluster: $[Fe_8S_8]$ and $[Fe_8S_7]$, as shown. The latter currently seems to be the preferred model.

The reaction process to produce NH_3 has been established (according to Rees and Howard, 2000) as:

(1) Formation of a complex between the reduced Fe-protein with two bound ATPs and the Mo-Fe protein.

(2) Electron transfer between the two proteins coupled to the hydrolysis of ATP.

(3) Dissociation of the Fe-protein accompanied by rereduction and exchange of ATP for ADP.

(4) Repetition of this cycle until sufficient numbers of electrons (and protons) have been accumulated so that a molecule of substrate can be reduced.

There seems to be a general agreement on the point that the Mo-Fe cofactor is the substrate reducing site. Note that the Fe-protein is a kind of ATPase. The ATPase activity is enhanced only when the Fe-protein binds with the Mo-Fe protein (Igarashi and Seefeldt, 2003). The reduction potential ($[Fe_4S_4]^{2+}/[Fe_4S_4]^{1+}$) of the Fe-protein is -0.30 V without MgATP, but changes to -0.43 V upon binding MgATP (see Igarashi and Seefeldt, 2003). It further changes to $-0.62\,V$ when the Fe-protein combines with the Mo-Fe protein (Igarashi and Seefeldt, 2003). This suggests that the binding of MgATP makes the $[Fe_4S_4]$ cluster more powerfully reducing, and this tendency is further enhanced when the Fe-protein combines with the Mo-Fe protein.

There are a number of studies that suggest that the Fe-protein's conformation changes significantly upon binding MgATP (Burgess and Lowe, 1996), including a suggestion that the $[Fe_4S_4]$ cluster becomes exposed. A recent x-ray structural determination (Tezcan *et al.*, 2005) of three states of the holo enzyme from *A. vinelandii* shed some light on this issue. Those three states are: (1) no MgATP on the Fe-protein, (2) with MgAMPPCP (ATP analogue but cannot be hydrolyzed) on the Fe-protein, and (3) with MgADP on the Fe-protein. The relative position between the Fe-protein and the Mo-Fe protein changes among these three states, and the distance between the $[Fe_4S_4]$ cluster and the Mo-Fe cluster becomes significantly shorter (by 5Å) when the Fe-protein is bound with the substrate analogue. This will certainly facilitate the electron transport.

There are a number of questions that are yet to be answered fully. First, (a) how is this potentiation of the $[Fe_4S_4]$ cluster affected; (b) how is the electron transfer to P-site (and then onto Mo-Fe) affected? Next, it seems that one cycle of two ATP's hydrolysis prompts transfer of a single electron; (c) would it go to the active site of the Mo-Fe or would it be deposited temporarily on the P-cluster, say, until two electrons are deposited there; (d) is the P-cluster simply another electron transfer agent to relay electrons; (e) is the N_2 binding site Mo or Fe of the Mo-Fe cluster; and (f) how is N_2 reduced?

Igarashi and Seefeldt (2003) provide a few tentative answers to some of these questions. First let us look at question (e). A detailed analysis of the electron density map at 1.16 Å resolution suggested that there is an atom at the center of the Mo-Fe cluster (Einsle *et al.*, 2002). This atom is identified as an N-atom by the authors. Theoretical studies of substrate binding to the Mo-Fe cluster suggested that the center of the cluster provides favorable interaction sites for N_2 (referred to in Einsle *et al.*, 2002). These results implicate to these authors (Einsle *et al.*, 2002) that N_2 fits into the vacant center of the Mo-Fe cluster. Durrant (2002), however, proposed a mechanism in which N_2 binds the Mo-atom displacing the homocitrate.

Durrant (2002)'s mechanism of N_2 reduction (regarding question (f)) assumes the following steps:

$$Mo-N\equiv N \rightarrow Mo-N=NH \rightarrow Mo=N-NH_2 \rightarrow$$
$$Mo\equiv N + NH_3 \rightarrow Mo + 2NH_3$$

Each step involves an addition of 1e and $1H^+$, and the last step is the addition of 3e and $3H^+$ to $N\equiv Mo$; this process may also consist of three one electron addition steps. The N-atom seen at the center of the Mo-Fe cluster (Einsle *et al.*, 2002) could be the $Mo\equiv N$ in the Durrant mechanism. The Durrant mechanism assumes that an electron is transferred (presumably through P-cluster) to the Mo-Fe cluster, which then adds that electron to the substrate one by one. So this is a proposed mechanism for question (f).

This reaction sequence is in contrast to the reaction sequence generally presumed; that is, two electrons addition at each step, as in:

$$N\equiv N \rightarrow HN=NH \rightarrow H_2N-NH_2 \rightarrow 2NH_3$$

Now, let's look at how electron transfer is affected. It has been observed that rates of electron transfer from the Fe protein to the

Mo-Fe protein significantly accelerate when MgATP hydrolysis on the Fe protein is taking place (Igarashi and Seefeldt, 2003). In other words, docking of the Fe protein (with MgATP) with the Mo-Fe protein likely promotes hydrolysis and the electron transfer. This implies that the energy released from the hydrolysis of ATP would enable the electron transfer. A likely pathway of electrons from the $[Fe_4S_4]$ cluster of the Fe protein to the P-cluster and that from the P-cluster to the Mo-Fe cluster has been mapped out (Igarashi and Seefeldt, 2003). However no reasonable model or explanation have been proposed for questions (a) through (d).

I tried to answer these questions in my 1977 book (Ochiai, 1977) based on the data and information then available. That model is modified here to accommodate the new information, and is mentioned here merely to provide an overall picture (see Fig. 8.4). It is to be emphasized that this is simply a speculation and needs to be verified or rejected by further studies. I am aware that the idea that the P-cluster comes close to the Mo site of the Mo-Fe cluster and acts as the direct reducing agent requires a special attention, because no indication to suggest such an action of the P-cluster has been obtained. Here are some hypotheses made in this mechanism for questions (a) through (f).

(a–b) Energy of ATP hydrolysis is used to change the conformation of the Fe-protein that forces a change in the $[Fe_4S_4]$ cluster in such a way that its reduction potential is significantly lowered as discussed in the 1977 book. The ATP hydrolysis does not necessarily initiate

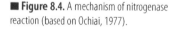

■ **Figure 8.4.** A mechanism of nitrogenase reaction (based on Ochiai, 1977).

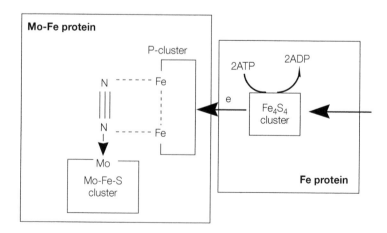

the electron transfer, though. The transfer of electron (to P-cluster) may require another conformational change due to docking of the Fe protein to the Mo-Fe protein, including a change in the reduction potential of the P-cluster and a change in the distances in the functional units.

(c–d) The P cluster is a special one and acts as the reducer of the substrate (N_2 and others), and the P cluster has an unusually low reduction potential, when activated. The reason for the requirement for the lower reduction potential of the P-cluster is that the special Fe—Fe in the P-cluster has to reduce N—N bond as depicted. This process, particularly the first step, requires a powerful reducing agent as discussed earlier.

(e) The Mo-atom is the N_2 binding site, and N_2 (as well as other substrates) binds to Mo as depicted. The reason for the rather unusual Mo-Fe-S cluster is that the low oxidation state of Mo required to bind N_2 is maintained by the reductive power of the Fe-S cluster part.

(f) In addition to N_2, nitrogenase reduces and splits the O—N bond of N_2O, N—N of N_3^-, C—N of RNC and acetylene. The electrons are to be added to π-antibonding orbital and σ-antibonding orbital. Hence it was assumed that electron(s) would be transferred from an antibonding Fe_2 unit, as shown in Figure 8.5 (Ochiai, 1977). The reduction potential of this unit can be lowered by diminishing the Fe—Fe distance, and presumably that is what is caused eventually by conformation changes by ATP hydrolysis. Here it is assumed that the

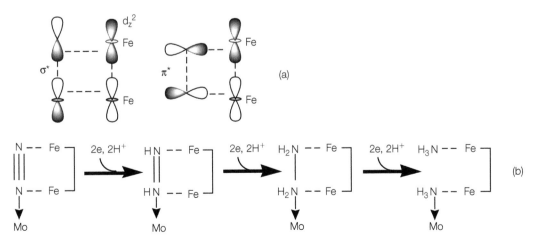

■ **Figure 8.5.** A molecular detail of mechanism of Figure 8.4 (based on Ochiai, 1977).

direct reducing unit is the special Fe—Fe pair in the P-cluster, which can become short. The Fes in each of the two cubes (of P-cluster) that is bridged by a cysteine (those in boldface in Fig. 8.3) could constitute this pair, and the distance between them may be shortened by a conformational change in the protein. Upon the conformational change, the pair could also come close to the N_2-binding site, as the intervening space is occupied mostly by water molecules (Rees and Howard, 2000). It is likely that each of two cubes in the P-cluster is one-electron deposit/conveyor, and so when two electrons are deposited on P-cluster, they are transferred to the substrate. In the mechanism shown in Figure 8.5, N_2 is supposed to bind from the outside of the Mo-Fe cluster (likely displacing homocitrate), as in Durrant mechanism (2002). One indication for this possibility is that a substrate such as isonitrile may not fit in the vacant space.

8.4. **OTHER NITROGENASES**

Nitrogenases that depend on V or Fe instead of Mo are known (Eady, 1996). As a matter of fact, the same *Azotobacter* species produces V-nitrogenase or Fe-only nitrogenase when sufficient Mo is not available (Eady, 1996; Rehder, 2000). V-dependent nitrogenase seems to contain a V-Fe-S cluster similar to that of Mo-Fe cluster (Rehder, 2000). The overall composition and structure of the V-nitrogenase is the same as that of the Mo-nitrogenase (Rehder, 2000). The existence of an Fe-only nitrogenase along with Mo- and V-dependent ones has suggested to many researchers that the common factor, Fe, is indeed the nitrogen binding/reducing site. However, as argued earlier regarding solid catalysis of NH_3 formation, several possible candidates for it exist; among them iron is the best overall, but Mo and Ni seem to be good candidates as well. Unfortunately data on V were unavailable, and hence the suitability of V cannot be ascertained. However, this argument suggests that several different metals may function as catalyst for ammonia formation. Therefore, it is just as likely that Mo, V, or Fe constitutes the nitrogen binding/reducing site, and that all the other Fe-S components act as electron conveyor, electron donor (to the substrate), or a mechanism to keep Mo (or V) in a reduced state, as argued for Mo-nitrogenase. If indeed only Fe is all that is needed as some researchers seem to believe, then there would be no reason for the biological systems to adopt Mo or V for the purpose, because, for one thing, Mo and V are much less abundant than Fe. Nonetheless, the fact is that Mo-nitrogenase is the most widespread among all nitrogen-fixing organisms.

Rehder (2000) discusses a number of model studies that show that V-complexes can bind N_2 and in certain cases can reduce N_2 to NH_3. This also is consistent with the idea that Mo, V, or Fe (in Fe-only case) is the N_2 binding and catalytic site.

REVIEW QUESTIONS

1. Estimate reaction enthalpies for the three steps from N_2 to $2NH_3$ as seen in the text, using bond dissociation enthalpies available in a general chemistry textbook. Compare your results with the data given in the text.

2. NO_3^- can be viewed as being reduced to, say, NH_3, or alternatively NO_3^- can be regarded as an oxidizing agent replacing O_2 in anaerobic condition. Chemically these two processes are equivalent, but in biological systems they take place in different manners, and play different roles. Discuss.

3. Discuss pros and cons for the two schemes proposed for reduction of N_2 to $2NH_3$ as the scheme of reduction in nitrogenase reaction.

PROBLEMS TO EXPLORE

1. Mo-nitrogenase contains three different unusual types of iron-sulfur clusters (Fe-Mo-S included). What is the implication of this fact? Discuss in detail.

2. Three different types of nitrogenase exist: Mo-Fe, V-Fe, and Fe-Fe. What would this suggest regarding the metal entity that binds N_2? Some ideas are mentioned in the text, but expand the argument on your own.

3. Nitrogenases require a large amount of energy (free energy) in the form of ATP in order to carry out the function (reduction of N_2 ammonia). Discuss the reason in detail.

4. An idea is proposed in the text that the reducing power (not reduction potential) of a Fe-Fe cluster may be increased by reducing the distance Fe-Fe. Discuss this idea.

Other Essential Elements

9.1. **INTRODUCTION**

First let's take another look at all the elements. The periodic table shown in Figure 9.1 includes some information on the degree of the biological essentiality of each element. This chapter deals briefly with some of those elements that are not touched on in other chapters; that is, N, S, P, B, Si, Se, V, Cr, Cl, Br, and I. This chapter gives some prominent features of biochemistry of the elements mentioned, and is not intended to be a comprehensive review of their biochemistry.

The essentiality of some other elements including F, Cd, Sn, and As has not been very well established, though an enzyme dependent exclusively on Cd (Cd-carbonic anhydrase) has been discovered recently (Lane *et al.*, 2005). Some aspects (particularly toxicity) of Cd, Sn, and As are treated in Chapter 11.

H																	he
li	be											B	C	N	O	F	ne
K	Mg											al	Si	P	S	Cl	ar
Na	Ca	sc	ti	V	Cr	Mn	Fe	Co	Ni	Cu	Zn	ga	ge	As	Se	Br	kr
rb	Sr	y	zr	nb	Mo	tc	ru	rh	pd	ag	Cd	in	Sn	sb	te	I	xe
cs	Ba	la*	hf	ta	W	re	os	ir	pt	au	hg	tl	pb	bi	po	at	rn
fr	ra	ac*															

*la = la ~ lu, ac = ac ~ lr

■ **Figure 9.1.** Elements found in the biosphere: (a) those in bold face (colored red) are macronutrients; (b) those in regular style (colored orange) are elements required in significant quantities for all the organisms; (c) those in italic (colored blue) are micronutrients, and many of them are universally required but some are essential to only a limited types of organisms; (d) those in italic (black) are suspected to be essential but have not been established so; (e) those in lowercase are not known to be essential to organisms.

9.2. **BIOCHEMISTRY OF NITROGEN COMPOUNDS**

Simpler nitrogen-containing compounds include NO_3^-, NO_2^-, NO, N_2, and NH_3. Nitrogen can be incorporated into various organic compounds only through NH_3. Such organic compounds include amino acids, flavin and its derivative, purine and pyrimidine bases of DNA and RNA, and various neuroactive compounds such as acetylcholine and dopamine. The incorporation of NH_3 into a biocompound is illustrated by the following reaction catalyzed by glutamate dehydrogenase.

$$\text{HOOCCH}_2\text{CH}_2-\overset{\overset{\displaystyle O}{\|}}{C}-\text{COOH} + NH_3 + NADH + H^+ \rightleftharpoons$$

$$\text{HOOCCH}_2\text{CH}_2-\overset{\overset{\displaystyle NH_2}{|}}{CH}-\text{COOH} + NAD^+ + H_2O$$

Hence processes that produce potentially NH_3 are critical. Two routes exist to produce NH_3; one is the reduction of N_2 (nitrogen fixation) and the other the reduction of NO_3^- (and other oxidized nitrogen compounds) all the way down to NH_3 (ammonification). An overall picture of the processes involved is discussed briefly in Chapter 8. It involves a number of metal-dependent enzymes. Reduction of NO_3^- can be an oxidation of another compound by NO_3^-, which can be reduced to NH_3 or any other intermediate stage. Hence there are two ways for use of NO_3^- (NO_2^-) reduction. One is to reduce it down to NH_3, which is incorporated into biocompounds; this reduction is then called assimilatory. The other is to use NO_3^- as an oxidizing agent (e.g., instead of O_2); it will not be reduced down to NH_3, and hence the resulting nitrogen compounds are not assimilated; this process is dissimilatory. One such process is the reduction of NO_2^- (or NO_3^-) to N_2; hence nitrogen will be returned to the atmosphere as a biologically inactive form. This process is termed dentrification. The organism uses the free energy released to produce ATP in these processes.

The reverse processes that oxidize NH_3 or other compounds to (NO_2^- to) NO_3^- are termed nitrification. In these processes the oxidizing agent is O_2. These will yield energy that will be used to produce ATP.

A biologically interesting nitrogen oxide is nitrogen monoxide NO. It is a free radical, and has been found to act as a second messenger in certain situations; for example, it is involved in dilation of an artery. Its formation and a biological sensor for it are discussed in Chapter 10.

9.3. BIOCHEMISTRY OF PHOSPHORUS

Oxidation and reduction do not play important roles in biotransformation of phosphorus compounds. The important reactions involving phosphorus include incorporation and release of phosphate groups in and from organic compounds and skeletal material (bones, etc.). The biocompounds containing phosphate entity are nucleotides, polynucleotides, and phospholipids (see the Introduction). Dead plants and animals are decomposed by fungi and bacteria, and the resulting phosphate is released to the environment. Phosphorus in terrestrial environment, particularly soil, is found typically bound with a carbohydrate-like compound (inositol); this is called phytic acid (as shown).

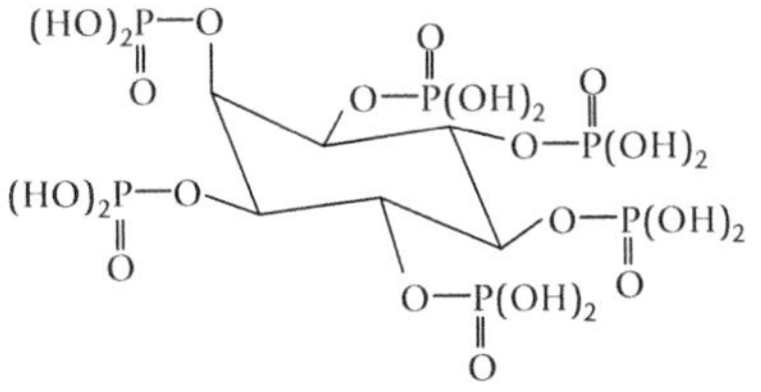

9.4. BIOCHEMISTRY OF SULFUR COMPOUNDS

9.4.1. Cellular Processes

Inorganic sulfur compounds including SO_2, SO_4^{2-}, SO_3^{2-}, $S_2O_3^{2-}$, and H_2S do not seem to be directly involved as such in biological systems. This does not suggest that these entities are not present as such in biological systems; many of them do. However, important S-containing biocompounds include amino acids, cysteine and methionine, glutathione and lipoic acid; all these compounds contain S nominally in the oxidation state of $-II$, the same as H_2S. There are, however, a number of sulfate esters known as chondroitin (sulfate ester of polysaccharide; Fig. I.5).

Some of the properties of inorganic S-compounds are given in Table 9.1 along with those of selenium. As indicated there, the reduction of SO_4^{2-} is rather difficult as suggested by the $E^{0\prime}$ value of -0.39 V (for SO_4^{2-}/SO_3^{2-}), and hence needs to be activated in order to be reduced. SO_4^{2-} first reacts with ATP forming adenosine-5'-phosphosulfate (APS). The enzyme for this reaction, APS-sulfurylase is found widely

Table 9.1. Difference between Sulfur and Selenium ($E^{0'} = E^0$ at pH 7)

Sulfur	Selenium
bond enthalpy S—H = 380 kJ/mol	bond enthalpy Se—H = 276 kJ/mol
H_2S: pK_1 = 6.9, pK_2 = 14.1	H_2Se: pK_1 = 3.9, pK_2 = 11
SO_4^{2-}/SO_3^{2-}: $E^{0'}$(pH 7) = −0.39 V	SeO_4^{2-}/SeO_3^{2-}: $E^{0'}$ = +0.44 V
SO_3^{2-}/S: $E^{0'}$ = −0.04 V	SeO_3^{2-}/Se: $E^{0'}$ = +0.26 V
S/HS^-: $E^{0'}$ = −0.27 V	Se/HSe^- (Se^{2-}): $E^{0'}$ = −0.7 V
$S_2^{2-}/2HS^-$: $E^{0'}$ = −0.48 V	$Se_2^{2-}/2Hse^-$: $E^{0'}$ = ?
radius (SO_4^{2-}) = 230 pm	radius (SeO_4^{2-}) = 234 pm

in animal tissues, baker's yeast, plant chloroplast, wild *Neurospora*, and such. In dissimilatory process, APS is reduced by APS-reductase. This process is employed by, for example, *Deuslfovibrio desulfuricans* and *D. nigrificans*. The reduction of APS is expressed as:

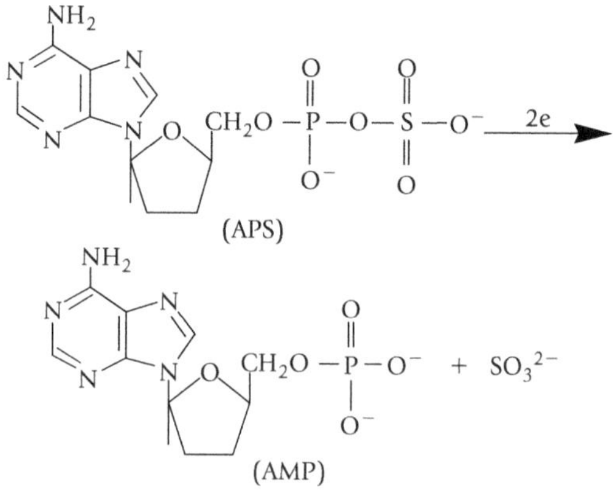

(APS)

(AMP)

This reaction can be regarded as:

$$APS + H_2O \longrightarrow AMP + 2H^+ + SO_4^{2-}$$
$$2H^+ + 2e^- + SO_4^{2-} \longrightarrow H_2O + SO_3^{2-}$$

Assuming that the free energy change for the first reaction is similar to that of hydrolysis of ADP (to AMP), the reduction potential of APS reaction is estimated to be raised by ca 0.16 V over the direct reduction of SO_4^{2-} (Ochiai, 1987).

In assimilatory process, APS is further phosphorylated (at 3′-C of ribose) by APS kinase to form 3′-phosphoadenosine-5′-phosphosulfate

(PAPS). This is reduced by PAPS reductase to SO_3^{2-}, which is then further reduced to H_2S by sulfite reductase containing FAD, FAMN, and iron-sulfur proteins. H_2S (HS^-) replaces OH of serine to form cysteine catalyzed by serine sulfhydrase as shown:

$$HOCH_2CH(NH_2)COOH + H_2S \rightarrow HSCH_2CH(NH_2)COOH + H_2O$$

The sulfhydryl group is readily subject to one-electron oxidation, often resulting in the formation of S—S bond. Hence such a compound plays the role of scavenging free radicals; examples include glutathione and lipoic acid.

9.4.2. **Marine Biogeochemical Cycling**

Sulfur emission, mostly in the form of dimethyl sulfide (DMS, $(CH_3)_2S$), from marine microorganisms play an important role in the global sulfur cycling (see Fig. A.2 and a recent review by Malin, 2006). DMS is oxidized to methyl sulfonic acid and sulfuric acid (and others); these compounds promote water condensation (hence rain). DMS sources include a number of phytoplankton (such as *Emiliania* (see Fig. 10.13), macroalgae, and coastal vascular plants (Gonzalez *et al.*, 1999). The precursor of DMS is dimethylsulfoniopropionate (DMSP), and is produced by these organisms by way of metabolizing sulfur. A reaction scheme of these sulfur-containing metabolites is shown in Figure 9.2. DMSP is presumed to play a role of osmolyte, predator deterrent and/or antioxidant (Howard *et al.*, 2006). DMSP is converted into DMS as shown:

$$(CH_3)_2S^+{-}CH_2CH_2COOH \rightarrow (CH_3)_2S + CH_2{=}CHCOOH + H^+$$

The enzyme for this reaction is DMSP lyase, which has not been well characterized yet (Steinke *et al.*, 1998).

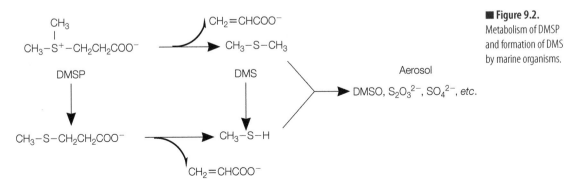

■ **Figure 9.2.**
Metabolism of DMSP and formation of DMS by marine organisms.

9.5. SELENIUM

Selenium is an extremely toxic element, and yet a large number of selenium-containing proteins/enzymes have been found in recent decades. It has been estimated that there may be up to 100 seleno-proteins (Burk and Hill, 1993). An extensive list of selenoproteins is given in a review by Birringer *et al.* (2002). In humans at least 25 selenoproteins have been identified, and selenium has been found to play some roles in preventing atherosclerosis, specific cancers, arthritis, diseases of accelerated aging, central nervous system pathologies, male infertility, and altered immunological functions (Patrick, 2004). Many of these functions of selenium seem to be due to the antioxidant function of the Se-associated proteins.

9.5.1. Chemistry of Selenium as Compared to That of Sulfur

Selenium participates in biochemistry as selenide (Se^{2-}), seleno-cysteine, and selenomethionine. That is, selenium takes the place of sulfur. The basic question is why selenium instead of sulfur is required. Sulfur and selenium behave similarly; both of them taking the oxidation states of $-II$ to $+VI$. Yet there are significant differences between them. Several comparative data are given in Table 9.1. Se—H bond is significantly weaker than S—H; this implies, for one thing, that hydrogen abstraction to form a free radical is easier with Se—H than with S—H. Selenide (Se^{2-}) is a weaker base than sulfide (S^{2-}). Selenate (SeO_4^{2-}) is thermodynamically more readily reduced than sulfate (SO_4^{2-}); though kinetically the reduction process is slow in both cases. HSe^- (or Se^{2-}) is a much stronger reductant than HS^- (S^{2-}). The sizes of sulfate and selenate ion are about the same; the organisms would have a difficulty in distinguishing them in, for example, uptake mechanism.

9.5.2. Glutathione and Selenium: Glutathione Peroxidase

The first well-established Se-enzyme is glutathione peroxidase (GPx). It turned out that there are several different types of glutathione peroxidase in human tissues; all are selenoproteins and act as an antioxidant. The first GPx is now known as classical and is found only in the cell cytosol. Other types of GPx are gastrointestinal GPx, phospholipid-hydroperoxide GPx (PHGPx), plasma GPx, and sperm nuclei GPx (Patrick, 2004). PHGPx is rather unique in that it can function inside cell membranes, though it is found also in cell nuclei, mitochondria, and cytosol. It is found also in high concentration in spermatozoa, and is involved in sperm maturation and the prevention of cellular apoptosis.

Glutathione (GSH) has the following formula, and GPx catalyzes the oxidation and S—S bond formation of GSH by H_2O_2 or ROOH.

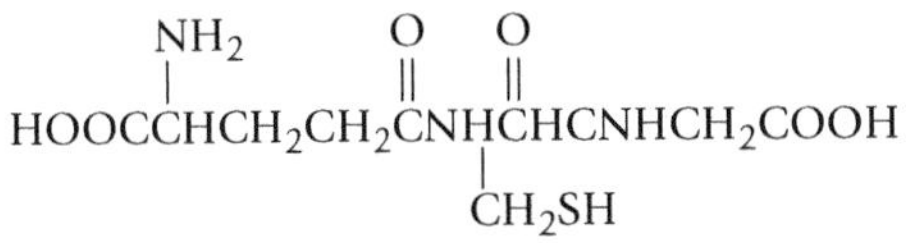

$$\begin{array}{c} \quad\;\; NH_2 \qquad\; O \quad\; O \\ \quad\;\; | \qquad\qquad || \quad\; || \\ HOOCCHCH_2CH_2CNHCHCNHCH_2COOH \\ \qquad\qquad\qquad\qquad\quad | \\ \qquad\qquad\qquad\qquad CH_2SH \end{array}$$

$$2GSH + ROOH \longrightarrow GS\text{—}SG + ROH + HOH$$

The catalytic site is a selenocysteine residue: (Cys)Se—H $(HSeCH_2CH(NH_2)COOH)$. The reaction mechanism is believed to be (Ganther *et al.*, 1974; Arteel and Sies, 2001):

(Cys)Se—H + ROOH $\longrightarrow$ (Cys)Se—OH + ROH (Step 1)
(Cys)Se—OH + GSH $\longrightarrow$ (Cys)Se—SG + HOH (Step 2)
(Cys)Se—SG + GSH $\longrightarrow$ (Cys)Se—H + GS—SG (Step 3)

This mechanism assumes a significant difference between H—S (of GSH) and H—Se (of RSeH). In other words, the rate-determining step (step 1) is assumed not to take place between GSH and ROOH; in other words, the following reaction does not happen:

G—S—H + ROOH $\longrightarrow$ G—S—OH + ROH (Step 1′)

If instead this reaction was possible (similarly step 2 is possible with G—S—OH as well) and the mechanism is correct, then there would be no necessity for selenocysteine. Is this assumption (i.e., step 1 is possible but step 1′ is impossible) reasonable? Why is it so?

In order to see if this might be so, a reasonable mechanism for step 1 needs to be found. Suppose that the following mechanism (i.e., acid-base type reaction) is reasonable; then it may be that the reaction assumed here may be much faster (perhaps by the order of 10^3–10^4) with Se-H than with S—H, because Se—H is a stronger acid than S—H.

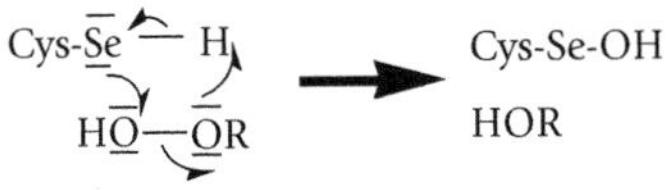

The next question is whether the Se in (Cys)SeOH is sufficiently electrophilic so that the S of GSH can attack it as presumed in step 2. The Se in (Cys)SeOH is nominally in 0 oxidation state whereas the Se in (Cys)SeH is nominally in $-II$ oxidation state. Therefore, the Se in (Cys)SeOH is certainly more electrophilic than that of (Cys)SeH, but is it sufficiently electrophilic? Similarly, is the assumption implicit in step 3 (i.e., acid-base type reaction) reasonable? In other words, the mechanism expressed in terms of steps 1 to 3 is based on acid-base type reaction, but is it reasonable?

9.5.3. **Thioredoxin Reductase**

Thioredoxin reductase from mammalian cells is another selenocysteine enzyme. Replacement of the selenocysteine residue by cysteine in rat cytosolic thioredoxin reductase using site-directed mutagenesis resulted in a functional mutant enzyme, but with only about one percent activity with thioredoxin as a substrate (Holmgren, 2000). For many selenocysteine-dependent enzymes, the S-analogues (S-cysteine) are functional but much less active than the Se-counterparts (Birringer et al., 2002).

Thioredoxin is a small protein containing Cys-X-X-Cys motif (where X is another residue), which undergoes oxidation-reduction. Its function is to reduce a disulfide (in a protein) to dithiol. The disulfide (oxidized) form of thioredoxin is reduced by NADPH catalyzed by thioredoxin reductase; that is:

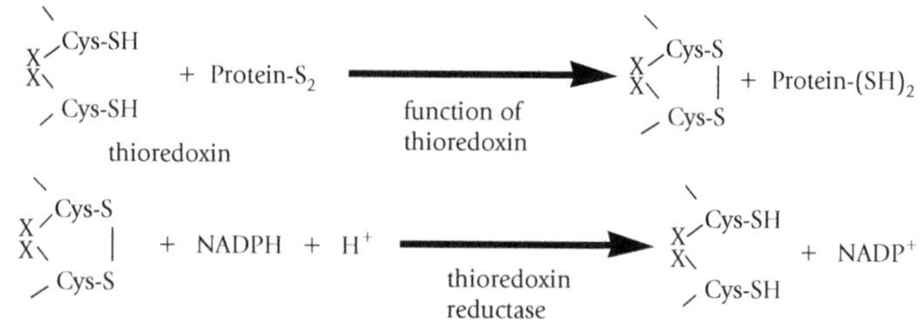

The mechanism is believed to involve the reduction of the disulfide of a special Cys-XX-Cys unit in the enzyme, and the resulting $(SH)_2$ then reduces the S-Se of CysS-SeCys. Finally, this $(SH)(Se^-)$ reduces the S—S of the thioredoxin (Birringer et al., 2002). Therefore, $Cys(SH)Cys(Se^-)$ must be a stronger reductant than $Cys(SH)Cys(SH)$. This may be reasonable in view of HSe^- being a stronger reductant than HS^- (see Table 9.1). The intermediary of the electron donor, NADPH, to the special CysS-SCys unit in the enzyme is FAD. FAD is capable of single electron transfer, rather than the transfer of a pair of electrons; that is, $FAD + H^+ + e \rightarrow FADH$, $FADH + H^+ + e \rightarrow FADH_2$. This may suggest that the reaction mechanism involves two single electron transfer steps.

9.5.4. **Other Selenium-Containing Proteins and Enzymes**

Other selenium-containing proteins and enzymes include glycine and other amino acid reductases, deiodinases (Birringer *et al.*, 2002), and hydrogenases and a formate dehydrogenase found in methane-forming archaea bacteria.

A selenocysteine-containing formate dehydrogenase H isolated from *E. coli* is a molybdenum enzyme, and its active site is a $Mo(pterin)_2$ (see Chapter 5). The selenocysteine Se is bound to the Mo in its oxidized state. The dehydrogenation of HCOOH (to CO_2) is believed to proceed in the regular manner as catalyzed by Mo-enzymes (Birringer *et al.*, 2002). The reason that a Se (Cys) is specifically required there is not understood. Another selenocysteine-containing enzyme, Ni-Fe hydrogenase from *Desulfomicrobium baculatum* (Garcin *et al.*, 1999), was discussed in Chapter 5. The specific reason for the presence of Se instead of S(Cys) is not well understood, except, perhaps, that Se will modulate the redox potential of the Ni entity.

Glycine reductase found in *Clostridial* bacteria was the selenoprotein, and subsequently many analogous amino acid reducing selenoenzymes have been found (Birringer *et al.*, 2002). Glycine reductase (and other amino acid reductase) generates ATP at the expense of glycine. A mechanism has been proposed by Arkowitz and Abeles (1991), but it is far from well established.

9.6. **BORON**

Boron is required only by plants. According to Blevins and Lukaszewski (1998), boron plays a structural role in plant cell walls (of mono- and di-cotyledons and gymnosperms), a role for some membrane functions, and a role in metabolic activities. It has been found that up to 90% of the cellular boron is localized in cell wall fraction. Borate bridges cell wall pectin polysaccharides. This accounts for the fact that the leaves of boron-deficient plants are brittle, whereas plants grown with an excess of boron produce leaves that are plastic or elastic. Borate forms the most stable diesters with *cis*-diols on furanoid ring (Blevins and Lukaszewski, 1998).

The web site "Plant Cell Walls" of Complex Carbohydrate Research Center, University of Georgia (O'Neill *et al.*, 2006; Selvendran and O'Neill, 1985) gives the details of chemistry of borate diester formations. The most important borate binding polysaccharide is

■ **Figure 9.3.** RGII, rhamnogalacturonan-II.

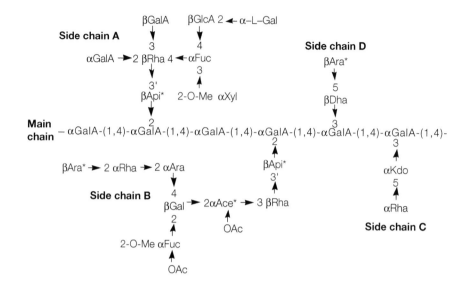

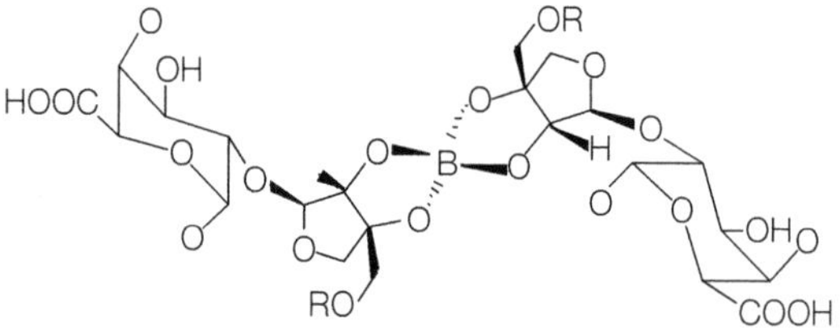

■ **Figure 9.4.** Borate ester cross-linking formation with RGII (after O'Neill *et al.*, 2006).

rhamno-galacturonan-II (RG-II) found in pectin. It is a relatively small polysaccharide, and consists of a main chain of 1,4-linked α-D-galacturonic acid (GalA) with several side chains of various oligosaccharides attached to the main chain. One such example is shown in Figure 9.3 (O'Neill *et al.*, 2006). It is believed that a borate cross-links two RG-IIs through diester formation with a pentose of furanoid type, likely apiosyl (β-Api(f)) residue in side chain A (see Fig. 9.4) (O'Neill *et al.*, 2006). Similar diols also are found in fucose and a few other hexose and pentose and their derivatives including arabinose and xylose, and borate would bind with these moieties as well.

Borate has been found to stabilize some pentoses, ribose, arabinose, xylose, and lyxose (Ricardo *et al.*, 2004). These compounds can form from formaldehyde through the formose reactions under an alkaline condition. However, they usually end up in an unidentifiable brown substance. For example, a mixture of formaldehyde and glycolaldehyde (condensation dimer of formaldehyde) formed a brown substance in the presence of $Ca(OH)_2$ at 25–40ºC. When the same mixture was allowed to react in the presence of borate mineral such as $NaCaB_5O_9 \cdot 8H_2O$, $Na_2B_4O_7$, or $CaB_6O_{11} \cdot 5H_2O$, it did not turn brown, and formed a mixture of borate-bound pentoses. The authors formulated a cross-linked (by borate) dimeric pentose like the one shown in Figure 9.4. The authors suggested that these observations may have some implication on the origin of life, particularly in view of the "RNA world" idea.

Other possible substances for borate cross-linking in primary cell walls include hydroxyproline-rich glycoproteins and proline rich proteins (Blevins and Lukaszewski, 1998). For example, the cell walls of boron-deficient bean root nodules have been observed to contain very low levels of hydroxyproline rich proteins, compared with the boron-sufficient controls.

An addition of boron to boron-deficient tissues has been observed to induce changes in cell membrane function (the following studies are quoted in Blevins and Lukaszewski, 1998). Phosphorus, chloride, and rubidium (K-analogue) uptake by boron-deficient roots of maize and *Vicia fava* was restored to about 40% of normal within 20 min after B was added. Boron deficiency was found to inhibit the vanadate-sensitive H^+-ATPase in microsomes isolated from sunflower roots. Addition of boric acid to low boron cells caused an instantaneous stimulation of the plasma membrane NADH oxidase.

It has been observed that the boron requirement for reproductive growth is much greater than for vegetative growth in most plants (quoted in Blevins and Lukaszewski, 1998). It also was observed that pollen tubes of petunia grew toward higher boron concentrations. The chemical/biochemical bases for these observations have not been elucidated.

9.7. SILICON

Silicon is found in all the organisms. Silicon is the second most abundant element in the Earth's crust, and it is not surprising that organisms contain it. The silicon concentration in organisms varies widely, reflecting the varying degree of exposure to silicon and its availability to organisms. Some organisms including diatoms, certain higher plants such as horsetail, grass, and bamboo, and animals do use silica as the structural material. Diatoms seem to require silicon even for some metabolic processes. It is assumed that silicon is essential to the majority of organisms, though the molecular mechanisms of the biological functions of silicon are not very clear (Perry and Keeling-Tucker, 1998; Exley, 1998).

9.7.1. **Chemistry of Silicon**

The simplest species in an aqueous medium is silicic acid formulated as $Si(OH)_4$. Silicic acid in natural waters is produced through weathering processes such as:

$$2SiO_2 \text{ (quartz)} + 4CO_2 + 12H_2O \rightarrow 2Si(OH)_4 + 4H_2CO_3$$

$$2KAlSi_3O_8 \text{ (K-feldspar)} + 2CO_2 + 11H_2O \rightarrow$$
$$4Si(OH)_4 + Al_2Si_2O_5 \text{ (kaolinite)} + 2KHCO_3$$

Silicic acid tends to condense and polymerize, resulting in amorphous silica:

$$2Si(OH)_4 \rightarrow (HO)_3SiOSi(OH)_3 + H_2O \rightarrow \rightarrow [SiO_n(OH)_{4-2n}]_m$$

The forward reaction (condensation) is faster by orders of magnitude than the reverse reaction (hydration) (Exley, 1998). In natural water, diatoms and other organisms facilitate this condensation. For example, frustules of diatoms are formed in a matter of hours to days, but dissolution of the frustules takes years to thousands of years or longer.

Another important reaction of silicic acid is the reaction with aluminum hydroxide to form hydroxyaluminosilicates (HAS):

$$Si(OH)_4 + Al(OH)_3 \rightarrow [(AlO)_{2n}(SiO)_n(OH)_{4n}] \text{ (HAS)}$$

This reaction is very facile because of the structural similarity of $[SiO_4]^{4-}$ and $[AlO_4]^{5-}$. This has an implication that silicic acid may remove toxic aluminum forming biologically inert HAS, and at the same time perhaps to release the useful phosphate that has been fixed as aluminum phosphate (Exley, 1998).

Reactions of silicic acid with organic compounds have been discussed but none has been unequivocally elucidated in living systems. Among them, the most interesting purported reaction is silicate ester formation with organic alcohols:

$$Si(OH)_4 + HOR \rightarrow Si(OH)_3(OR) + H_2C$$

Ester formation can be effected in lab by reacting silicic acid with an alkyl halide; for example,

$$Si(OH)_4 + RCl \rightarrow Si(OH)_3(OR) + HCl$$

Whether such an esterification is possible in biological systems has not been determined. However, there are a number of indications that suggest its possibility. For example, a silicon complex was isolated from the conifer *Thuja plicata* and was shown to be 1:3 complex of silicon and a tropolone, thujaplicine (thpl): $[Si(thpl)_3]^+$ (Perry and Keeling-Tucker, 1998). Model compounds such as Si-catecholate complex, $[Si(C_6H_4O_2)_3]^{2-}$ have been synthesized. There is also a strong possibility that $Si(OH)_4$ interacts through hydrogen-bond with alcoholic OHs in organic compounds.

9.7.2. **Frustules of Diatoms**

A prominent use of silicon, as amorphous silica, is found in diatoms. Diatoms, microscopic unicellular photosynthetic algae, in both freshwater and saltwater, wrap themselves with cell walls made of amorphous silica called frustules. Many frustules are beautifully shaped, as illustrated in Figure 9.5. This silica is hydrated silicon dioxide, $SiO_2 \cdot nH_2O$; its composition is variable. Diatoms use silicic acid as the source of silicon, and apparently facilitate deposition of amorphous silica on their cell walls. The proteins of the cell wall were found to contain more serine, threonine, and glycine, and less glutamic acid, aspartic acid, aromatic amino acids, and sulfur-containing amino acids than is found in the proteins within the cell (Hecky *et al.*, 1973).

More recently, Kröger and coworkers (2001) isolated and identified two small polypeptides, sillafin-1A$_1$ and sillafin-1A$_2$, from the cell wall of a diatom *Cylindrotheca fusiformis*. Both peptides were shown *in vitro* to precipitate nanospheres of silica within seconds when

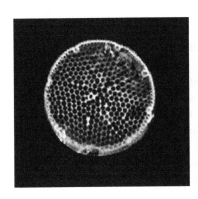
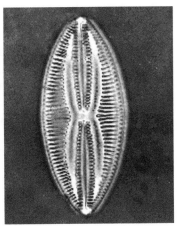

■ **Figure 9.5.** Examples of microscopic diatoms, *Lyrella lyra* and *Coscinodiscus radiatus*, from University College London, Micropalaeontology web site.

added to a silicic acid solution. They must provide a nucleation site for silica formation. They are 15 and 18 amino acids long, and contain seven and eight serine residues, respectively, and additionally, four lysine residues. Interestingly, all the lysine residues are post-translationally modified; that is, the ε-amino groups are linked to linear polypropylamines (5–11 units). One of those is the terminal of an ε-N,N,N-trimethyl-δ-hydroxylysine in sillafin-1A$_1$. Because both proteins are equally effective in forming silica, it was suggested that the polypropylamine chains are essential for silica precipitation. It has been shown that similar synthetic polypeptides without poly-amines attached could not precipitate silica at pH $<$ 7. The silica formation in diatoms takes place in an acidic intracellular compart-ment. Hence it is obvious that the amines are all protonated and that abundance of positive charges acts as a catalyst to polymerize silicic acid and form the nanocrystals of silica, though how it works is not clear. The function of abundantly present (up to 50%) alcoholic resi-dues (serine and threonine) is not clear. The authors (Kröger *et al.*, 2001) suggest that they are O-glycosilated.

9.7.3. **Spicules in Sponge**

Spicules, the skeletal elements in sponges *Demospongiae* and *Hexactinellida*, are made of amorphous silica. Cha *et al.* (1999) found that a marine sponge *Tethya aurantia* produces a large amount of silica spicules, which constitute three-fourths of the dry weight of the organism. Each spicule, a filamentous element 1 to 2 mm long and 30 μm in diameter, contains an axial filamentous protein. This protein consists of three subunits, and has been named *silicatein*. Silicatein α has been demonstrated *in vitro* to catalyze the polymer-ization of silicon alkoxides to form polymerized silica at neutral pH (Cha *et al.*, 1999).

The major components of silicatein α are Gly (15.2%), Ser (13), Asx (12.5), Ala (9.6), Glx (8.8), Leu (7), and Thr (6.7) (Shimizu *et al.*, 1998). The alcoholic amino acids (Ser, Thr, and Tyr) amount to about 27% of all the amino acids. This is similar to the other pro-teins involved in the formation of amorphous silica as discussed earlier. The amino acid sequence of silicatein α deduced from cDNA turned out to be very similar to a protease, cathepsin L (Shimizu *et al.*, 1998). From this congruency between two proteins, the authors (Cha *et al.*, 1999) suggested that the two proteins function similarly. The catalytic triad (His, Asn, Cys) in cathepsin L corresponds to a triad (His, Asn, Ser) in silicatein.

Both silicatein (in sponge) and sillafin (in diatom) function similarly; they induce the polymerization of silicic acid to form amorphous silica. Yet, they are quite different in structure and functional groups, as the respective authors suggest. A common feature in both the proteins is the preponderance of alcoholic amino acid residues. It seems more reasonable to assume that this common feature has something to do with the amorphous silica formation. In other words, the serine and threonine (and tyrosine) residues in these proteins play the role of polymerization or nucleation site of silica formation, as suggested by Hecky *et al.* (1973), and the rest of each protein is, perhaps, involved in regulating the overall crystal shape, fibrous, globular, or otherwise.

9.7.4. **Other Biological Functions of Silicon**

Silicon has been shown to be essential for normal growth overall and of skeletons in particular in rats, chicks, and other vertebrates (Perry and Keeling-Tucker, 1998). Silicon is found at the growth front in young bones of mice and rats and facilitates mineralization of bones (approximately hydroxy calcium phosphate = apatite mineral). An observation was reported that an amorphous silica layer was the site on which nucleation and growth of the apatitic phase takes place. In mammals, silicon is located along with calcium in an osteoid tissue. Low silicon also has been shown to cause abnormalities in connective tissues, collagen, and cartilage (Perry and Keeling-Tucker, 1998).

Silica deposition in plants occurs mainly in the cell wall. As discussed earlier, plant cell walls contain a high level of pectin, a polysaccharide rich in galacturonic acid. The pectic substance is bound with $Ca(II)$, but it also contains a significant amount of silicon in high silica-accumulating plants. How silica deposits on pectic substances is not known; no specific silica-forming agent has been identified.

The interaction of silicon with aluminum in terms of HAS formation mentioned earlier appears to be one of the reasons for essentiality of silicon in biological systems. It has been observed that the presence of silicon reduced the gastrointestinal absorption of aluminum (Perry and Keeling-Tucker, 1998). In studies on aluminum toxicity on plants, it was found that addition of silicon species to the growth medium reduced the extent of aluminum-induced root damage, and that aluminum was located with silica deposits in the roots. Silicon added to soils was also reported to increase the uptake of phosphate by plants; silicon presumably binds aluminum (to form HAS) and releases the phosphate fixed as aluminum phosphate (Exley, 1998).

9.8. **VANADIUM**

Vanadium has been known to accumulate in many sea squirts (ascidians), and also in *Amanita muscaria*, a poisonous mushroom. Vanadium also constitutes the catalytic site of haloperoxidases in seaweed and others.

9.8.1. **Vanabins**

Some sea squirts, Ascidians, are known to accumulate a high concentration of vanadium in blood. For example, *Ascidia gemmata* accumulates vanadium in its blood cells at as high a concentration as 0.35 M. Other organisms that accumulate vanadium in blood cells include *A. ahodori*, *A. sydeneiensis samea*, and *Phallusia mammillata*. A polychaete, *Pseudopotamilla accelata*, also is known to accumulate vanadium selectively (Yoshihara *et al.*, 2005).

The blood cell called vanadocyte of *A. gemmata* has a vacuole that contains a highly acidic (pH 1.9 and 0.5 M SO_4^{2-}) vanadium (III) solution. A correlation seems to exist between the vanadium concentration and the pH of the vacuole solution. Vanadium (III), like Fe(III) or Al(III), is highly acidic and precipitates at pH higher than about 4.5. The low pH of the vacuole seems to be maintained through a H^+-ATPase, which injects H^+ into the vacuole. Vanadium enters the blood cells likely as V(V), and then is reduced to V(IV) nonenzymatically by NADPH, and further to V(III) in the vacuole. It has not been elucidated what form the V(III) species takes in the vacuole, nor the role that it plays. Ascidians are the most primitive chordate, and their blood vessels are supposed to circulate oxygen. Hence one of the roles hypothesized is an oxygen carrier. However, there are other ascidians and similar organisms that do not accumulate vanadium.

Several proteins have been identified to bind vanadium in the cytoplasm of vanadocytes of *A. sydeneiensis samea*. They are named as vanabin; vanabin 1~4. Another protein, vanabinP was found not in blood cells but in blood plasma (Yoshihara *et al.*, 2005). All these proteins have secondary structures similar to each other, particularly in regard to the position of cysteine residues, and bind not a single but tens of vanadium (IV). They are rich in lysine and cysteines, the latter of which are spaced fairly regularly. VanabinP has been found to bind as many as 13 V(IV) ions per mole with a dissociation constant of 2.8×10^{-5} (Yoshihara *et al.*, 2005). These proteins are proposed to be vanadium carriers.

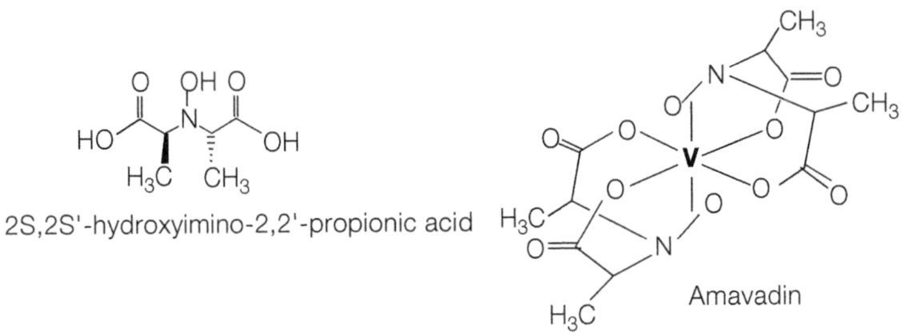

■ Figure 9.6. Hydroxyimino-2,2'-propionic acid and its vanadium complex (amavadin) (after Gamer *et al.*, 2000).

2S,2S'-hydroxyimino-2,2'-propionic acid

Amavadin

9.8.2. **Amavadin**

Amanita muscaria, a poisonous mushroom, contains a significant level of vanadium. It has been shown that one of the vanadium-containing compounds, amavadin is $[(hidp)_2V^{IV}]^{2-}$ (Hubregtse *et al.*, 2005), where hidp = 2S,2S'-hydroxyimino-2,2'-dipropionic acid (see Fig. 9.6). The vanadium complex is chiral and has the structure shown in Figure 9.6 (based on the x-ray crystal structure reported by Gamer *et al.* (2000). However, the biological function of amavadin is unknown.

9.8.3. **Haloperoxidases**

Vanadium-dependent haloperoxidases have been found in various organisms; many brown algae, some red algae, and one green alga, a lichen *Xantora parientina*, and some fungi (Almeida *et al.*, 2001). It seems that different peroxidases exist separately for chlorine, bromine, and iodine. Chloroperoxidase (ClPO) is found only in terrestrial organisms, and bromo (BrPO) and iodoperoxidases (IPO) are found mostly in the marine organisms (Almeida *et al.*, 2001). A ClPO from *Curvularia inaequalis* (a fungus) has exactly the same V-entity at the active site (Messerschmidt *et al.*, 1997) as that of BrPO from *Ascophyllum nodosum* (Weyand *et al.*, 1999). Hence the halogen-specificity seems to be dependent on the rather small difference in the overall structure of active site. For example, the halide specificity of vanadium-dependent BrPO from a marine alga *Corallina pilulifera* was altered by a single amino acid substitution (Ohshiro *et al.*, 2004). That is, the mutant enzymes R397W and R397F showed a significant ClPO activity as well as BrPO activity.

BrPO catalyzes halide oxidation by hydroperoxide. In marine algae, the enzyme is involved in synthesis of halogenated marine bioproducts, such as halogenated indoles, terpenes, and acetogenins, and volatile halogenated hydrocarbons (Almeida *et al.*, 2001; Carter *et al.*, 2002; Colin *et al.*, 2003).

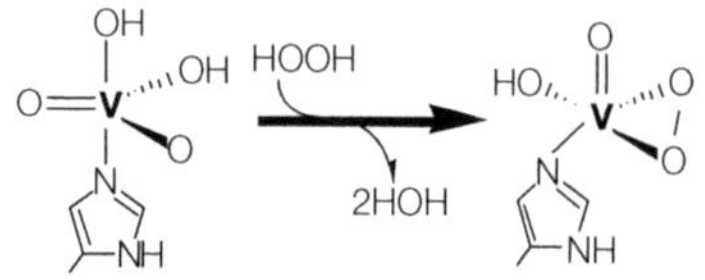

Figure 9.7. The structure of the catalytic (V) site in V–haloperoxidase (after Rehder et al., 2003).

The catalytic active site of BrPO from *Ascophyllum nodosum* is a trigonal bipyramidal V(V) entity (Weyand *et al.*, 1999). The equatorial positions are occupied by three O^{2-} or OH (V—O distances range from 1.5 to 1.6 Å), and another O^{2-} (OH) (at a distance of 1.8 Å) and a histidine N (at 2.1 Å) occupy the axial positions; it seems that vanadate ($[VO_4]^{3-}$) is bound to a histidine N. The structure around the V in a fungal ClPO has been found to be the same (Messerschmidt *et al.*, 1997).

The major form is vanadate $[V^VO_4]^{3-}$ when V_2O_5 is dissolved in a highly alkaline solution, whereas it is $[VO_2]^+$ at low pH (<2) (Greenwood and Earnshaw, 1997). Various dimeric and oligomeric forms do exist in the intermediate pHs. H_2O_2 added to an aqueous vanadate solution displaces O^{2-}s and forms a peroxo complex: $[(O)V(O_2)(H_2O)_4]^+$. Further reactions will produce a tetra (peroxo) species: $[V(O_2)_4]^{3-}$. For example, deep violet crystals of $k_3[V(O_2)_4]$ have been isolated (Greenwood and Earnshaw, 1997). That is:

$$(H_2O)_xV(=O)_2 + 4H_2O_2 \rightarrow (O=)V(O_2)(H_2O)_y \rightarrow$$
$$(O=)V(O_2)_2(H_2O)_z \rightarrow \rightarrow [V(O_2)_4]^{3-}$$

In view of this kind of chemistry of V(V), the V(V) of the active site of haloperoxidase is believed to form a peroxo complex when HOOH is added. Indeed, the x-ray crystallography of the peroxo form (Messerschmidt *et al.*, 1997; Rehder *et al.*, 2003) showed that the V-complex is a tetragonal pyramid with a peroxo, N(his) and O in the base and another O at the apical position (see Fig. 9.7).

It is known that peroxo-V(V) is an effective oxidant of many inorganic and organic compounds, including halides (Bortolini and Conte, 2005). Examples of such reactions include:

$$X^- \text{ (Halogen)} \rightarrow HOX, X_2; \quad R—HC(OH)—R' \rightarrow R—C(=O)—R';$$
$$C=C \rightarrow \text{epoxide}$$

$$RH \text{ (aliphatic or aromatic)} \rightarrow ROH; \quad SO_3^{2-} \rightarrow SO_4^{2-};$$
$$R—S—R' \rightarrow R—SO—R'$$

Hence, it seems reasonable to assume that haloperoxidases produces $V(O_2)$, which then oxidizes X^- to XOH or a species equivalent to X^+. A reasonable mechanism can be written as in Figure 9.8.

Figure 9.8. A mechanism of V-peroxidase reaction (Ochiai).

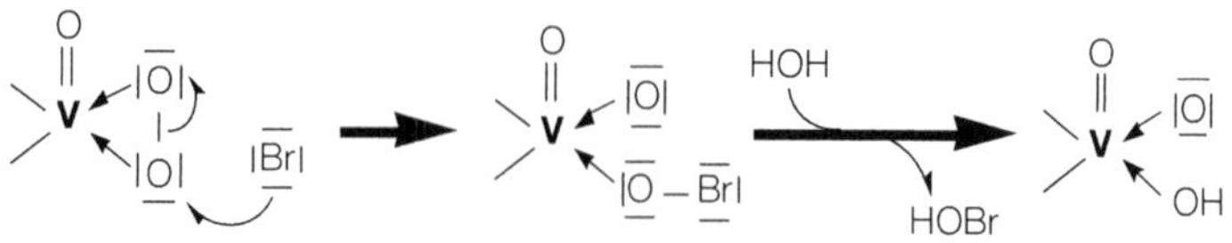

9.9. **CHROMIUM**

Chromium has been suspected to be essential to many organisms, but its biochemistry is poorly understood. The most intensely studied is the chromium involved in GTF (glucose tolerance factor); that is, it is somehow involved in the function of insulin.

A low-molecular-weight chromium binding substance was isolated from livers of some mammalian species (Yamamoto *et al.*, 1984). This compound is now dubbed as chromodulin (CMD) (Vincent, 2001, 2004). It is an oligopeptide of molecular weight of 1438, and is composed of glycine, cysteine, aspartate, and glutamate with the acidic amino acids comprising more than half. One mole of this compound binds four moles of Cr(III) very tightly. The binding constant is 10^{21} M^{-4} for CMD + 4Cr^{III} → CMD(Cr)$_4$. When apochromodulin binds Cr(III)s that are brought by transferrins, it changes the peptide conformation, and this holopeptide then binds to the insulin receptor, and amplifies the receptor's tyrosine kinase activity (Vincent, 2001, 2004). This manifests as the hormonal action of insulin. The chromiums in the holochromodulin were found to exist in the form of an antiferromagnetically coupled trinuclear assembly that is interacting with a fourth Cr(III). The Cr(III)s are octahedrally coordinated by Os of carboxylate groups of the oligopeptide (Jacquamet *et al.*, 2003).

9.10. **HALOGENS AND THE LIKE**

Chloride Cl$^-$ is omnipresent and a crucial counter anion. Iodine is a component of hormone thyroxine in mammals. Bromide and iodide, Br$^-$ and I$^-$, are found widely in all living organisms. In order for halides and analogues (Cl$^-$, Br$^-$, I$^-$, SCN$^-$, etc.) to be incorporated into organic compounds or made to react with relatively inert organic compounds, they have to be made reactive. Cl$_2$ and Br$_2$ are reactive enough but often require light in order to react with organic compounds, through free radical reactions that are difficult to control. Another reactive form is formally X$^+$, which can be in the form of XO$^-$ or HOX. The formation of HOX (hypohalous acid) is carried out by peroxidases dependent on hydrogen peroxide. HOXs are also oxidants strong enough to kill invading bacteria.

There are three different types of peroxidases present in living organisms. One is vanadium-dependent ones found in seaweeds as discussed earlier. There are two different families of heme-dependent peroxidases (Furtmüller *et al.*, 2006). One is plant, fungal, and (archae) bacterial peroxidases, and the other is a group of peroxidases of a similar structure (but different from those of plants and fungi) found in mammals. Some examples of the first type are discussed in Chapter 5. Peroxidases of the second type in mammals are also called "Du-ox" (dual oxidase). However, peroxidases of primary and tertiary structures similar to those in mammals recently have been found in lower animals such as arthropods, mollusks, and even a worm *Caenorhabditis elegans* and fruit fly *Drosophila* (Furtmüller *et al.*, 2006). Hence these are now characterized as animal peroxidases, in contrast to the plant peroxidases earlier.

9.10.1. Formation of Volatile Halocarbons in Macroalgae

Seaweeds produce a number of halogen-containing compounds that are held within algal membrane-bound vesicles, and smaller ones are released into the atmosphere (Ekdahl *et al.*, 1998). These compounds are thought to act as hormones or repellants in biological defense mechanism (Almeida *et al.*, 2001). V-dependent halo-peroxidases in macroalgae (brown kelp, etc.) oxidize halide, Cl^-, Br^-, and I^-, and form HOCl, HOBr, and HOI, as discussed earlier. These compounds are believed to halogenize hydrocarbons to form volatile halogenated compounds of hydrocarbon such as CH_3I, C_2H_5I, CH_2ICl, CH_2IBr, CH_2I_2, $CHBr_3$, CH_2Br_2, $CHBr_2Cl$, CH_3Br, and C_2H_5Br (Carpenter *et al.*, 2000) in reactions of the type:

$$R—H + XOH \rightarrow R—X + H_2O$$

The mechanism for this halogenation is not well understood. However, it is true that these volatile halocarbon compounds are present in a significant amount in the atmosphere above the sea area where brown algae and other seaweeds abound.

9.10.2. HOX Formation in Mammals and Others

9.10.2.1. *Formation of HOX by a Fungal Chloroperoxidase*

A chloroperoxidase from a fungus *Caldariomyces fumago* has been well characterized (Kühnel *et al.*, 2006). It is a P-450-like enzyme with a heme protoporphyrin-IX of which Fe is bound to a cysteine S^-, and catalyzes chlorination of cyclopentadienone in the biosynthesis of antibiotic caldariomycin (Kühnel *et al.*, 2006). It is capable of bromination and iodination as well. Unlike P-450, though, it uses HOOH (instead of O_2) to oxidize halides. These authors and others presume that the reaction between Fe(III)-heme and HOOH produces compound I (CpI, O = Fe(IV) plus porphyrin π-radical, see Chapter 5). As discussed in Chapter 6, an alternative mechanism is that the reaction leads to O = Fe(V) superferryl, as in the case of P-450. The superferryl O = Fe(V) seems to behave as O(Fe(III)). That is, this O (in O(Fe(III))) is like the first excited singlet state of O-atom, and acts as an acceptor of a pair of electrons. Hence,

$$|\overline{X}| \overset{\frown}{} \overline{O}| \to Fe^{III} (\leftrightarrow |\overline{O}| \to Fe^{V}) \longrightarrow |\overline{X} - \overline{O}|^{-} \ + \ Fe^{III}$$

The substrate specificity (in chlorination reaction) of this enzyme turned out to be rather low, suggesting that the subsequent chlorination reaction of a substrate is independent of the active site and may take place between the substrate and XO^- or HOX formed without the intervention of the active site (Kühnel *et al.*, 2006).

9.10.2.2. *Formation of HOX by Mammalian Peroxidases*

Several different peroxidases are found in mammals; the major ones are thyroid peroxidase, lactoperoxidase, myeloperoxidase, and eosinophil peroxidase (Furtmüller *et al.*, 2006). They belong phylogenetically to the same gene pool, and hence the overall 3D structures are similar to each other, but there are differences in details.

Phagocytic cells, leukocytes, and macrophage in mammals produce a number of reactive oxygen species including O_2^- and HOOH through NAD(P)H oxidase (see Chapter 11 and Fig. 11.1). These reactive oxygen species (ROS) are used to kill invading bacteria. It turned out, however, that the HOOH produced is used to oxidize halides and the like, such as SCN^- (Wang and Slungaard, 2006; Senthilmohan and Kettle, 2006). The reaction is described by:

$$HOOH + X^- + H^+ \longrightarrow HOX + H_2O \ (X = Cl, Br, I, SCN)$$

These HOXs (hypohalous acid) are stronger and more specific toxic agents against invading bacteria.

White blood cells, leukocytes, are a part of the nonspecific immune system, and consist of neutrophils, eosinophils, basophils, natural killer cells, macrophages, and others. The major enzyme of neutrophils is myeloperoxidase. Eosinophils seem to be specialized in eliminating invasive parasites (Wang and Slungaard, 2006).

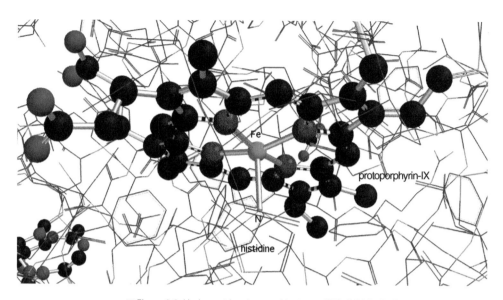

■ **Figure 9.9.** Myeloperoxidase-heme and its vicinity. (PDB 1D2V: Fiedler, T.J., Davey, C.A., Fenna, R.E. 2000. X-ray crystal structure and characterization of halide-binding sites of human myeloperoxidase at 1.8 Å resolution. *J. Biol. Chem.* **275**, 11964–11971.)

The structure of myeloperoxidase has been determined by x-ray crystallography (Fiedler *et al.*, 2000). The enzyme consists of two identical halves, each of which is made of two polypeptides; these units are connected by a single (Cys)S—S(Cys) bond. Each half contains a heme group, the Fe of which is bound to N^- of histidine imidazole (see Fig. 9.9); this binding of imidazole is the same as that in horseradish peroxidase (see Chapter 5). The protoporphyrin-IX is covalently bound to the protein through three positions. It is believed, as in horseradish peroxidase, that the reaction of H_2O_2 with Fe(III) form of myeloperoxidase results in the formation of CpI; O^{2-}=Fe(IV) (ferryl) plus porphyrin π-radical. This is two steps higher than the oxidation state of the resting state Fe(III), and is believed to remove two electrons from X^- to form X^+ (in the form of XO^-). In view of the fifth ligand of the heme group, the formation of superferryl may not be feasible in this enzyme system (refer to Chapter 6). But then, how CpI oxidizes the single entity X^- by removing two electrons is not understood, because the electron accepting centers are different; one is on the Fe and the other on the porphyrin ring.

In this connection is an interesting report that showed that Br_2 and BrCl (and probably Cl_2) are direct products in myeloperoxidase-H_2O_2-X^- (Spalteholz *et al.*, 2006). In other words, perhaps, CpI's

Fe(IV) and π-radical accept an electron each separately from X^- to form two $X\cdot$ species that combine to form X_2 or XY species. Whether X_2 or XY rather than HOX is the halogenation agent is not clear. It is possible that the intermediate $X\cdot$ free radical is the halogenating agent. It is possible that $X\cdot$ is further oxidized by the Fe(IV) center to X^+. In other words, a two-step oxidation of X^- (to $X\cdot$ and then X^+) may be compatible with formation of both X_2 and HOX.

Hypochlorous acid HOCl and hypothiocyanous acid HOSCN are the major products of myeloperoxidase, whereas HOBr becomes predominant at higher pH even in the presence of physiological levels of chloride (Senthilmohan and Kettle, 2006). On the other hand HOSCN and HOBr are the major products in the case of eosinophil peroxidase (Senthilmohan and Kettle, 2006). These products are used to kill invading parasites and bacteria. However, eosinophil peroxidase exhibits also a deleterious effect in pathogenesis of various human allergic diseases including asthma (Wang and Slungaard, 2006), very likely by promoting tissue damages due to the oxidative agents.

The substrate specificity of eosinophil peroxidase *in vivo* has been found to be in the order of $SCN^- > NO_2^- > Br^- >> Cl^-$ (Wang and Slungaard, 2006). Indeed HOSCN is the primary product of this peroxidase as well as lactoperoxidase. It has been shown that NO_2^- is oxidized by a single electron to $NO_2\cdot$. The mechanism of this peroxidase has not been elucidated, but is believed to involve the formation of CpI. That is, it seems to involve two separate steps of single electron oxidation.

Thyroxine is a prominent hormone that contains iodine (see below for the structure). Two iodated tyrosine molecules couple to form it. Iodination of tyrosine is catalyzed by thyroid peroxidase, which belongs to the animal peroxidase family. It requires I^-, H_2O_2, and Tg (thyroglobulin) (Ruf and Carayon, 2006). A likely mechanism involves the formation of I^+ or (HOI), which substitutes the aromatic H in an electrophilic manner at the *ortho*-position(s) of tyrosine. The coupling between two iodated tyrosine molecules in Tg leads to the formation of thyroxine (T_4). The mechanism of this part is obscure, but it might involve a free radical coupling.

REVIEW QUESTIONS

1. Phosphorus P is a congener of nitrogen N. N has an extensive biochemistry involving oxidation-reduction reactions as outlined. On the contrary, the biochemistry of P does not seem to involve oxidation-reduction reactions. Why? What is the chemical basis for this difference?

2. Selenium Se is a congener of S. However, though S is an essential element, Se is quite toxic even at low level. Then why do organisms (including humans) require Se for certain functions? List as many reasons as possible.

3. In view of the functions of B (boron) discussed in the text, suggest why B may not be vital to animals.

4. List the main functions of silicon in various organisms.

5. Vanadium seems to be found in various organisms unrelated to each other. What are they, and what is vanadium doing in them?

6. Explore literature for the function(s) of V in the poisonous mushroom, *Amanita muscaria*.

7. It seems that oxidation of halides and the like (Cl^-, Br^-, I^-, and NCS^-) is widespread among living organisms. What roles do the oxidized products play in organisms?

PROBLEMS TO EXPLORE

1. A project is suggested: list in the form of a table all the biochemical roles and functions of all the essential elements (including not only those in this chapter but also all those discussed in other chapters as well), because no such table is provided in this book.

2. A list of the elements essential to organisms is given in Figure 9.1. There are some indications that elements such as Sn and As might be essential. These elements are toxic and this aspect is dealt with in Chapter 11. Explore literature for possible positive biological effects of these elements, and discuss whether their essentiality is reasonable.

3. The role Cr plays in enhancing insulin's function is far from well understood. In view of chemistry of chromium, does the discussion in this chapter make sense? Explore further and discuss.

4. One of the major functions of vanadium is that of haloperoxidases; that is, halide oxidizing enzymes using hydrogen peroxide. There are two types of haloperoxidases; one is V-enzymes and the other is the more widespread heme-enzymes. What is the implication of two completely different types of enzymes that catalyze the same type of reaction?

5. Explore the similarities and differences between haloperoxidase and cytochrome P-450 dependent monooxygenases.

6. Many seaweed species produce halogenated hydrocarbons (of small molecules). Explore what significance the production of these compounds has to them.

Chapter 10

Metal-Related Physiology

Although all the inorganic elements' interactions with biological systems take place at the molecular level, the results of such interactions are manifested as physiological processes. The biochemical reactions discussed in the previous chapters (see Chapters 4–9) are indeed reflected in physiological processes at higher levels. Such a physiology is indeed a part of the entire biology, and will be omitted here. Only a few selected physiological processes that directly involve inorganic elements are discussed in this chapter.

First, any inorganic element has to be dealt with by the biological systems. To begin with, it has to be incorporated into the system (uptake/absorption). It will likely be transported from the point of uptake to other tissues where the element is required or stored; this requires transportation. The element then has to be unloaded from the transporter. Any excess may need to be stored away or removed. Some of these processes need to be regulated (regulators). Let us call all these processes collectively "metabolism of inorganic elements."

Some physiological processes are carried out by metallic elements. Ca(II) is involved in a great variety of cellular processes, which manifest themselves as physiology. The examples of such processes include muscle contraction, blood coagulation, and release of neurotransmitters at a synapse. A whole variety of hormone actions are mediated by Ca(II), a second messenger. Metal containing entities are involved in sensing/monitoring some small molecules such as O_2, CO, NO, and H_2. Unequal distributions of Na(I) and K(I) in and out of the cell create an electric potential across the cell membrane. This is the basis of electric signaling.

The skeletal structures are not exactly physiology, but are included here. A number of essentially inorganic substances constitute some solid structures in biology. Examples include mammalian bones, teeth, shellfish shell, bird eggshell, frustules (of diatoms), and spikes in sea squirts. Some bacteria and probably some migrating birds use magnetite (Fe_3O_4) or others as a compass needle to navigate their movements.

Some metallic elements may interfere with some physiological processes such as brain function. Toxicity of metallic elements in general can be considered to belong to this category. The bulk of toxicity issues will be discussed in Chapter 11.

10.1. METABOLISM OF METALLIC ELEMENTS

All compounds, both inorganic and organic, are metabolized in the widest sense by organisms. It is impossible to discuss metabolisms of all the essential and nonessential elements, and besides, the details of those processes for many elements have not been elucidated. The overall picture of even iron, the most important and most intensely studied element, is not very clear yet. Only a few elements will be discussed here, including iron, copper, and zinc. The metabolism of some toxic elements will be dealt with in Chapter 11.

10.1.1. Iron Metabolism (in Mammals)

As mentioned repeatedly in this treatment, the most characteristic feature of iron is a relatively facile conversion between Fe(II) and Fe(III). And Fe(II) and Fe(III) have quite different solution chemistry. That is, Fe(II) is stable in aqueous medium of pH up to 7.5, but Fe(III) precipitates as the hydroxide above pH 2. Fe(III) forms chelates with O-ligands much more strongly than Fe(II). For example, the formation constant of Fe(III)(EDTA) is 10^{25} whereas that of Fe(II)(EDTA) is 10^{14} (where EDTA = ethylenediamine tetraacetate). These differences are reflected in how iron is processed in metabolism.

The overall picture of iron metabolism has been fairly well established; only the molecular details have not been elucidated until recently (Arredondo and Núñez, 2005). Figure 10.1 gives an overall picture of iron metabolism in a human (mammalian) body, indicating some of the recognized agents involved, such as transferrin (Tf), ferritin, divalent metal transporter (DMT), duodenal ferric reductase (Dcytb), ferroportin (Fpn), ferroxidase (Fox, ceruloplasmin), and hephaestin (Heph). This is not a complete picture; it is likely that more

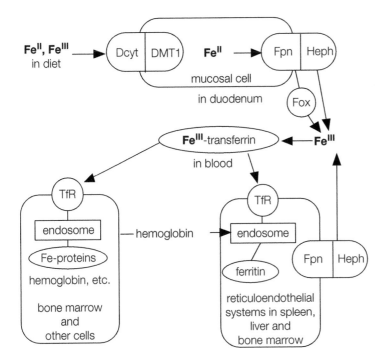

■ **Figure 10.1.** Metabolism of iron in mammals. Dcyt = duodenal cyt b (ferric reductase), DMT1 = divalent metal transporter 1, Fpn = ferroportin, Fox = ferroxidase (ceruloplasmin), Heph = hephaestin (ferroxidase), TfR = transferrin receptor.

agents (relevant proteins) and more details are yet to be elucidated. In addition to these agents, there is an elaborate system to regulate body iron, because either deficiency or excess can lead to diseased states. Some of these agents involved will be discussed at molecular levels.

10.1.1.1. *Ferric Reductase*

The absorption of iron takes place mainly in the duodenum (the upper portion of the small intestine) in mammals. The iron contained in the foodstuff can be either in the Fe(II) or Fe(III) state. The major uptake mechanism located in the duodenum is DMT1, which carries any divalent metal cations, but mainly Fe(II). An absorption pathway for Fe(III) does exist, but the amount of uptake of iron through this route is minor. Therefore, the important first reaction that iron should undergo is the reduction of Fe(III) to Fe(II). The enzyme for such a reaction, ferrireductase, seems to be expressed on the duodenal brush border membrane. It has been found that it has a strong homology to cytochrome b_{561}, and hence has been dubbed Dcytb (McKie *et al.*, 2002).

Several genes for ferrireductases have also been identified: *fre1* and *fre2*. The gene sequences suggest similarities to b-type cytochromes (Aisen *et al.*, 2001).

10.1.1.2. *Divalent Metal Transporter (DMT1)*

The major carrier of Fe(II) into the mucosal cells of duodenum has been identified as DMT1. This protein is also known as DCT1 (divalent cation transporter) or Nramp2 (natural resistance-associated macrophage protein 2). It mediates import of divalent metals including Fe(II), Mn(II), Co(II), Ni(II), Cu(II), Zn(II), Cd(II), and Pb(II), though the primary physiological role is the uptake of Fe(II). DMT1 is a single chain of 561 amino acid residues and molar mass of 61 kDa (Aisen *et al.*, 2001). This protein contains twelve transmembrane domains (TMDs) with both N- and C-terminals exposed to the cytoplasm. How Fe(II) is transported across membrane by DMT1 has not been elucidated, except that the conduit created by several TDMs is believed to conduct a divalent cation; usually 6TMDs seem to create a conduit. Ca-ATPase is of the same type and is discussed in more detail in a later section. A similar DMT2 is also known but is yet to be studied in detail. Similar transporters have been identified in yeast (*Saccharomyces cerevisiae*) as well. The iron thus imported enters into a labile iron pool. It can remain as free Fe(II) or binds to intracellular transporters.

10.1.1.3. *Ferroxidase*

For the iron to be transported throughout the body, it needs to be bound with a protein transferrin (Tf). Iron has to be Fe(III) in order to bind to Tf. So the iron imported in the form of Fe(II) needs to be oxidized to Fe(III). This reaction is catalyzed by ferroxidase, which is also known as ceruloplasmin in blood serum, and a similar membrane enzyme dubbed as hephaestin (Heph). Ceruloplasmin has been discussed in Chapter 5. Hephaestin (see Fig. 10.1) is also a multicopper enzyme (Petrak and Vyoral, 2005).

10.1.1.4. *Transferrin (Tf) and Transferrin Receptor (TfR)*

Fe(III) bound with Tf is carried by blood to the target tissues. Vertebrate Tfs are glycoproteins of 80 kDa. The single chain polypeptide is segmented into two lobes, that of the N-terminal half and that of the C-terminal half. Each lobe binds one Fe(III). Each lobe consists of two similar domains N1, N2 and C1, C2. An x-ray crystallographic study (Thakurta *et al.*, 2004) on the apo-form of hen's serum Tf showed that both lobes are open; it seems that the N1 and N2 domain are separate from each other with a connecting hinge; the same with C1 and C2. In this open structure the (Fe-) coordinating residues, one histidine, two tyrosines, and one aspartate are all exposed. In addition two lysine residues (209 and 301) are far apart (8.2 Å). When two Fe(III)s are loaded, the overall structure becomes closed; see Figure 10.2 for a comparison of the two forms in the case of human transferrin. The aforementioned ligands and CO_3^{2-} coordinate to each Fe(III) and the two lysine residues are now closer (2.5 Å) (Thakurta *et al.*, 2003). The structure about Fe(III) is a five-coordinate tetragonal pyramid: the monodentate CO_3^{2-} and a histidine N are at trans to each other; a tyrosine O and a monodentate aspartate are another set of trans-ligands; and a tyrosine O is at the apical position (PDB 1D4N). In another Fe(III)-bound Tf, CO_3^{2-} seems to be a bidentate ligand (PDB 1H76). The extraneous CO_3^{2-} is an obligatory ligand for strong binding.

The two lysine residues (positively charged) mentioned might be involved in binding to the carbonate anion, and hence closing the structure of apo-Tf. However they do not seem to be involved in binding the carbonate, though they come very close to each other in the closed structure. Instead of these residues, a positively charged arginine and a peptide NH (of serine) seem to provide binding to the carbonate ligand through ionic and hydrogen-bonding interactions.

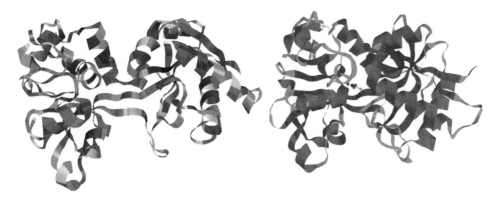

■ **Figure 10.2.** The structures of the N-lobe portion of human transferring; (a) apo-form and (b) Fe(III)-bound form. ((a) PDB 1BTJ: Jeffrey, P.D., Bewley, M.C., MacGillivray, R.T., Mason, A.B., Woodworth, R.C., Baker, E.N. 1998. Ligand-induced conformational change in transferrins: Crystal structure of the open form of the N-terminal half-molecule of human transferrin. *Biochemistry* **37**, 13978–13986; (b) PDB 1A8F: MacGillivray, R.T., Moore, S.A., Chen, J., Anderson, B.F., Baker, H., Luo, Y., Bewley, M., Smith, C.A., Murphy, M.E., Wang, Y., Mason, A.B., Woodworth, R.C., Brayer, G.D., Baker, E.N. 1998. Two high-resolution crystal structures of the recombinant N-lobe of human transferrin reveal a structural change implicated in iron release. *Biochemistry* **37**, 7919–7928.)

No matter which the case might be, CO_3^{2-} certainly contributes to bringing two lobes closer, enhancing the binding of Fe(III) to the site.

The Fe(III) of the holoprotein (Tf) is released upon lowering pH. Release starts at pH about 6.5 and is complete at pH 4 *in vitro*, but it seems to be so *in vivo* as well (Thakurta *et al.*, 2004). Lowering pH is a factor in releasing Fe(III) from Tf, but this factor alone may not be sufficient to allow the facile release (Aisen *et al.*, 2001).

Tf is not incorporated into a cell as such; it must be bound to a receptor (TfR). The Tf-TfR complex is then incorporated into a cell by endocytosis (Aisen *et al.*, 2001). Another similar receptor, TfR2 is known, but TfR is the predominant one.

10.1.1.5. *Ferritin*

Ferritin is widely distributed from bacteria to human. Mammalian ferritins consist of 24 subunits of H and L types. H-type is heavier (with 21 KDa) than L (19.5 KDa). The 24 subunits form a nearly spherical structure with a cavity of 70 to 80 Å at the center (see Fig. 10.3). This cavity accommodates as many as 4500 iron atoms in approximately hydroxide form (Aisen *et al.*, 2001). Mineralogically the iron

■ **Figure 10.3.** An overall shape of ferritin. (PDB 1IER: Granier, T., Gallois, B., Dautant, A., Estaintot, B.L.D., Precigoux, G. 1997. Comparison of the structures of the cubic and tetragonal forms of horse-spleen apoferritin. *Acta Crystallog. Sect. D* **53**, 580–587.)

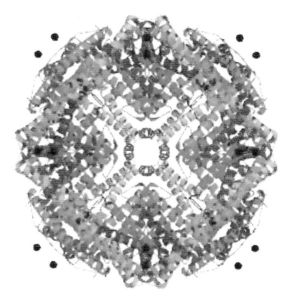

core looks like ferrihydrite $5Fe_2O_3 \cdot 9H_2O$ with magnetite Fe_3O_4 or hematite Fe_2O_3 as minor components. The iron that is to be sequestered into ferritin is either in Fe(II) or Fe(III) state, but it is incorporated as Fe(III)-hydroxide/oxide. Therefore, Fe(II) needs to be oxidized.

The oxidation and incorporation seems to take place as follows (Aisen *et al.*, 2001). The H-polypeptide is supposed to have a ferroxidase activity. (Whether ferroxidase activity is necessary is questionable.) Two Fe(II)s are bound to a H-ferroxidase center, and reacts with O_2 to form $Fe(III)(O_2^{2-})Fe(III)$. It is likely to further react with two more Fe(II)s to end up with $2O^{2-}(2H_2O)$ and 4Fe(III). The reaction may not be as straightforward, and may produce some polynuclear Fe(III) clusters bridged by Os. In fact, an x-ray crystallographic picture shows several different Fe species (mononuclear, dinuclear, trinuclear, pentanuclear, etc.) bound to the polypeptides of a bacterial ferritin (Zeth *et al.*, 2004). It seems that the degree of polymerization increases as the iron entities move toward the interior (PDB 1TKP). As more Fe(III)s enter the cavity, nucleation and growth of minerals of approximate composition $Fe_5O_3(OH)_9$ (ferrihydrite) takes place.

Release of iron from ferritin is not as well understood. *In vitro*, addition of a reducing agent along with a Fe(II)-specific chelating agent captures the iron readily released from ferritin. Such a mechanism is not known *in vivo*. Instead, it seems necessary to degrade the ferritin by lysosomal or proteosomal degradation to release the iron (Aisen *et al.*, 2001).

10.1.1.6. *Ferroportin (Fpn)/Hepcidin*

The basolateral membrane of duodenal mucosal cells and other cells export iron. This is carried out by a protein named ferroportin (also known as Ireg1 and Mtp1). It is a protein with multiple TMDs (Pignatti *et al.*, 2006); otherwise not much is known yet.

It turned out that Fpn is the cellular receptor for an iron-regulatory hormone hepcidin. When hepcidin is added to the cultured cells expressing Fpn, the Fpn is internalized and degraded. In other words, hepcidin reduces iron-export. Hepcidin is a 25-amino acid peptide produced by liver cells. Its synthesis is stimulated by inflammation or by iron-overload (Ganz, 2003).

10.1.1.7. *Regulation of Ferritin and Transferrin*

A main iron regulatory system works on ferritin and TfR (transferrin receptor), not directly on Tf. It is effected through respective mRNA in conjunction with IRPs (iron regulatory protein). Each RNA has on it an iron-responsive element (IRE); on 5′-portion on mRNA of ferritin and 3′-portion of mRNA of TfR.

IRP1 is a cytosolic counterpart of mitochondrial aconitase (see Chapter 4). Therefore, given enough Fe, IRP1 has a full-fledged iron-sulfur cluster and acts as aconitase, and cannot bind to the IRE of mRNA. The result is production of ferritin, which sequesters away the excess iron. Without the iron-sulfur cluster (in an iron-poor condition), IRP1 binds to the IRE of mRNA, and blocks the expression of ferritin gene. Hence, no incorporation of iron into ferritin occurs.

IRP2 has not been well characterized, but it is known to protect the mRNA from degradation. At high intracellular levels of iron, IRP2 is oxidized and degraded, and hence no protection for the mRNA. Therefore, the mRNA of TfR is rapidly decomposed, and little TfR is produced. At low iron levels, IRP2 protects the mRNA of TfR, and hence TfR will be produced, enhancing availability of iron.

10.1.1.8. *Iron Metabolism in Bacteria, Fungi, and Plants*

Note: There is no set designation system for naming these various proteins, and as a result, confusingly, different notations have been used by different researchers.

Autotrophs and fungi have to obtain iron from their environments. Iron in today's environment is rarely in soluble Fe(II) form on the Earth. Fe(III) exists usually as insoluble hydroxide, oxides, or mixtures. However, these organisms have to be prepared to use iron in any form: Fe(II), free Fe(III), or Fe(III)-insoluble form. Hence, for example, a fungus *Saccharomyces cerevisiae* has all three different systems: Fe(II)-permease (transporter), Fe(III)-permease, and Fe(III)-siderophore dependent facilitator (Kosman, 2003). Fe is oxidized by ferroxidase (Fet3; Aisen *et al.*, 2001) or reduced by ferric reductase (Fre1, Fre2) located on the cell membrane, depending on the condition the iron is found in.

Fe(II) is directly transported by Fe(II)-transporters, one of which is designated as Fet4 with six TMDs and very similar to DMT1 mentioned earlier (Kosman, 2003). Fe(II) or Fe(II) reduced by Fre is then oxidized by an oxidase Fet3, which is similar to the multicopper enzyme ceruloplasmin. And the resulting Fe(III) is transported into the cell by a transporter Ftr1, which has six TMDs also.

Bacteria, fungi, and plants produce small molecules to capture Fe(III), which is rather inaccessible or in short supply. These small molecules are called *siderophores* (iron-carrier). Fe(III) prefers O^--ions as ligands over N or other ligand atoms. Several different types of

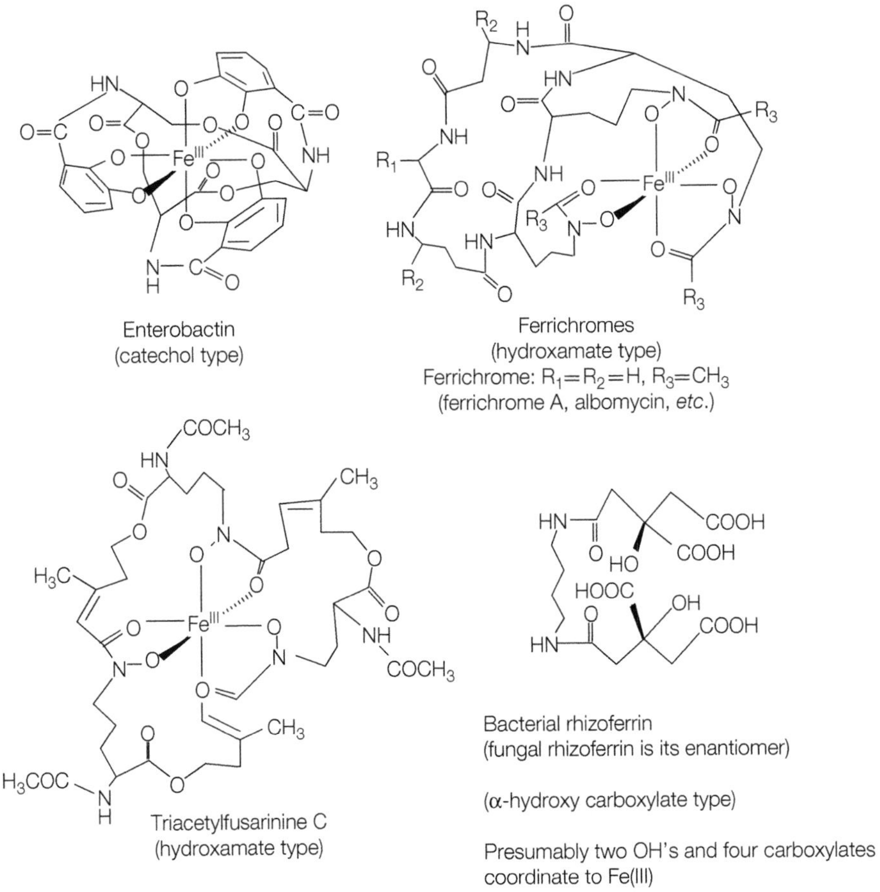

■ **Figure 10.4.** Examples of siderophores and their Fe(III)-complexes.

Enterobactin
(catechol type)

Ferrichromes
(hydroxamate type)
Ferrichrome: $R_1=R_2=H$, $R_3=CH_3$
(ferrichrome A, albomycin, *etc.*)

Triacetylfusarinine C
(hydroxamate type)

Bacterial rhizoferrin
(fungal rhizoferrin is its enantiomer)

(α-hydroxy carboxylate type)

Presumably two OH's and four carboxylates coordinate to Fe(III)

compounds are known as siderophores: hydroxamate, catecholate, or α-hydroxy carboxylate. All of these have six Fe(III)-coordinating O^--ions; that is, hexadentate ligand. A few examples of Fe(III)-siderophore complexes are shown in Figure 10.4. These siderophores form very stable Fe(III)-chelates; formation constants range from 10^{29} to 10^{32}.

E. coli picks up ferrichrome (Fe(III)) through its outer membrane by an active iron transporter protein called FhuA (formerly TonA) (Braun and Braun, 2002). Ferrichrome is then transported into cytoplasm by FhuB. FhuA is energized by an energy-transmuting system located in the cytoplasmic membrane. FhuA has a rather wide selectivity, and transports other antibiotic compounds and acts as a receptor for a few small polypeptide toxins. FhuB is associated with FhuC that is an ATPase. Another similar protein FepA located in the outer-membrane transports enterobactin (Fe(III)) (Braun and Braun, 2002).

10.1.2. **Copper Metabolism**

10.1.2.1. *Outline of Copper Metabolism in Mammals*

The metabolism of copper is not as well understood as that of iron. One difference between iron and copper is that copper is much more cytotoxic, and hence the free copper level in cytoplasm is regulated at lower levels. A number of proteins classed as copper chaperone may be playing this role to keep a free copper level low. An outline of copper homeostasis is given in Figure 10.5.

A Cu-specific transporter protein (Ctr1) has been identified in yeast, humans, and mice. It is believed to be the main mechanism of Cu-import in the brush border of intestinal cells. The DMT1 mentioned earlier for iron-transport also seems to be able to import Cu(I) (Arredondo and Núñez, 2005).

The main exporter of copper in most cells is MNK protein. A defect in this gene is responsible for Menkes disease. The same function is carried by WND in hepatic cells, a lack of which causes Wilson disease. These proteins are ATPase of P-type and transport copper actively by using ATP. MNK (also known as ATPase 7A) and WND (ATPase 7B) are involved also in the export/import in the intracellular vesicles such as trans Golgi network, where Cu is incorporated into Cu-proteins. Both MNK and WND have multi TMDs, and a number

■ **Figure 10.5.** Metabolism of copper in mammals (based on Arredondo and Núñez, 2005). Ctr1 = Copper transporter 1; DMT1 = Divalent metal transporter 1; Hah1, Atox1, Ccs, Cox17 = Copper chaperones; MNK = Menkes ATPase; MT = Metallothionein; WND = Wilson ATPase.

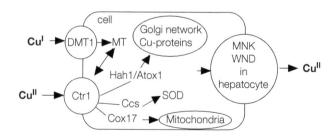

of repeats of the same motif: Gly-Met-Thr-Cys-X-X-Cys (Andrews, 2002). This Cys-X-X-Cys is supposed to be the Cu-binding site. One such structure in which Cu(I) is coordinated linearly by two cysteine Ss has been captured by NMR spectroscopy (Banci *et al.*, 2005).

Metallothionein, a small protein that strongly binds several metallic ions, including Cu(I), Zn(II), Cd(II), and Hg(II). In the case of Cu, MT is considered to be a Cu-storage, particularly in hepatic cells. Metallothionein will be discussed in some detail in Chapter 11. The Cu-transporter in the circulatory system is believed to be serum albumin.

10.1.2.2. *Copper Metabolism in Bacteria and Plants*

Plants require copper, especially in photosynthetic systems. Some bacteria, but not all, also require copper. In the *Arabidopsis* plant, COPT1 in the plasma membrane imports copper, which is carried by a copper chaperon CCH and then is pumped into a vesicle or Golgi apparatus by Cpx-ATPase, PAA1 (Williams *et al.*, 2000). Cpx-ATPase is similar to other ATPases of P-type such as Ca(II)-ATPase, and MNK and WND mentioned earlier. COPT1 is supposed to be similar to Ctr1 and DMT1.

Copper metabolism in Gram-negative bacteria such as *E. coli* is summarized in Finney and O'Halloran (2003). Copper efflux from *E. coli* is carried out by proteins CopA, CusA (and associated CusB/C), and PcoB. In periplasmic space (between outer membrane and cytoplasmic membrane), a few enzymes are oxidizing and reducing copper (Cu(I)-Cu(II)). The expression of these proteins is regulated by a number of gene-regulators, depending on the copper-stress. The first line of control is done by a protein CueR. This protein is quite unusual in that the induction of CopA through CueR takes place at an incredibly low free Cu(I) concentration, 10^{-21} mol/L (Changela *et al.*, 2003), it is very sensitive to the presence of Cu(I). X-ray crystallographic study showed that Cu(I) is coordinated by two cysteine S's in a linear manner. This unusual structure is believed to be mainly responsible for the high sensitivity (Changela *et al.*, 2003).

A small molecule coined as methanobactin has been found that serves as a chelator for copper, like siderophores for iron. It is produced by methane-oxidizing bacterium, *Methylosinus trichosporium* OB3b,

Note: 10^{-21} mol/L is a very low value; it corresponds to only 600 atoms of free Cu(I) ions in L. If one cell is 10μ in diameter, the volume of one cell is 10^{-12} L. Then, 600 atoms/L means 6×10^{-10} atom/cell; practically zero.

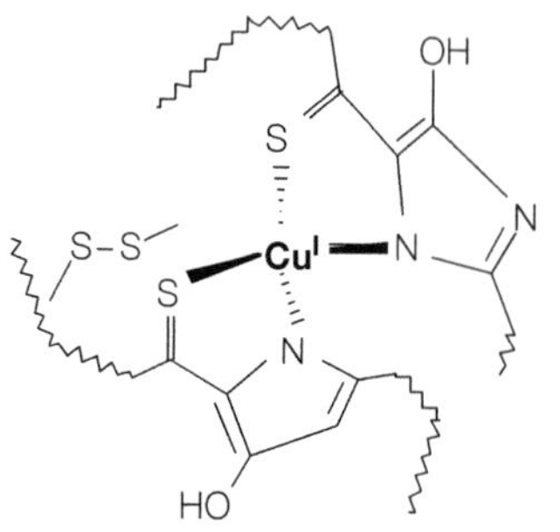

■ Figure 10.6. Cu-binding site of Cu-methanobactin (see Kim *et al.*, 2004 for details).

and composed of a tetrapeptide, tripeptide, and several other moieties and has a composition of $C_{45}H_{62}N_{12}O_{14}Cu$ (Kim *et al.*, 2004). It has an unusual copper binding entity, 4-hydroxy-5-thionyl imidazole, which binds through 2Ns and 2Ss to Cu(I) in an approximately tetrahedral manner (see Fig. 10.6).

10.1.3. **Zinc Metabolism**

10.1.3.1. *In Mammals*

Zn is also universally required for all sorts of cell functions. The pathological signs of Zn-deficiency include growth failure, impaired parturition, neuropathy, decreased and cyclic food intake, diarrhea, dermatitis, hair loss, tendency to bleed, hypotension, and hypothermia (Tapiero and Tew, 2003). Anorexia, and alterations of gastrointestinal, central nervous, immune, skeletal, and reproductive systems are associated with Zn-deficiency states. In view of the wide-ranging roles, it is not surprising to find a fairly elaborate system of Zn-metabolism (homeostasis).

Eighty-five percent of the whole body Zn is found in brain, muscle, and bone; 11% is found in skin and the liver. Almost all of it is present in cells; 50% in cytoplasm, organelles, and vesicles; 30 to 40% in the nucleus; and the remainder in membranes (Tapiero and Tew, 2003).

Transport of Zn is much simpler than those of iron and copper, as there is no oxidation-reduction involved. Zn-transports across membranes in mammals are carried out by proteins of two gene families: *znt* genes and *zip* genes. The expressed proteins are ZnT1~9 and Zip1~4 proteins (Zrt- or Irt-like proteins). ZnT and Zip proteins seem to have opposite roles. ZnTs reduce cytoplasmic Zn-level by enhancing Zn-efflux from cells or sequestering into intracellular vesicles. On the other hand, Zip transporters increase intracellular Zn level by exactly reverse processes; promoting the transport of Zn into cytoplasm from the extracellular space and/or the intracellular vesicles (Liuzzi and Cousins, 2004).

ZnTs typically have six TMDs (transmembrane domains), though some (e.g., ZnT5) have 12 TMDs. This structure (with multi TMDs) is common in all transmembrane transporters. ZnTs have a long histidine-rich loop between TMD IV and V. This loop is considered to be Zn(II)-binding site (Liuzzi and Cousins, 2004). Zip transporters are predicted to have eight TMDs, and a histidine-rich loop of variable

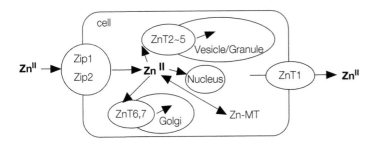

■ **Figure 10.7.** Metabolism of zinc in mammals (based on Zalewski *et al.*, 2005; see the text for details).

length between TMD III and IV (Liuzzi and Cousins, 2004). An outline of the Zn homeostasis involving ZnTs and Zips in mammals is given in Figure 10.7 (based on Zalewski *et al.*, 2005).

In addition, two pathways of Zn-uptake exist. One is a saturable receptor-mediated transport of Zn-albumin by transcytotic vesicles, and the other is a nonsaturable cotransport with albumin or histidine as ligands through intercellular junctions. There is also a suggestion that Zn enters cells complexed with cysteine or histidine through a Na/amino acid co-transport mechanism (Tapiero and Tew, 2003). In the neuronal cells, gated channels also seem to be involved in Zn-uptake. Examples are Na(I)/Zn(II) exchangers, N-methyl-D-aspartate receptor-gated channel, and voltage-gated L-type Ca(II)-channel. In all these processes the molecular details are yet to be determined.

A storage for Zn(II) in cells is metallothionein (MT). It is a cysteine-rich, small protein, and several variations are known. MT1 and MT2 are expressed in almost all tissues, but MT3 and MT4 are tissue-specific; MT3 is expressed in brain. MTs are involved in the regulation of cellular metabolism of Zn and Cu, in detoxification of heavy metals such as Cd, Cu, and Hg, and in protection against reactive oxygen species and alkylating agents. MTs will be discussed in Chapter 11.

10.1.3.2. *In* E. coli

Zn homeostasis in *E. coli* is carried out by Zn transporters, ZnuABC, which imports Zn(II) into the cell; and ZntA, which exports Zn(II). (*Note*: ABC stands for ATP-Binding Cassette.) These proteins are controlled by metal-sensitive regulatory proteins Zur and ZntR, respectively (Finney and O'Halloran, 2003). The structure of ZntA has been determined by x-ray crystallography (see Finney and O'Halloran,

2003); Zn(II) is bound by two cysteines, one aspartate and another undetermined ligand arranged in a tetrahedral manner.

Another Zn(II) exporter in the inner membrane of *E. coli*, YiiP, has recently had its structure determined (Lu and Fu, 2007). It is a homodimer, and each subunit consists of an intracellular segment with four Zn(II) sites and six TMDs. When an excess Zn(II) is present in the cell, Zn(II) ions bind to the intracellular domain and it causes an active dimer formation. The active dimer transports Zn(II) out of the cell. When the cellular level of Zn(II) returns to the normal level, the transport becomes inactive and ceases transportation of Zn(II). This control mechanism seems to be faster and finer than that provided by gene expression.

10.1.4. **A Mg(II) Transporter**

CorA is another divalent cation transporter. It is considered to be the primary transporter of Mg(II) in bacteria and archaea, but distributed widely in eukaryotes as well. It has been demonstrated that it transports Mg(II) in both directions (Eshaghi *et al.*, 2006). The structure of CorA from *Thermotoga maritime* has been obtained by x-ray crystallography. It is a pentamer with a cone shape and 10 TMDs. The N-terminal domain seems to provide the initial Mg(II) binding sites. One of the binding sites M1 provides two Asps and one Glu to bind Mg(II) and it seems that three water molecules also bind. The M2 binding site seems to bind not bare Mg(II) but a hydrated Mg(II). There is, as in other metal ion transporters, a metal-conducting pore formed by the TMDs. This pore is lined by O-ligands: glycine's carbonyl O and Tyr (Eshaghi *et al.*, 2006).

10.2. **PHYSIOLOGICAL ROLES PLAYED BY METALLIC ELEMENTS**

Inorganic elements do play direct physiological roles. Some examples will be discussed here.

10.2.1. **Na/K-ATPase and Ca-ATPase**

10.2.1.1. *Mechanism*

Neuronal electric signal transduction is carried out by electrically charged entities. The electrical potential across membrane is maintained, for example, by Na/K-ATPase. This potential is transiently reversed by, for example, a sudden opening of a Na-channel. This manifests as an electrical signal in the neuronal system. Na/K-ATPase is a typical transmembrane cation transporter (of P-type) energized by ATP hydrolysis. A part (N-domain) of the structure has been determined by x-ray crystallography (Hakansson, 2003). The overall structure was determined only at 11 Å resolution by electron micros-copy (Rice *et al.*, 2001) and was shown to be very similar to the struc-ture of Ca-ATPase, which plays the role of pumping out Ca(II) from cytoplasm.

The structure of Ca-ATPase in various states has been determined by x-ray crystallography and is discussed in detail by Møller *et al.* (2005). The structure of 2Ca(II)-bound ATPase was first x-ray crys-tallographically obtained by Toyoshima *et al.* (2000). It consists of several domains, 8TMDs = Ca-conduit, NBD (nucleotide-binding domain or ABC = ATP-binding cassette or ATPase domain) stick-ing out into cytoplasm and a random coil protruding into the peri-plasm. The two Ca(II)s are bound in the TMD portions and each ion is coordinated by seven Os of various kinds. Structures in other states have also been determined and have led a tentative mechanism as shown in Figure 10.8 (Møller *et al.*, 2005). The free energy released from the hydrolysis of ATP is supposed to cause conformational changes that affect extrusion of Ca(II), but the molecular details have not been delineated.

10.2.1.2. *Ion Selectivity in Metal Ion Transporters and Channels—A General Discussion*

The metabolism of metal ions and pumping of various ions involve a number of transporters (pumps) and channels as outlined earlier. Ion pumps (ion transporters) build concentration gradients across mem-brane (though some ion pumps seem to simply remove extra ions). This gradient (negative free energy) is then used as an energy source to pump nutrients into cells (nutrient transporters), generate electri-cal signals, regulate cell volume, and secretes electrolytes (through ion exchangers). Ion pumps energized by ATP generally are termed P-type and include H^+-ATPase, Ca-ATPase, and Na/K-ATPase. Transporters of iron, copper, and zinc of P-type mentioned earlier are all of this type. The mechanism of energy transduction of ATP-hydrolysis has been discussed in the previous section in regard with Na/K-ATPase and Ca-ATPase. Similar structures have been found with and similar mech-anisms have been presumed for the other transporters mentioned

■ **Figure 10.8.** A schematic description of the transmembrane Ca(II) movement by Ca-ATPase (Møller et al., 2005).

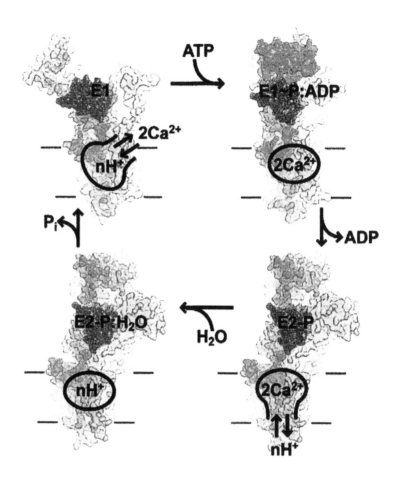

in earlier sections as well. The transports by all these proteins are active processes, against the concentration gradients.

Channels allow translocation of ions (and others) along the gradient, and hence are passive processes. In both the active and the passive processes, the proteins have to work on only a specific entity (ion or others). That is, one of the crucial issues is how selectivity is attained. This will be discussed briefly here. However, not much has been learned yet about the molecular details of transporters and channels for iron, copper, zinc, and others. Only the cases of better understood ions, such as Na(I), K(I), Ca(II), and Cl(-I) will be mentioned (Gouaux and MacKinnon, 2005). Typically a constriction in a pump or a channel that specifically binds an ion plays the role of a selective gateway.

Ions are hydrated in an aqueous medium, including periplasmic space and cytoplasm. For an ion to bind to a specific site, it has to strip coordinating water molecules fully or partially. The discriminating factors available for the specific binding include the electric charge and the size on the part of ion, and the nature of binding ligands on the part of the protein. In the case of alkali metal ions, Na(I) and K(I), the distinguishing factor is facility and ease of stripping of hydrating molecules, and the ionic size. The magnitude of hydration energy is much greater with Na(I) than K(I). Hence a K(I)-binding site can use relatively weaker binding than a Na(I)-binding site. This manifests also in the form of the distance of metal cation to the ligand atoms, because the hydration energy is largely determined by the cation ionic size. In other words, the ionic size seems to be the crucial discriminating factor.

One of the two Na(I)-binding sites in Na(I)-dependent leucine transporter consists of carbonyl Os (of the peptide bond), OHs of side chain, and one carboxylate being arranged in an approximate octahedron. The other site has five ligands, carbonyl Os and hydroxyls. The average Na(I)-O distance is 2.28 Å (Gouaux and MacKinnon, 2005). The constriction of K(I)-channel consists of four consecutive K(I)-binding sites. Each K(I) is bound by eight partially negative ligands (carbonyl Os and side-chain OHs). And the mean distance of K(I)-O is 2.84 Å. Thus, the discriminating constriction of these proteins is commensurate with the respective ion radius; the radius of Na(I) is 0.95 Å and that of K(I) is 1.33 Å.

There are some indications that the discriminating factors for cations such as Fe(II, III), Cu(I, II), and Zn(II) are not the size but specific ligands. Fe(III) binds preferentially O$^-$ ligands such as tyrosyl, and hence the iron transporters use such ligands for binding. Histidine N is a ligand favored by Zn(II) or Cu(II), and cysteine S seems to be a good choice to bind Cu(I) or Zn(II), though not exclusively. This is consistent with the inorganic chemistry of these cations.

10.2.2. Ca(II)—Second Messenger and Other Functions

Ca(II) is the most widely employed second messenger and is involved in a vast array of cell physiology, including muscle contraction, neurotransmitter release and other secretion processes, cell proliferation, and cell death. A review by Berridge *et al.* (2000) provides a good overall view of signaling by Ca(II).

In eukaryotic multicellular organisms, the cytoplasmic Ca(II) concentration ($[Ca^{II}]_i$) is of the order of 10^{-7}M in the resting state and rises quickly to $10^{-6} \sim 10^{-5}$ M in an activated cell. The typical Ca(II) concentration in the extracellular lumen is of the order of 10^{-3} M, and that in the Ca(II)-storing vacuoles such as endoplasmic reticulum (ER) and sarcoplasmic reticulum (SR) is of the order of 10^{-4} M. The flow of Ca(II) is regulated through voltage-, receptor-, and store-operated channels. The important intracellular Ca(II) storages include ER, SR, and mitochondrion (Brini, 2003).

10.2.2.1. *Control of Cytoplasmic Ca(II) Concentration*

Ca(II) as a second messenger causes a huge array of physiological processes, and the critical step for this is the fast increase of the $[Ca(II)]_i$ in cytoplasm. $[Ca(II)]_i$ in the resting cell is quite low, and the $[Ca(II)]$ in everywhere else (extracellular and the vacuolar storages) is much higher than that. Hence, what is required is to provide conduits for Ca(II) to enter the cytoplasm. This would not require energy, as the process is along the concentration gradient. Once $[Ca(II)]_i$ becomes high in the cytoplasm and the physiological processes have been elicited, Ca(II) has to be removed. This removal process is to be conducted against the concentration gradient and hence requires active processes, typically using ATP hydrolysis free energy. One of the agents for this last process is Ca(II)-ATPase, which has been discussed earlier. A general overall $[Ca(II)]_i$ control scheme is given in a very lucid review article by Berridge *et al.* (2000). It is simplified and shown in Figure 10.9.

Ca-signals are initiated by stimulus (electrical signal, hormone, or otherwise). These signals open up various Ca(II)-channels, the plasma membrane Ca(II)-channel first. One of such channels is the same as IP$_3$R described later. The number of membranous IP$_3$R sites is small, several per cell. This channel imports Ca(II) very rapidly. The major channel is Orial1, which is distributed widely in the plasma membrane, but its rate of Ca(II) imports is slower (100 times slower) than the scarce IP$_3$R channel (Gill *et al.*, 2006; Dellis *et al.*, 2006). Ca(II), which entered in the cytoplasm, then elicits actions of other second messengers: inositol-1,4,5-triphosphate (IP$_3$), ryanodine (RY), and others. These chemical substances bind to the respective receptors (IP$_3$R and RYR), which act as Ca(II)-channels in ER and SR. The result is the release of Ca(II) into the cytoplasm. As a further flexible control of $[Ca(II)]_i$, a number of buffers exist in the

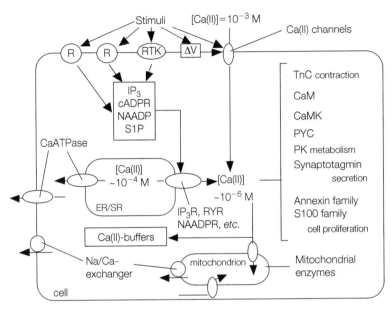

■ **Figure 10.9.** Ca(II) transport and signaling system and effects on physiology (based on Berridge *et al.*, 2000). cADPR = Cyclic ADP ribose; CaM = Calmodulin; CaMK = Calmodulin-dependent protein kinase; $\triangle V$ = Depolarization; IP$_3$ = Inositol-1,4,5-triphosphate; IP$_3$R = IP$_3$ receptor; NAADP = Nicotinic acid dinucleotide phosphate; NAADPR = NAADP receptor; PK = Phosphorylase kinase; R = Receptor; RTK = Receptor tyrosine kinase; RYR = Ryanodine receptor; S1P = Sphingosine 1-phosphate; TnC = Troponin C.

cell; proteins such as parvalbumin, calbindin-D$_{28K}$, and calcetinin are examples.

10.2.2.2. *Basic Mechanisms of Ca(II)—Physiology*

Many of the cellular effects of Ca(II) are mediated by the Ca(II)-binding proteins; two classics examples of such Ca(II)-binding proteins are calmodulin (CaM) and troponin C (TnC); these can be considered as Ca-sensors. Upon binding up to four Ca(II) ions, CaM undergoes a significant conformational change (see Fig. 10.10), which enables it to bind to specific proteins eliciting specific cellular responses. The Ca(II) binding residues are different among the four. Three of the four are coordinated by five O-ligands in approximately tetrahedral pyramid shapes. The majority of the O-ligands are aspartate O$^-$; the others include the carbonyl C=O of peptide bond, threonine OH, and water. One Ca(II) seems to be coordinated by only three ligands. Calmodulin kinase II (CaMKII) is a major target of the Ca-second messenger system. Once bound to Ca(II)-CaM, the CaMKII is activated, which then phosphorylates proteins/enzymes. This is involved in cell proliferation, fertilization, learning, and memory. Troponin C has a similar structure to CaM, and is a control factor on the interaction between actin and myosin in muscle.

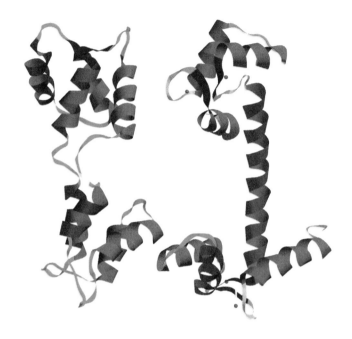

■ **Figure 10.10.** Calmodulin without Ca(II) (the figure on the left: PDB 1CFD: Kuboniwa, H., Tjandra, N., Grzesiek, S., Ren, H., Klee, C.B., Bax, A. 1995. Solution structure of calcium-free calmodulin. *Nat. Struct. Biol.* **2**, 768–776), and that on the right with Ca(II) bound with the four Ca(II)'s: pink dots (PDB 1EXR: Wilson, M.A., Brunger, A.T. 2000. The 1.0 Å crystal structure of Ca(2+)-bound calmodulin: An analysis of disorder and implications for functionally relevant plasticity. *J. Mol. Biol.* **301**, 1237–1256).

Calcineurin (Cn) is another major target protein that also is activated by Ca(II)-CaM. Cn is a serine-threonine phosphatase, and upon activation, it directly binds to, and dephosphorylates nuclear factor of activated T cells (NFAT) transcription factors, and hence is involved in the regulation of gene expression (Zayzahoon, 2005). Other Ca(II)-induced physiological processes are summarized concisely in the review by Berridge *et al.* (2000, 2003).

10.2.2.3. *Synaptotagmin, an Example of Physiology Mediated by Ca(II)*

The neurosignal transmittance is mediated by the release of a neurotransmitter from synaptic vesicles. The exocytosis of the synaptic vesicle is known to be one of the fastest cellular/physiological processes; it occurs on the micro- to millisecond scale (Chapman, 2002). The process is outlined in Figure 10.11; the fusion of the vesicle membrane and the synaptic cell membrane is triggered by Ca(II). As seen in the figure, the vesicle docks to the active site through a protein complex called SNARE, which is an acronym of soluble N-ethylmaleimide-sensitive fusion protein (NSF) attachment protein receptor, and consists of vesicle protein synaptobrevin (i.e., associated

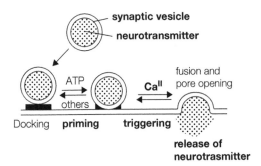

synaptic vesicle

neurotransmitter

ATP

others

Ca^{II}

fusion and
pore opening

Docking   **priming**   **triggering**

release of
neurotrasmitter

■ **Figure 10.11.** A schematic representation of release of neurotransmitter from synapsis (based on Chapman, 2002).

with the vesicle membrane) and two plasma (of neuron) membrane proteins: syntaxin and synaptosome-associated protein of 25 kDa (SNAP-25). This complex seems to pull the two membranes together with ATP and other factors (priming process); in fact it leads to fusion but at a much lower rate. The fast fusion requires another protein synaptotagmin, which triggers the fusion process when it binds Ca(II) (Chapman, 2002).

The N-terminal of synaptotagmin anchors in the vesicle membrane and the C-terminal divides into two domains, C2A and C2B. C2A binds three Ca(II)s and C2B binds two (see Fig. 10.12). These bindings are not very strong with $K = 10^3 \sim 2 \times 10^4$ for C2A and $2 \sim 3 \times 10^3$ for C2B in the free state. The one side of the Ca(II)'s coordination sphere is open, suggesting possible further binding of ligands. Indeed the heads of C2A and C2B penetrate partway into the outer lipid bilayer, and Ca(II) binding becomes very strong. It has been demonstrated that Ca(II)s bind the anion heads of a phospholipid, phosphatidylserine of the membrane. This promotes squeezing of the two membranes and twisting of the SNARE, and the fusion of bilayers of the vesicle and the plasma membrane ensues (see Fig. 4 of the article by Chapman, 2002, for the details). By the way, Sr(II) and Ba(II) can function as a substitute for Ca(II) in triggering, but Mg(II) was found to be ineffective. A recent report (Martens *et al.*, 2007) gives more details of mechanism of synaptotagmin.

10.2.2.4. *Why Calcium(II)?*

Ca(II)'s physiological functions or signaling are diverse. It is involved throughout the life history of an organism: in fertilization, embryonic pattern formation, cell differentiation, transcription factor activation, apoptosis; and in other physiological processes including

■ **Figure 10.12.** Ca(II) (orange ball) binding to synaptotagmin C2A and C2B (from Chapman, 2002).

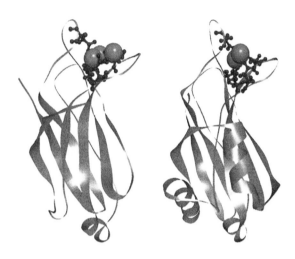

neural transmission, muscle contraction, and blood coagulation to name a few.

Is there any common feature among all these diverse processes involving Ca(II)? First this agent (Ca(II)) has to be able to flow rapidly; that is, the local concentration has to change rapidly. Second, it should be able to bind rapidly with the proteins that it is supposed to exert its effects on, and with some strength. In this second stage, it has to effect enough conformational changes, perhaps, in the protein bound with it, so that the protein affected then can exert its function—the protein is switched on. Besides, the agent has to be readily available to begin with. Assuming that these are required characters for the agent, a question can then be asked, why Ca(II)? This issue was discussed by this author (Ochiai, 1987, 1991).

Ca(II) is one of the most abundant and readily available elements on Earth, and it is soluble in water. Alkali metals (Na(I), K(I)), and Mg(II), also have the same characters. The ionic potential (electric charge/ionic radius) is 5.5 (nm^{-1}) for Na(I), 4.6 for K(I), 12.2 for Mg(II), and 10.4 for Ca(II). The hydration energy is -412 (kJ/mol) for Na(I), -336 for K(I), -1980 for Mg(II), -1650 for Ca(II). The ionic potential could represent the electric field effect to cause conformational change in a protein when a cation is bound. Na(I) and K(I) are not very effective in this sense. Mg(II) could be a little more effective than Ca(II), but Ca(II) is strong enough. By the way, other divalent cations including Fe(II), Zn(II), and Cu(II) are much more effective in this sense (see Chapter 4), but these are much less available.

Ca(II), as a matter of fact, binds fairly strongly with a wide variety of biomolecules. Important among such compounds is phosphate, which is vital to many cellular functions. If Ca(II) is allowed to bind free phosphate or phosphate derivatives in a cell, it forms insoluble precipitates that hamper cellular processes, and the available phosphate will be reduced. Hence, the free calcium concentration in the cytoplasm has to be kept low (typically $[Ca(II)] = 10^{-7}$ mol/L in the resting cell), by way of some specific Ca(II)-binding proteins. The $[Ca(II)]$ in the extracellular milieu is of the order of 10^{-3} mol/L. The intracellular concentration of Ca(II) needs to be changed very rapidly by up to 100-fold in order to be able to play the role of a second messenger. The $[Mg(II)]$ in the resting cell is relatively high $(10^{-4}\sim10^{-3}$ mol/L), and a change of concentration of 10- to 100-fold will require a significantly larger (likely by a factor of 3~4) quantity of Mg(II) ions being moved than in the case of Ca(II). This will not be easy to accomplish. A ligand (protein included) binding and substitution rate is known to be much faster with Ca(II) than with Mg(II), partly because of the lower ionic potential of Ca(II). Ca(II) is thus preferable to Mg(II) in these regards.

The electric potential effect on protein conformation of Ca(II) may not be as strong as that of Mg(II), but it would be enough, as illustrated for the case of calmodulin (see Fig. 10.9). Another character that may be useful with Ca(II) is the fact that the coordination of ligands about Ca(II) is rather flexible, in terms of both the number of ligands to be bound and the coordination structure. This flexibility may make Ca(II) more widely acceptable for a variety of situations. Lastly the ligands that Ca(II) prefer to bind are O-donor atoms, carboxylate O^-, alcoholic O^- or OH of serine and threonine, O of the peptide bond carbonyl, and phosphate O^-. These entities are more preponderant among the naturally occurring compounds than N-atom ligands.

By the way, the effectors of the second messenger Ca(II) are associated mostly with phosphate addition or removal. Phosphate PO_4^{3-} (or $-OPO_3^{2-}$), if it replaces the H of $-OH$ or other residues, will change significantly the local electric potential distribution in a protein, causing significant changes in its conformation.

10.2.3. **Zinc-Enriched Neuron (ZEN)**

Zinc is most abundant in the brain among all the tissues and organs. The total zinc concentration in the brain is about 1.5×10^{-4} M on average, but the Zn content is quite high in certain areas of brain. For example, the Zn content in the synaptic vesicles of some neurons in the forebrain has been found to be $>10^{-3}$ M (Mocchegiani et al., 2005). Neurons that contain free zinc in the vesicles in presynaptic regions are present also in other brain areas, and are called zinc-enriched neurons (ZEN). ZENs have been found in various brain tissues, but the Zn-content is higher in gray than white matter.

The physiological role of synaptic Zn is not very well understood, but is believed to be mainly involved in the modulation and release of some neurotransmitters including glutamate and γ-amino butyric acid (GABA). It has been shown that Zn released along with glutamate reduces the ability of glutamate to activate postsynaptic NMDA receptors and favors synaptic non-NMDA receptor activation (Mocchegiani et al., 2005); NMDA is N-methyl D-aspartate.

It also has been suggested that Zn(II) does indeed bind to the postsynaptic receptor and that it even enters the postsynaptic neurons through a Ca-A/K-channel. As a result, Zn(II) exerts a number of physiological functions such as intracellular signal (Mocchegiani et al., 2005). Though many other physiological manifestations of Zn(II) in neurons have been observed, no significant progress has been made to delineate the molecular mechanisms of such actions of Zn(II) (see also Chapter 11).

An exception is metallothionein (MT). MT-III is brain-specific, and it regulates the intracellular Zn(II), which should be fairly low, as the higher level of free Zn(II) is cytotoxic. MT-III binds Zn(II) strongly; formation constant $K = 3 \times 10^{13}$. As MT is also a detoxifier for Hg(II), Cd(II) and the like, the details of MT will be discussed in Chapter 11.

10.2.4. **Sensors for Small Molecules**

Organisms may need to monitor and regulate the level of small molecules such as H_2, O_2, CO, NO, and C_2H_4, depending on their needs. A number of receptors for O_2 (hemoglobin, etc.) already have been mentioned, and the ethylene receptor will be discussed in the next section. These receptors under certain circumstances may function as sensor/monitors. However, specific molecules seem to have been developed that monitor the presence or level of H_2, O_2, NO, and CO to induce necessary actions. The iron atom, particularly that of hemes, would bind O_2, CO, or NO and are indeed involved in the sensing processes of these molecules.

10.2.4.1. *Oxygen Sensors*

Oxygen, being an essential entity for aerobic organisms including humans, is closely monitored and, if not sufficient, organisms attempt to increase the supply of oxygen. In humans, the major site for this monitoring of O_2 is located in the carotid artery near the heart. A small body called a carotid body in the artery is made of two types of cells, sustentacular and glomus cells, and the latter is the major site that senses O_2 level (Lahiri *et al.*, 2006). Two different mechanisms seem to exist that sense O_2 and influence mechanisms to increase O_2 supply. One is fast in response and electrical in nature, and the other is histological and biochemical in nature and involves expressions of genes associated (Lahiri *et al.*, 2006).

A cytochrome-450 enzyme (see Chapter 6), hemoxygenase-2, seems to be involved in fast sensing/effecting. It breaks down heme to iron, biliverdin, and CO. CO produced activates a K(I)-channel (BK channel), and hence depolarizes the membrane potential under normal conditions (with sufficient $P(O_2)$). Under a hypoxic condition (low $P(O_2)$), less CO is produced and hence diminishes the excitatory influence on the BK channel. This results in polarization of the membrane potential and initiates a nerve impulse and releases neurotransmitters such as dopamine, and these processes lead eventually to increased breathing and others to increase oxygen supply (Lahiri *et al.*, 2006).

The key factor in the histological process is HIF-1 (hypoxia inducible factor). HIF-1 consists of HIF-1α and HIF-1β. HIF-1β is constitutively expressed and unaffected by changes in $P(O_2)$. HIF-1α on the other hand is continuously synthesized and decomposed under normal $P(O_2)$ (Lahiri *et al.*, 2006). It seems that HIF-1α has prolyl hydroxylase activity and hydroxylates its own prolyl residues. The thus modified HIF-1α is degraded by a proteosomal system. The prolyl hydroxylation activity requires the presence of O_2, Fe(II), 2-oxoglutarate (and ascorbic acid). This combination suggests that this enzymatic activity is 2-oxo-glutarate dependent monooxygenase (hydroxylase) reaction as discussed in Chapter 6. Indeed, the Fe(II) octahedral coordination environment consists of two histidine and one aspartate residues and an OH with the remaining two being occupied by O and N of an inhibitor as found in the x-ray crystallographic structure (McDonough *et al.*, 2006). In hypoxia where $P(O_2)$ is low, obviously the prolyl hydroxylase activity is suppressed and HIF-1α accumulates, combines with HIF-1β, and forms holoprotein HIF-1. This then binds other factors termed as coactivators CBP and P300, and these factors activate RNA polymerase and other transcription factors. This will produce necessary proteins that diminish the hypoxia condition (Lahiri *et al.*, 2006).

A rhyzobial oxygen sensor is FixL (Gong *et al.*, 1998). *Rhyzobium* is a nitrogen-fixing bacterium under anaerobic conditions, and has to monitor the $P(O_2)$. FixL is a histidine kinase. It has a heme group, and autophosphorylates at a histidine residue with ATP when it is devoid of oxygen (deoxy form); that is, $P(O_2)$ controls the kinase activity. The phosphoryl group is then transferred to the transcription factor FixJ and triggers a cascade of gene expression to produce necessary proteins for nitrogen fixation. In the presence of O_2, the heme binds O_2 and the kinase activity is suppressed, and hence no expression of nitrogenase is carried out. Likewise, the CN^- bound with the met form of the heme suppresses the kinase activity.

10.2.4.2. *CO Sensors*

CO, though toxic, has been found to play a variety of roles in cell physiology, and constitutes ligands for Ni- or Ni/Fe-containing hydrogenases and CO dehydrogenase (CODH; see Chapter 5).

The best-known CO-sensor is CooA, found in a facultative photosynthetic bacterium *Rhodospirillum rubrum*, and regulates the CODH activity (Roberts *et al.*, 2004). This is the prototype of such CO-sensors found in a number of bacteria. CooA is a homodimeric heme-containing protein. CO binds to the Fe(II) form of the heme. The reduction potential of the heme is quite low, $-0.3\,V$, which matches the reduction potential of CODH.

The fifth and sixth ligand of the heme are histidine and proline, respectively (Lanzilotta *et al.*, 2000). CO is believed to replace the proline ligand, and to cause a substantial conformational change in the DNA binding domain of CooA, allowing the binding. In the oxidized form, Fe(III) binds a cysteine residue instead of a histidine in its fifth coordination site. It is likely that this replacement reduces the reduction potential, and that once reduced it will be replaced by a nearby histidine. Only CO can replace the proline and can cause the necessary conformational change when it is bound with the heme.

Neuronal PAS domain 2 (NPAS2) monomer (a mammalian transcription factor) contains two heme prosthetic groups as CO-binding sites, and appears to act as another CO-sensor (Roberts *et al.*, 2004).

10.2.4.3. *NO-Sensors*

NO is important in the regulation of blood flow and pressure and also inhibiting the activation of blood platelets, and it is recognized as a neurotransmitter in certain types of nerves. NO is also used as a primary defense mechanism against microorganisms.

NO is synthesized from arginine through a reaction shown below, catalyzed by NOS (NO-synthase). NOS is a cytochrome P-450 enzyme. Isoforms NOSI and NOSIII are constitutively expressed, and NOSII is induced by external stimulus such as bacterial lipopolysaccharides (Bruckdorfer, 2005).

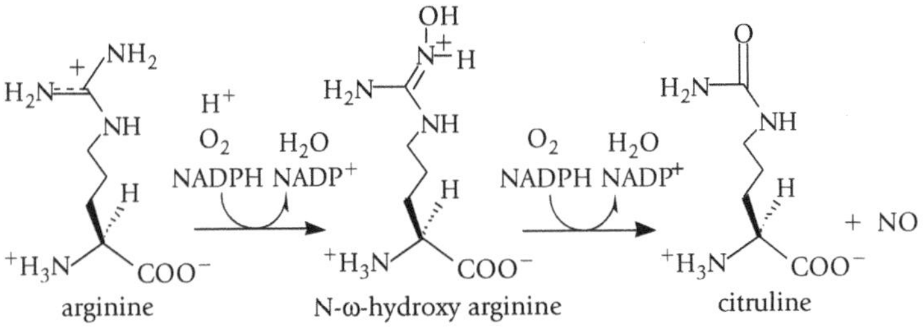

The enzyme consists of an oxygenase and a reductase domain. The oxygenase domain has binding sites for cyt P-450 and tetrahydrobiopterin, and is linked to the reductase domain through a binding site for calmodulin. The reductase domain contains binding sites for FAD, FMN, and NADPH (Bruckdorfer, 2005).

A NO-sensor is soluble guanyl cyclase (sGC). It is a heterodimeric (α/β) heme protein, and is expressed in almost all mammals and many vertebrates. The N-terminal region of the β-subunit binds the heme. Binding of NO to the Fe of the heme triggers a conformational change in sGC through cleavage of the proximal histidine-to-Fe bond, forming a five-coordinate high-spin protoporphyrin IX-NO adduct (Roberts *et al.*, 2004). This causes a several fold enhancement of its GTP cyclase activity. The cGMP is the second messenger and produces its physiological effects.

10.2.4.4. *H_2-Sensors*

Ralstonia eutropha, a facultative chemolithotropic proteobacterium, oxidizes H_2 using two Ni—Fe hydrogenases (see Chapter 5): a membrane-bound (MBH) and a cytoplasmic NAD-reducing one (SH) (Bernhard *et al.*, 2001). Hydrogenase gene transcription is controlled by a regulatory system consisting of proteins HoxA, HoxB, HoxC, and HoxJ. Sensing H_2 requires HoxBC protein (now called regulatory hydrogenase (RH)), in addition to HoxA and HoxJ. RH contains a typical Ni—Fe hydrogenase cluster, and indeed showed some hydrogenase activity, though about 200 times lower than the Ni—Fe enzyme. RH is insensitive to the presence of O_2, CO, or C_2H_2. HoxJ has a histidine kinase activity (Bernhard *et al.*, 2001). This is reminiscent of FixL discussed earlier. The mechanism of H_2 is yet to be determined. A tentative mechanism includes (a) binding of H_2 to the Fe site of the Ni-Fe cluster, (b) modifying the RH conformation by the two electrons, (c) affecting the HoxJ kinase activity, and (d) causing a cascade of gene expressions.

10.2.4.5. *Redox Sensors*

Cells are constantly under an oxidative stress due to ROS (reactive oxygen species) such as superoxide radical $^{\cdot}O_2^{-}$, $^{\cdot}OH$ radical, and H_2O_2. This necessitates redox sensors; two such sensors that have been relatively well characterized are SoxR and OxyR, found in bacteria such as *E. coli* (Pomposiello and Demple, 2001).

SoxR is a polypeptide with CX_2CXCX_2C (C = Cys, X = others) in the C-terminal region, and these cysteine residues are the ligands for a $[Fe_2S_2]$ cluster in the monomer. The presence of the cluster is not required to maintain the protein conformation so as to bind to a specific DNA sequence. SoxR has a strong transcription-activating capacity when the cluster is in the oxidized form [Fe^{III}—Fe^{III}], but it loses its capacity when it is reduced to [Fe^{II}—Fe^{III}]. Thus, SoxR in the SoxR/DNA/RNA polymerase complex activates the RNA polymerase to express genes under high oxidative stress due to superoxide free radicals (so that the cluster is oxidized to $[Fe^{III}—Fe^{III}]$) (Pomposiello and Demple, 2001). By the way, NO can also activate SoxR, and this activation is independent of the presence of O_2.

E. coli contains also OxyR, a sensing protein of hydrogen peroxide, and at least 15 other bacteria contain OxyR. OxyR seems to activate in its oxidized form such genes as *dps* (DNA and iron-binding protein), *gorA* (GSH reductase), *grxA* (glutaredoxin), *katG* (peroxidase), and *fur* (iron-binding repressor of iron-transport). OxyR forms a tetramer in solution; the monomer has a molar mass of 34 KDa. When OxyR is exposed to $100{\sim}1000\,\mu M$ H_2O_2, an intramolecular disulfide bond forms between Cys199 and Cys208 (Pomposiello and Demple, 2001). OxyR in its oxidized form binds to the promoter regions of the target genes and activates transcription. GSH-glutaredoxin restores OxyR to the reduced state (Pomposiello and Demple, 2001).

10.2.5. **Plant Hormone Ethylene and Copper**

Ethylene is a plant hormone, involved in a number of physiological processes including seed germination, cell elongation, fertilization, and fruit ripening (Alonso and Stepanova, 2004). A model plant, *Arabidopsis*, has been found to have several ethylene receptors, ETR1, ETR2, EIN4, ERS1, and ERS2 (Alonso and Stepanova, 2004). It has been shown that ETR1 expressed in yeast increases its ethylene binding ability by 10- to 20-fold when $300\,\mu M$ $CuSO_4$ is added (Rodriguez *et al.*, 1999). Other transition metals except for Ag(I) did not have any enhancing effect on the ethylene-binding of ETR1. Ag(I) and Cu(I) are known to bind olefins; hence it is reasonable to presume that Cu(I) is the ethylene-binding entity in ETR1, and probably in the other ethylene receptors as well. The amino acid sequence of ETR1, as compared with a few other similar proteins suggested to Rodriguez *et al.* (1999) that cysteine (65) and histidine (69) are involved in binding Cu(I). A mutant with Ser in place of Cys and Ala in place of His showed no ethylene-binding activity (Rodriguez *et al.*, 1999).

10.2.6. **Magnetic Navigation**

Some bacteria that use magnets as their navigation tool are known as magnetotactic. Such bacteria include *Magnetospirillum magnetotacticum* and *M. magneticum*. They are aquatic motile organisms and can navigate along the geomagnetic field lines (Bazylinski and Franke,

2004). It has been demonstrated that they contain magnetosomes, organelles that house magnets, small crystals of magnetite (Fe_3O_4), or greigite (Fe_3S_4). The organelle wraps a nanocrystal of the iron oxide or sulfide and a series of such organelles are lined just inside the membrane. It turned out that the organelle is formed as an invagination of membrane, and the magnetosomes are connected by a network of filaments like cytoskeletal actin (Komeili *et al.*, 2006). It is interesting to note that such bacteria are found mostly at the oxygenic-anoxic interface in the aquatic environment (Bazylinski and Franke, 2004). That is, they live in an area where the oxygen tension is relatively low but not too low. Very likely this is because a high oxygen pressure makes it difficult to form the iron oxide of intermediate oxidation state, magnetite (nominally mixture of Fe(II) and Fe(III)).

There is some evidence that some birds, for example, pigeons and some other migrating birds, also use magnetite as a compass needle (Mora *et al.*, 2004).

10.2.7. **Radiation Shields**

Transition metals can easily change their oxidation states. This implies that such metallic entities may function not only as free radical quenchers as have been discussed, but also may function as a radiation shield. Indeed such a function has been found. A bacterium *Deinococcus radiodurans*, as the name implies, can grow under chronic γ-radiation or recover from acute high doses of γ-radiation. It has been shown to accumulate high intracellular Mn(II), 0.4–4 nmol/mg protein (Daly *et al.*, 2004), and the intracellular Mn/Fe ratio ranges from 0.2 to 2.5. This ratio in ordinary radiation-sensitive bacteria is as low as 0.007 to 0.0001. Other bacteria that are radiation-resistant bacteria such as some species of *Enterococcus*, *Lactobacillus*, and cyanobacteria also have been shown to accumulate Mn(II). It has been demonstrated that damages by ionizing radiation were not prevented by other metals such as Fe, Co, or Mo. The radiation damage is caused partially by reactive oxygen species such as $O_2^{\cdot-}$ and $OH^{\cdot}$. Mn(II) may play roles to remove these reactive species, but apparently not by way of Mn—SOD alone. It is considered to be unlikely that Mn(II) directly protects radiation damage of DNA.

10.3. **BIOLOGICAL SKELETONS (BIOMINERALS)**

All organisms require some physical means to protect their bodies, maintain their physical postures, and grind food. These require some physical strength. Organic compounds may not provide enough mechanical strength in certain situations. Hence inorganic solid materials are used to fill such needs. Silicon compounds, silica and amorphous silica, are used as frustule in diatoms, spicules in sea squirts, grasses, bamboo, and such; these are discussed in Chapter 9. Therefore, only calcium compounds will be dealt with here. Calcium carbonate (either calcite or aragonite) is the extra skeleton covering the organisms. On the contrary, the internal skeleton is made mainly of calcium phosphate. Other calcium compounds including calcium oxalate and some strontium compounds also are used for similar purposes in some organisms.

10.3.1. **Calcium Carbonate**

The basic mechanism of formation of $CaCO_3$ biomineral has been known for years. That is, the organic matrix materials, acidic proteins, glycoproteins, or polysaccharides provide nucleation site for calcium carbonate (calcite or aragonite).

A recent study identified three proteins in the shell formation of a Mediterranean mussel, *Pinna nobilis* L (Marin and Luquet, 2005). They are named caspartin (17 KDa, in the prism matrix), calprismin (38 KDa), and mucoperlin. Caspartin is rich in acidic aspartate, and the anionic carboxylate groups of the aspartate are believed to provide nucleation site for calcium carbonate. Calprismin is also acidic but not so much as caspartin. Mucoperlin is a glycosylated mucin-like protein in the nacre. [A typical sea shell consists of three layers: the outermost periostracum, thin protective layer; the middle portion; the major prismatic $CaCO_3$ layer; and the nacre layer, mother of pearl lining, consisting of a thin layer of $CaCO_3$.]

The eggshells of birds are also $CaCO_3$, and its rate of formation is remarkably high. A chicken produces 5 g of the mineral phase in 24 hours. The calcified shell consists of 95% of $CaCO_3$ and 5% of organic phase (Lakshiminarayanan *et al.*, 2003). The basic structure and the mechanism of formation of avian shell seem to be the same irrespective of species, though the details differ. Lakshiminarayanan *et al.* isolated a protein, ansocalcin from goose egg matrix, and determined its amino sequence. It has a strong homology to ovocleidin-17 from chicken egg matrix. They showed that ansocalcin catalyzed crystallization of calcite just as the entire organic matrix extract does. Ansocalcin is not particularly acidic, with 11 glutamates and 6 aspartates among 132 residues. The mechanism for crystallization is not known.

The coccoliths of minute pelagic plankton coccolithophores might be the greatest source of biological calcium carbonate (Westbroek, 1991). A number of coccoliths cover a single cell. The coccolith of *Emiliania huxleyi*, for example, is an oval scale consisting of a radial array of elaborate crystals of calcite. The shapes of these coccoliths are aesthetically appealing (Hagino and Okada, 2006, for photographs of some examples; see Fig. 10.13). Another group of microorganisms that form aesthetically beautiful shells of calcium carbonate are radiolarians.

A coccolith is formed inside the cell and is then exuded after crystallization is terminated (Westbroek, 1991). Three acidic polysaccharides PS1~3 have been identified in another coccolithophore, *Pleurochrysis carterae* (Marsh, 2003). The acidic polysaccahrides and base plates (of calcite) are formed in *medial* to *trans* Golgi cisternae, and then are transferred to the mineralizing vesicle. PS2 is a linear polysaccahride with a repeating unit consisting of D-glucuronic, *meso* tartaric, and glyoxylic acids (see Fig. 10.14), has a high capacity to bind Ca(II), and facilitates calcite nucleation. It also works as a buffer of Ca(II). The growth of calcite crystal requires PS3, polygalacturonomannan. *Emilinania huxleyi* lacks acidic polysaccharides equivalent to PS1 and PS2, but has CP that is equivalent PS3. The inner membrane surface of mineralizing vesicle seems to provide nucleation site in the *E. huxleyi* cell.

Another calcified biomineral is that of corals. It covers about 2.8×10^6 km^2 of the earth, and calcification rate is estimated to be $2-6$ kg $(CaCO_3)$/m^2/y (Allemand *et al.*, 2004). The mechanism of $CaCO_3$ deposition seems to be basically the same as in other calcified biominerals as mentioned earlier, but the details are yet to be revealed (Allemand *et al.*, 2004).

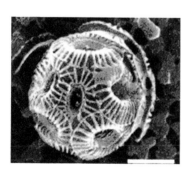

■ **Figure 10.13.** Coccoliths (Hagino and Okada, 2006).

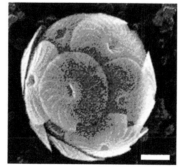

■ **Figure 10.14.** SP2 (after Marsh, 2003).

10.3.2. **Calcium Oxalate**

Calcium oxalate (CaC_2O_4) is distributed among all the photosynthetic organisms from small algae to angiosperms and gymnosperms (Franceschi and Nakata, 2005). The function of Ca oxalate seems to be high-capacity calcium regulation and protection against herbivores; that is, it acts as an unpalatable/toxic substance when eaten. The formation of calcium oxalate crystals in these organisms seem to be regulated genetically, and to be specific in shape and size.

10.3.3. **Calcium Phosphate**

The minerals of vertebrates' bone and dentin are essentially hydroxyapatite ($3Ca_3(PO_4)_2Ca(OH)_2$). These minerals usually contain a little (4–6%) carbonate, and hence they should be called carbonated hydroxyapatite (Wilt, 2005); however, here hydroxyapatite is used. Calcified bone contains approximately 25% of organic matrix (including cells (osteocyte) themselves), 5% of water, and 70% of hydroxyapatite (Sommerfeldt and Rubin, 2001). Two different bone formation mechanisms are known (Sommerfeldt and Rubin, 2001; Provot and Schipani, 2005). One involves mesenchymal cells differentiating directly into osteoblasts, which form bone. This type of membranous or intramembranous bone formation is found during skull formation and others.

If mesenchymal cells first differentiate into chondrocytes, a different kind of bone forms—skeletal bone. Chondrocytes produce cartilaginous templates, which develop by interstitial and appositional growth. Chondrocytes are replaced by osteoblasts, which then lay down extracellular matrix on which calcification proceeds. Eventually osteoblasts turn into osteocytes, which are embedded in calcified structure (bone).

The freshly produced matrix prior to mineralization (by osteoblasts) consists of about 94% collagen. Other proteins include osteocalcin, which may control bone formation, and those involved in mineralization process (osteonectin, osteoputin, sialoprotein). These noncollagenous proteins are proteoglycans and involved in defining the spatial organization of the extracellular matrix. For example, sialoprotein is an acidic phosphorylated (ser/thr) and sulfated (tyr) glycoprotein, and takes part in bone formation by binding to hydroxyapatite through two to three stretches of polyglutamic acid.

The mineralization is poorly understood, but is believed to be initiated by a vesicle that buds from osteoblast or chondrocyte. The small matrix vesicle (30–100 nm in diameter) contains Ca(II) ions, phospholipids, and alkaline phosphatases. The membrane of the matrix vesicle contains more sphingomyelin and phosphatidyl serine than the membrane from which it buds out. When the ionic product of [Ca(II)] and [PO_4^{3-}] exceeds the critical value, mineralization commences, assisted by the outer surface of the vesicle, and eventually the vesicles disintegrate (see Fig. 10.15). The mineral has an approximate composition of $Ca_{10}(PO_4)_6(OH)_2$ ($3Ca_3(PO_4)_2 \cdot Ca(OH)_2$; hydroxyapatite) and takes a plate-like form 20–80 nm long and 2–5 nm thick (Sommerfeldt and Rubin, 2001). These crystals are much smaller than the naturally occurring apatites and less perfect in structures, and hence more reactive and more readily decomposed.

Bone provides not only mechanical strength but also the source of calcium. Hence the calcified minerals are being formed and resorbed depending on the demand. The latter is carried out by osteoclasts. The osteoclasts secrete lytic enzymes and pump out H^+ through proton pumps making the pH 2–4 in the subosteoclastic space. The low pH and the secreted enzymes degrade the matrix and free Ca(II). An activated osteoclast resorbs $2 \times 10^5\,\mu m^3$/day; this amount of bone is formed by 7 to 10 generations of osteoblasts with an average life span of 15 to 20 days (Sommerfeldt and Rubin, 2001). In other words, resorption is very rapid.

The outermost layer of tooth is enamel, which is the hardest part of body, is 90% hydroxyapatite, and covers the main body of tooth, dentin. Dentin is a little less mineralized than enamel, and a little more flexible. Ameloblasts secrete amelogenin and other proteins. Amelogenin is known to be involved in the formation of enamel (Wilt, 2005). Odontoblasts secrete dentin sialoprotein and dentin

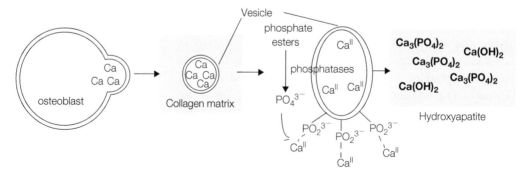

■ **Figure 10.15.** A schematic description of bone (hydroxyapatite) formation process (adapted from Ochiai, 1987).

phosphoprotein, both of which are tooth-specific; but the major parts of proteins, collagens, osteopontin, osteonectin, and such are common to both bone and dentin (Papagerakis *et al.*, 2002).

REVIEW QUESTIONS

1. Why do most organisms have particularly elaborate systems of securing and controlling the level of iron in their bodies?

2. Figure out a general mechanism by which the level of a metal cation in a cell is regulated.

3. Distinguish an active transmembrane transport and a nonactive transmembrane transport.

4. Siderophores often are used as antibiotics. Why do they function as such?

5. Summarize the reasons that Ca(II) plays many vital roles.

6. O_2 is an essential substance for aerobic organisms including humans, and hence elaborate systems to regulate O_2 level have been developed in the organisms. List such mechanisms.

7. It is interesting that some organisms have developed mechanisms to defend themselves against radiation. How could transition metals be employed for this purpose?

8. In transport of iron, iron often is reduced to Fe(II) and then oxidized back to Fe(III). Why do the physiological systems have to do this: oxidizing and reducing in dealing with iron?

9. List all the discriminating factors for cations at the transporting site (binding site). You might list some factors other than those mentioned in the text.

PROBLEMS TO EXPLORE

1. DMT1 (Divalent Metal Transport) is a typical transmembrane transporter of cations. Explore in detail the structural bases for cation transport by DMT (and the like) in literature.

2. Explore a general mechanism of ATP-energized transmembrane transport.

3. No description is given in the text for the physiological functions of NO (nitrogen monoxide), though NO-synthesis

and NO-sensors are mentioned. Explore the mechanisms of the physiological functions of NO.

4. The interior (endo)skeleton in many organisms, such as bones and teeth, is made basically of calcium phosphate, whereas the outer (exo)skeleton (egg shell, sea shells, etc.) is made of mostly calcium carbonate. Discuss why this might be so.

5. Alkali metals, Na and K, are widely used in all organisms. Yet, many plants do not seem to use much of Na(I). That is why many herbivores (including humans) need to take NaCl as a separate nutrient. In fact, ash (of burnt plants) contains a lot of K_2CO_3 but not many Na(I) compounds. Why don't terrestrial plants use much of Na(I)?

Chapter 11

Environmental Bioinorganic Chemistry

11.1. GENERAL CONSIDERATIONS

I published an article with this title in 1974 (Ochiai, 1974), and developed the idea further in a later book (Ochiai, 1977). It was suggested that the environmental issues are a legitimate part of bioinorganic chemistry. The organizers of the Gordon Research Conference initiated a biannual conference of "Environmental Bioinorganic Chemistry" in 2002.

Chapter 1 of this book is devoted to the biogeochemical cycling of elements (see also Ochiai, 1997, 2004); that is, in what way and how fast are elements cycled through the four spheres—atmosphere, hydrosphere, lithosphere, and biosphere. The environmental issues are essentially due to the extra rates of cycling or emission (above the natural rates) of material due to human activities. An estimate of the extra rate of emission to the environment due to the anthropogenic activities was given for several elements in the biogeochemical cycling diagrams in Chapter 1 and in the Appendix. A similar data set is given in Table 11.1 (Ochiai, 2004). The ratio of total [anthropogenic + natural] emission/natural emission (the last column) represents the degree of human activities compared to the natural one. This ratio of unity represents the background (natural) level, and the value (higher than one) indicates the possible severity of pollution that an element may cause. Hence the value is coined as "relative extra burden"; the burden that is imposed on the biosphere, individual organisms as well as collectively. It must be pointed out that these numbers are the planetary overall and average values, and that local conditions vary enormously.

As we just implied, the human activities have imposed an enormous burden on the biosphere in terms of quantities of these elements.

Table 11.1. Anthropogenic Emission of Selected Elements and Its Comparison with Natural Emission (kg/y)

Element	Anthropogenic emission to atmosphere	Anthropogenic emission to hydrosphere and soil	Natural emission to atmosphere (total)	Natural emission[a] to hydrosphere and soil	Extra burden[b] in atmosphere, hydrosphere
Al	7×10^9		4.9×10^{10}	1.5×10^{12}	1.1
As	1.9×10^7	1.4×10^8	1.2×10^7	1×10^8	2.6, 2.4
Cd	7.6×10^6	3.7×10^7	1.3×10^6	1×10^7	6.8, 4.7
Cr	3×10^7	1.4×10^8	4.4×10^7	1×10^9	1.7, 1.1
Cu	3×10^7	1.2×10^9	2.8×10^7	7×10^8	2.1, 2.7
Fe	1.1×10^{10}		2.8×10^{10}	7.4×10^{11}	1.4
Hg	3.6×10^6	1.5×10^7	2.5×10^6		2.4
Mn	3.8×10^6	2.6×10^8	3.2×10^8	2×10^{10}	1.01, 1.01
Mo	3.3×10^6	1.1×10^8	3×10^6	6×10^7	2.1, 2.8
Ni	5.6×10^7	5×10^8	3×10^7	1×10^9	2.9, 1.5
Pb	3.3×10^8	2.0×10^9	1.2×10^7	3×10^8	29, 7.7
Se	4×10^6	8×10^7	9×10^6	1×10^7	1.4, 9
V	8.6×10^7	1.6×10^8	2.8×10^7	3×10^9	4.1, 1.05
Zn	1.3×10^8	3.9×10^9	4.5×10^7	2.6×10^9	3.9, 2.7

Sources: Nriagu, J. O. and Pacyna, J. M. 1988. Nature **333**, 334–339. Nriagu, J. O. 1989. Nature **338**, 47–49 (for the year 1983). Lantzy, R. J. and Mackenzie, F. T. 1979. Geochim. et Cosmochim. **43**, 511–525.

[a]The data for Al, Fe, Cu, Pb, and Zn: from Schlesinger, W. H., Biogeochemistry, Academic Press (1997); otherwise estimates based on the estimate of the total flux of particulates (carried by river) of 1.5×10^{13} kg/y and the concentration data from Martin, J-M. and Whitfield, M., Trace Metals in Sea Water (eds. Wong, C. S., Burton, J. D., and Goldberg, E. D., 1983), Plenum Press, 265–296.

[b]Relative extra burden = Total emission (natural and anthropogenic)/Natural emission.

We could but would not extend this idea to synthetic organic compounds that are another category of material threatening the biosphere. However, metals and others may be involved in rendering harmless these organic, particularly halogenated, compounds. This chapter will not focus on this issue, as reactions and interactions of such compounds with metals or metalloenzymes are discussed throughout this book.

Three issues that are relevant to bioinorganic chemistry will be considered:

(1) The mechanisms of adverse effects of these elements on organisms; i.e., the molecular basis of pollution/toxicity.

(2) How the organisms cope with the adverse effects of elements.

(3) How to reduce such effects or remove the toxic material by biological means (aside from physicochemical means); that is, bioremediation of metals.

11.2. **TOXICITY OF INORGANIC COMPOUNDS**

11.2.1. **Abundance and Toxicity**

Let us look at Figure 1.4. It shows some correlation between the concentration of elements in a human body and that in seawater (environment). It is to be noted that the majority of universally essential elements are located at the upper right portion in the diagram. The exceptions are relatively few: Co, I, Se, and Mo. Boron and nickel are essential only to certain types of organisms. This suggests that the organisms would preferably make use of the more abundant elements, and that certain elements are necessary even though they are not abundantly available in the environment. The organisms have not found more abundantly available alternatives for these exceptional elements: Co, I, Se, Mo, Ni, and so on (Ochiai, 1995a). On the other hand, there are elements that are abundantly available and yet organisms have found no use for; for example, Al.

The elements in the lower left portion of the diagram tend to be toxic. It can be suggested generally that the organisms would have difficulty in dealing with those elements that are not abundant because the organisms would encounter them only rarely; therefore the organisms have not learned how to deal with them. Such elements tend to be toxic (Ochiai, 1995a). However, certain kinds of organisms may have had enough contact with, for example, mercury, because of its habitat, and some of them should have acquired means to cope with its toxicity during the long evolution process; otherwise they would not have survived. That is, they have developed defense mechanisms against the toxicity of the element they had encountered often. It must be pointed out, of course, that any element when present in excess in an organism can have an adverse effect, whether it is essential or not. Organisms must have at least a certain limited capacity to control the levels of elements in their bodies and tissues. This issue was discussed further in Chapter 10.

11.2.2. **Toxicity of Reactive Oxygen Species, and Defense Mechanisms Against Them**

Oxygen O_2 is abundant in today's atmosphere, but it has a high oxidative potential and is universally toxic to organisms. The reactivity of free O_2, however, is not quite high, because O_2 is in a triplet ground state, and its reactions with ordinary substances in the singlet state are forbidden, and hence are kinetically slow. O_2 can be converted relatively easily to other more toxic, reactive forms such as HOOH, $O_2{}^{\cdot-}$ (superoxide radical), $^{\cdot}OH$ (hydroxyl radical), and a singlet oxygen ($^1\Delta_g$) (see Chapter 6). These chemical entities often are referred to as reactive oxygen species (ROS).

These substances are toxic themselves, but more importantly their productions and conversions to other forms are often intimately associated with metallic entities, particularly those of transition metals. The basic chemistry of reactions involving O_2 and ROS is discussed in Chapter 6. Some important types of reactions will be restated here.

(a) One of the two unpaired electrons on O_2 couples with an odd electron:

$$O_2 + e \longrightarrow \ OO^{\cdot-}$$
$$O_2 + {}^{\cdot}R \longrightarrow \ {}^{\cdot}OOR$$
$$O_2 + M^{+n} \longrightarrow \ {}^{\cdot\cdot}OO-M^{+(n+1)} \ (\longrightarrow \ O_2{}^{\cdot-} + M^{+(n+1)})$$

(b) The resulting entities have an odd electron that will react with another odd electron or an entity with an odd electron:

$$^{\cdot}OO^-(X) + {}^{\cdot}Y \longrightarrow Y(OO^{2-})X$$
$$R^{\cdot} + {}^{\cdot}OOR \longrightarrow ROOR$$

(c) The free radical entities listed (and others) may abstract a hydrogen atom from a C-H bond; for example:

$$^{\cdot}OOR + H-C(L) \longrightarrow HOOR + {}^{\cdot}C(L)$$
$$^{\cdot}OH(R) + H-C(L) \longrightarrow HOH(R) + {}^{\cdot}C(L)$$

(d) Reaction of HOOH(R) with another odd electron breaks the O-O bond:

$$HOOR + e \longrightarrow HO^- + {}^{\cdot}OR \ (or \ HO^{\cdot} + {}^-OR)$$
$$HOOR + M^{+n} \longrightarrow HO^- + {}^{\cdot}OR \ (or \ HO^{\cdot} + {}^-OR) + M^{+(n+1)}$$

On the other hand, ROOH can also react with a metal in a higher oxidation state:

$$HOOR + M^{+(n+1)} \longrightarrow H^+ + {}\cdot OOR + M^{+n}$$

In these processes, a transition metal ion with variable oxidation states can create reactive free radical intermediates and hence enhance such a chain reaction.

(e) If the reaction of a free radical ($\cdot$OH, $\cdot$OOR, and others) produces a relatively stable or stabilizable entity (instead of reactive $\cdot$C(L) as earlier), the damaging effects of the free radical will be quenched; examples are:

$$RO\cdot + (HO)C_6H_4(OH) \longrightarrow ROH + (\cdot O)C_6H_4(OH) \text{ (semiqinone)}$$
$$2(\cdot O)C_6H_4(OH) \longrightarrow (HO)C_6H_4(OH) + O{=}C_6H_4{=}O \text{ (quinone)}$$
$$RO\cdot + HS(L) \longrightarrow ROH + {}\cdot S(L); \ 2\cdot S(L) \longrightarrow (L)S{-}S(L)$$

That is, polyphenols and analogues (flavins) and thiols can act as quenchers for free radicals, in particular ROSs with free radical characters.

(f) Singlet state $O_2(^1\Delta_g)$ can be produced chemically or photochemically:

$$O_2(^3\textstyle\sum_g) + h\nu \longrightarrow O_2(^1\Delta_g)$$
$$HOOH + NaClO \longrightarrow O_2(^1\Delta_g) + H_2O + NaCl$$
$$O_2{}^{\cdot-} + {}\cdot OH \longrightarrow {}^-OH + O_2(^1\Delta_g)$$

Phagocytic cells such as leukocytes and macrophages ingest invading microorganisms and kill them. They utilize the ROSs. A scheme used to produce several ROSs in macrophage is shown in Figure 11.1. In the process of NAD(P)H oxidase, superoxide ($O_2{}^{\cdot-}$) may be an intermediate for the formation of HOOH. Similar reactions may produce unnecessary ROSs in other cells. They exert their damaging effects through reaction schemes (c) and (d), particularly on the cell membrane's hydrocarbon and/or DNA/RNA. The halogen derivatives (ClO^- or the like) seen in Figure 11.1 are discussed in Chapter 9.

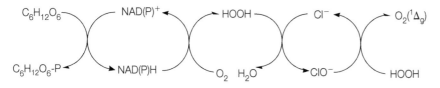

■ **Figure 11.1.** One process (in macrophage) of production of several ROSs (reactive oxygen species).

A number of devices have been developed in the organisms to counter the toxic effects of ROSs. They include catalase (decomposition of HOOH, see Chapter 5), glutathione peroxidase (decomposition of ROOH, see Chapter 9), superoxide dismutase (rendering $O_2^{\cdot-}$ into HOOH and O_2, see Chapter 5), and free radical quenchers such as ascorbic acid (vitamin C), α-tocopherol (vitamin E), flavins, polyphenols (e.g., coenzyme Q_{10}), thiol-containing compounds such as glutathione, lipoic acid, and Se-analogues. Quenching of the singlet oxygen is not a chemical reaction but rather requires an energy transfer. Carotene is known to be an effective singlet oxygen quencher. Many of these compounds are known as antioxidants.

11.3. MOLECULAR MECHANISMS OF TOXICITY OF INORGANIC COMPOUNDS

11.3.1. Discrimination of Elements by Organisms—General Considerations

In order for an element (its compound) to exert its toxic effects, it has to be incorporated in the body of an organism in the first place. If the body of an organism can discriminate a toxic element from nontoxic elements, the problematic element may be prevented from entering the body. A general question is then how an organism incorporates an element (or its compound) and how selectively it can do so. This issue is discussed from a somewhat different perspective in Chapter 10.

A number of mechanisms to import/export (actively or passively through channels) and transport (carry) specific elements through cell membranes or through the body fluid have been discussed in the previous chapter. In a multicellular organism, there are a number of barriers to go through for an element to reach a target cell. In all these processes, the basic mechanism depends on the binding of an element to a specific binding site (receptor/transporter). The question is how specific it is. The binding is governed by physicochemical factors, and can be expressed in terms of binding strength (thermodynamic factors) and binding rate (kinetic factors). The latter factor often is governed by a catalytic entity. For example, the incorporation of Fe(II) into protoporphyrin is catalyzed by iron chelatase, which is specific for iron. The selectivity is hence being carried out by not the porphyrin itself, but by an entity that enhances the process of chelation. In a way, this is a multistep process, and the whole process is governed kinetically.

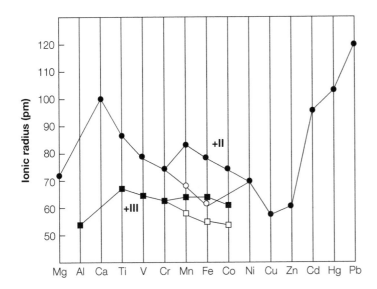

■ **Figure 11.2.** Ionic radii: $+$II $=$ divalent cations in high-spin state (black circle) or in low-spin state (white circle); $+$III $=$ trivalent cations in high-spin state (black square) or in low-spin state (white square).

Metallic elements can be taken up either as a cation or as an oxy-anion. Exceptions for this general rule include the absorption of a metal-chelate as a whole, as in the case of Fe(III)-siderophore and likely Fe(II, III)-heme. Many channels and receptor sites have structural discriminating factors: (a) size of the accepting site, (b) binding ligands, and (c) their arrangement (coordination structure).

The ionic radii of several relevant metal cations are shown in Figure 11.2. The divalent cations, V(II), Cr(II), Mn(II), Fe(II), Co(II), Ni(II), and Mg(II) have similar sizes, 77 ± 6 pm. Hence, if the size alone is the discriminating factor, for example, an Fe(II)-transport site would not be very specific for Fe(II) alone, and allow these cations to bind. The sizes of Ca(II), Cd(II), Hg(II), and Pb(II) are also sufficiently similar to each other so that a Ca(II) binding/transporting site may bind/transport Cd(II), Hg(II), and Pb(II) as well.

The ionic radii of anions are given graphically in Figure 11.3. The doubly charged tetrahedral oxyanions, SO_4^{2-}, SeO_4^{2-}, CrO_4^{2-}, and MoO_4^{2-} have similar sizes within $\pm5\%$. Hence a SO_4^{2-} (essential) uptake/transport site would very likely do the same with these other oxyanions. Likewise, a PO_4^{3-} absorption/transport mechanism would absorb/transport AsO_4^{3-}.

Another discriminating factor is the matching between the metal cation and the ligands. Acid-base interaction (i.e., metal cation and ligand interaction) is regarded to consist of ionic interaction and

■ **Figure 11.3.** Radii of anions and oxyanions.

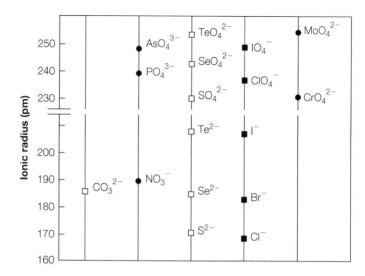

covalent interaction. The ionic interaction portion of cation acid strength can be described by the ionic potential (z_{eff}/r, where z_{eff} is the effective nuclear charge and r the ionic radius) as discussed in Chapter 4.

The covalent portion is due to the atomic orbital overlap, which is associated with the polarizabilities of the entities involved. The overall acidity and basicity are then expressed in terms of inherent character and hardness-softness. The latter is related to the polarizability/covalency. The metal cations tend to become softer as the number of electrons increases; that is, as the oxidation state becomes lower, and/or the atomic number becomes higher for the same electric charge of cations. And the softer cations prefer to bind softer bases. Softer acids (cations) prefer to bind N-ligands over O-ligands, P-ligands over N-ligands, or S-ligands over O-ligands. Among the common cations, Mg(II), Ca(II), Mn(II), and Fe(III) are hard acids and these cations prefer to bind O-ligands. Fe(II), Co(II), Ni(II), and Cu(II) are a little softer than those in the previous sentence, and tend to bind N-ligands preferably. Zn(II) and Cu(I) are softer and prefer to bind S-ligands. Cd(II), Hg(II), and Pb(II) are much softer and strongly bind S-ligands.

In addition, a metal has a preferable coordination structure. For example, Zn(II) and Co(II) (in a high-spin complex) prefer a tetrahedral structure, whereas Cu(II)'s preference is square planar, and so forth (see Chapter 2). These factors (preferred ligands and preferred structure) are

apparent in the metal coordination site structure of all the metalloproteins and metalloenzymes discussed throughout this book. The binding sites for metal cation absorption/transport proteins likewise utilize
these factors to increase their binding specificity.

The specificity of binding of a cation or anion to a protein or other
biomolecules is far from perfect in general, and hence wrong elements may be taken up. This is the situation of a single metal cation
at a binding site at a time. However, the reality is much more complicated. Let's suppose that there are two elements, M_1 and M_2, and
two binding sites, L_1 and L_2, in the same locale. L_1 is meant to be the
binding site for M_1 (the binding constant $K_{L1}(M_1)$ but it also binds
M_2 with $K_{L1}(M_2)$, and also with L_2. That is:

$$M_1 + L_1 \Leftrightarrow M_1L_1{:}K_{L1}(M_1); \; M_1 + L_2 \Leftrightarrow M_1L_2{:}K_{L2}(M_1), \text{ etc.}$$

Further if it is assumed that $[M_1]_{total} \ll [L_1]_{total}$, $[M_2]_{total} \ll [L_2]_{total}$,
and $K_{Li}(M_i) > K_{Lj}(M_i)$, the apparent $K'_{L1}(M_1)$ as defined here will be
smaller:

$$K'_{L1}(M_1) = [M_1L_1]/\{([M_1]_{total} - [M_1L_1]) \times ([L_1]_{total} - [M_1L_1])\} =$$
$$K_{L1}(M_1)/\alpha_{M1}\alpha_{L1} \quad \text{where } \alpha_{M1} = 1 + K_{L2}(M_1)[L_2] \text{ and } \alpha_{L1} =$$

$$1 + K_{L1}(M_2)[M_2]$$

This clearly shows that the apparent binding constant is smaller than
the real binding constant, implying that the specificity is reduced in
the presence of competing entities.

In a multicellular organism an element may have to go through several steps (sites) before reaching its target. This situation can improve
the specificity. Let's suppose that two metals M_1 and M_2 go through
two transfer steps L_1 and L_2. Before they go through the first transfer step L_1, their concentration ratio is $[M_1]/[M_2]$. After the first step,
this ratio will become $K_{L1}(M_1)[M_1]/K_{L1}(M_2)[M_2] = (K_{L1}(M_1)/K_{L1}(M_2))$
$([M_1]/[M_2])$. After the second step L_2, it will be $(K_{L1}(M_1)K_{L2}(M_1)/$
$K_{L1}(M_2)K_{L2}(M_2))([M_1]/[M_2])$. If at each step the selection factor
$K_{Li}(M_1)/K_{Li}(M_2)$ is assumed to be 10 to 1, after the second step, the
ratio $[M_1]/[M_2]$ would be 100 times higher than the original. Hence,
even if the selectivity at each step may not be very high, a multistep
selection process would increase the specificity.

The important factors in the uptake of anions are dehydration of the
anion and hence the hydrophobicity of the binding site. These factors may be efficient in distinguishing vastly different anions (e.g.,

between NO_3^- and SO_4^{2-}), but would not be sufficient to distinguish between, for example, SeO_4^{2-} and SO_4^{2-} or AsO_4^{3-} and PO_4^{3-}.

11.3.2. Oxidative Stress and Metals and As— General Effects

Toxicity of transition metals with variable valence is associated with oxidative stress. Fe, Cu, Cr, Mn, and Co could be involved in the formation of ROSs, exchanging electron(s) with O_2, O_2^-, ROOH, and others as outlined earlier. The presence of an excessive level of ROS species in a cell is termed oxidative stress. ROSs can damage a cell membrane's hydrocarbons, particularly unsaturated ones, and DNA (Eaton and Qian, 2002; Gaetke and Chow, 2003; Papanikolaou and Pantopoulos, 2005; Valko *et al.*, 2005). Lipid peroxides formed from the free radical attack on polyunsaturated fatty acid (of phospholipids) react further with redox metals and may eventually produce mutagenic and carcinogenic malonaldehyde, 4-hydroxy-nonenal, and other exocyclic DNA adducts (Valko *et al.*, 2005).

For example, organ damage caused by chronic overload of iron manifests in slow insidious onset of dysfunction in a wide range of organs, particularly liver, heart, and pancreatic beta cells (Eaton and Qian, 2002). A disease, hemochromatosis, is caused by a defect in iron metabolism leading to iron overload. Iron overload is particularly severe in mitochondria, and is believed to damage the mitochondrial membrane and DNA, and that leads to mitochondrial dysfunction (Eaton and Qian, 2002). It was observed that mitochondrial lipid peroxidation occurred at hepatic Fe-concentrations one-half to one-third of those necessary to trigger microsomal lipid peroxidation (Eaton and Qian, 2002). DNA is quite inert toward oxidants such as HOOH, but its scission occurs in the presence of added iron. It was demonstrated that this effect could be countered by Fe-chelators such as desferrioxamine. Damage on DNA also includes oxidative modification of bases in DNA and formation of DNA-protein cross-links. It is known that mitochondrial DNA is more fragile than nuclear DNA.

Chronic Cu toxicity primarily affects the liver, resulting in liver cirrhosis with hemolysis. Cu-toxicity also manifests in renal tubules (kidney) and brains (Gaetke and Chow, 2003). It can progress to coma, hepatic necrosis, vascular collapse, and death (Gaetke and Chow, 2003). Menkes disorder is caused by Cu-deficiency; it causes mental retardation and neuronal degeneration due to deficiency of Cu-enzymes necessary for brain development. Wilson's disease is caused by an excessive presence of Cu in the liver and brain. As shown in Figure 10.4, Wilson ATPase is the Cu-exporter out of hepatic cells, and hence its deficiency can result in an accumulation of Cu. It then leads to cirrhosis and the extra Cu is also deposited in the brain and cornea of the eye. Evidence has been found to indicate that Wilson's disease is associated with lipid peroxidation in liver mitochondria and reduction of antioxidants such as vitamin E (Gaetke and Chow, 2003). As Cu constitutes an antioxidant enzyme, superoxide dismutase, its deficiency can also lead to an oxidative stress to hepatic and other cells. Other redox metals such as Cr, Mn, V, and Co may also be involved in bringing about the oxidative stress.

The toxicity of some nonredox metals also has been shown to be associated with oxidative stress. Hg(II), Cd(II), Pb(II), and Ni(II) can bind thiol groups, such as glutathione, lipoic acid, and cysteine-proteins, resulting in a depletion of free radical scavengers (i.e., antioxidants) (Valko et al., 2005). Zn(II) is also toxic in this sense, but Zn(II) is essential for so many biofunctions that its toxicity is rather subtle, depending on the exact localization, its speciation, and local concentration. For example, Zn(II) can be neurotoxic (see later), and yet it is essential and can be neuroprotective under certain circumstances (Chen and Liao, 2003).

A characteristic feature of chronic arsenic toxicity is chromosomal damage. It can lead to mutagenesis and cancer. One possibility of toxicity is the oxidative stress. The most commonly occurring As is arsenate, AsO_4^{3-}, which is readily reduced in mammalian cells by glutathione. Hence, it reduces the availability of the antioxidant, glutathione. Besides, the resulting arsenite with $As^{III} = O$ is believed to bind two thiols in the manner such as:

$$HO-As^{III}=O + 2HSR \longrightarrow HO-As^{III}\begin{array}{c} SR \\ \diagup \\ \diagdown \\ SR \end{array} + H_2O$$

For example, pyruvate oxidase is known to be sensitive to As(III), because the lipoic acid in the enzyme system is trapped in this manner. In general, arsenite seems to bind thiol groups. For example, it has been demonstrated that the differences in the number of cysteine residues in human and mouse hemoglobin are responsible for the greater accumulation of arsenic species in mouse blood (Aposhian and Aposhian, 2006). Reactions of the same type, binding to the thiol groups, would diminish the quantity of the thiol antioxidants, hence heightening the oxidative stress. As discussed later, arsenic is methylated in mammals and others. The methylated species (MMA, DMA) are better removed in urine than arsenate, but, particularly in the As(III) forms, they are more toxic (Hirano *et al.*, 2004).

11.3.3. **Individual Element's (Acute) Toxicity**

In addition to the general toxic effects through ROSs as described before, individual elements exert their own specific, often acute toxic effects on organisms. A few prominent examples of such toxicity will be mentioned here; an exhaustive discussion is not feasible. Some of the general mechanisms for toxicity are: (a) a toxic element replacing the functional element and (b) an element blocking the active site, catalytic or otherwise. There are other mechanisms as well, though.

11.3.3.1. *Cd(II) and Hg(II)*

Cd(II) and Hg(II) being congeners of Zn(II) can occupy the Zn-binding sites in many Zn enzymes and proteins. Yet, Cd(II) and Hg(II) are much larger than Zn(II), and Cd(II) or Hg(II)-proteins formed would not function as well as the corresponding Zn(II)-proteins, or the activity may be lost altogether. Cd(II), Hg(II), or other cations can bind to the active site, which consists of appropriate ligands, especially cysteines and/or histidine. Hg(II) binds especially cysteine residues, and hence inhibits enzymes whose active site contains cysteine; such enzymes are numerous. Cd also was the cause of *Itai-itai* (ouch-ouch) disease in Japan; this is characterized as extremely fragile bones.

11.3.3.2. *Pb(II)*

Pb(II) has been shown recently to be rather special in this regard. Two of the important molecular targets of Pb-toxic effects have been identified as synaptotagmin (Ca-binding) and ALAD (δ-aminolevulinic acid dehydratase) (Godwin, 2001). The latter, also known as porphobilinogen synthase, is a Zn-enzyme, and is involved in the formation of porphyrins; hence one of the main symptoms of Pb-toxicity is anemia (Jaffe *et al.*, 2001). The Zn-ligands in ALAD are cysteinyl S^-s. The intriguing question was why Pb interferes specifically with ALAD while there are many Zn(II)-enzymes with cysteines as ligands. It turned out that there are two types of Zn-binding sites: ZnA and ZnB in ALAD (Jaffe *et al.*, 2001). The catalytic ZnB site is rather unusual, with only three cysteine residues binding in a trigonal pyramid manner (Godwin, 2001; PDB 1H7P; Erskine *et al.*, 2001). Pb(II) also binds the three cysteines in the same fashion fairly tightly (PDB 1QNV; Erskine *et al.*, 2000), but Pb(II) is less acidic than Zn(II) and may not function as a proper catalyst. In a study with smaller Pb(II)-coordination compounds, it was shown that Pb(II) prefers to avoid four-coordinated structures, and instead binds three S ligands in trigonal pyramidal fashion (Magyar *et al.*, 2005).

Another target of Pb(II) is zinc-finger transcription factors (Zawia *et al.*, 2000). If Pb(II) binds a zinc finger protein, it very likely changes its conformation and, hence, its interaction with the DNA, and could result in disruption of the transcription.

Pb(II) causes neurological problems. It is believed that Pb(II) interferes with the ability of Ca(II) to trigger exocytosis of neurotransmitters in neuronal cells (see Chapter 10). In concert with this notion was an observation that picomolar concentrations of Pb(II) activated Ca(II)-dependent PKC (protein kinase C). PKC has a C2 domain similar to those in synaptotagmin (see Chapter 2), and the C2 domain contains a multi-Ca(II) binding site and also binds phospholipids in a Ca(II)-dependent manner as in synaptotagmin. Pb(II) promotes phospholipid binding more effectively than Ca(II), suggesting that Pb(II) can bind the Ca(II)-signal protein, synaptotagmin. This could be one of the main reasons for the neurotoxicity of Pb(II) (Chapman, 2002). The similarity of the size of Pb(II) 1.17 Å to that of Ca(II) 1.00 Å may be the main reason for this successful substitution, but with the disruption of the Ca(II) signaling process.

11.3.3.3. *Organometallic Compounds*

One factor that is not a fundamental issue but of practical importance is the chemical speciation. Hg(II) implies free Hg(II) ion, though it is hydrated in reality. Inorganic Hg(II) compound HgX_2 typically is present in the dissociated form: $Hg(II) + 2X^-$. Cell membranes in general have low permeability toward Hg(II) or any other cations, and hence metal cations have difficulty in reaching the intracellular target molecules. Hence, often Hg(II) can go through the gastrointestinal system without being absorbed much. The result will be diarrhea but not very serious. CH_3HgX (monomethyl mercury) is the form of Hg found often in the environment; CH_3Hg^+, being less ionic and having some affinity toward the membrane because of the methyl group, can be absorbed relatively easily through membrane. $(CH_3)_2Hg$ would go through cell membranes more readily. Hence they can easily reach their targets and exhibit toxic effects, although $(CH_3)_2Hg$ has to get rid of at least one of the methyl groups. Other organometallic compounds including $Pb(C_2H_5)_4$ tend to be absorbed readily into cells for the same reason, and then if, for example, degraded into $Pb(C_2H_5)_3{}^+$ or others, can affect its toxicity more acutely and severely than the inorganic counterparts. The oxidation state of Pb in these compounds is +IV, which is more acidic than Pb(II), and Pb(IV) would bind ligands more strongly than Pb(II), though the acidity is moderated by the alkyl groups.

11.3.3.4. *Organotin Compounds*

Inorganic Sn-compounds generally are regarded as innocuous (Hoch, 2001). The major forms of inorganic Sn are of Sn(II). Sn(II) is a congener of Pb(II) and would be toxic like Pb(II), if it is allowed to interact with thiol and other substances. A major difference between Sn(II) and Pb(II) is their acidity. Sn(II) is much more acidic and as a result, Sn(II) tends to precipitate as $Sn(OH)_2$ above pH 2. [Pb(II) is stable up to pH 7.5.] This implies that Sn is not available for affecting toxicity. The tendency to form hydroxide is much stronger with Sn(IV) than with Sn(II); hence free Sn(IV) would not be present much in aqueous media. This might be the reason for a relative nontoxicity of inorganic Sn compounds.

Organotin compounds are a different story. Organotin compounds have been developed for a number of industrial uses such as stabilizer for polyvinyl chloride, and biocides including the use for antifouling paint for boats (Hoch, 2001). The fact that organotin compounds are used for biocide purposes implies that these are toxic. Trialkyl tin ($R_3Sn^{IV}X$) turned out to be the most toxic, expected from the previous discussion. It is interesting to note that tin with different alkyl groups acts most effectively for different types of organisms. For example, trimethyl tin (R = CH_3) is specific to insects; triethyl tin, particularly triethyl tin acetate ($(C_2H_5)_3Sn(CH_3COO)$) is the most poisonous to mammals; R = C_4H_9 is specific for fish, algae, mollusks; R = phenyl is specific for fungi, mollusks, and fish (Hoch, 2001). Tributyl tin (R = C_4H_9) is the most widely used component of antifouling paint.

11.3.3.5. *Be(II), Al(III)*

Be(II), a congener of Mg(II), replaces Mg(II) in Mg(II)-dependent enzymes, but would not exhibit the enzymatic activity, as Be(II) binds usually much more strongly than Mg(II). Al(III) likewise would bind too strongly with catalytic residues such as carboxylate and serine/threonin-OH, and replace Mg(II) inhibiting Mg(II)-dependent enzymes.

11.3.3.6. *Tl(I)*

The acute toxicity of thallium Tl(I) is known to be more severe than those of Hg(II), Cd(II), Pb(II), and Cu(II) in mammals. Tl(I) is similar to K(I) in electric charge and size, and hence interferes with K(I)-related enzymes and transport systems including Na/K-ATPase and likely K(I)-channels. Tl(I) also binds strongly with thiol groups, and inhibits enzymes/proteins, which have sulfhydryls at the active sites (Peter and Varraghavan, 2005).

11.3.3.7. *Cr*

Cr(III) is known not to be very toxic, perhaps because Cr(III) binds ligands only very slowly. $Cr^{VI}O_4^{2-}$, however, is quite toxic. It is strongly oxidative and likely oxidizes most organic compounds; it degrades important biocompounds. The toxicity of Cr(VI) is widespread. The details of the mechanisms have not been studied well, but include the formation of ROSs in the process of the oxidation of organic compounds. The toxicity of chromium to higher plants and microorganisms has been reviewed by Cervantes *et al.* (2001) and Shanker *et al.* (2005). CrO_4^{2-} is believed to be taken up by a SO_4^{2-} absorption system in the plant roots. It is reduced readily in the root, likely by an enzyme, chromate reductase, and the resulting Cr(III) is less toxic.

11.3.3.8. *Ni(II)*

The high cancer incident rate among workers dealing with nickel was first reported in 1933, and it is prominent for the respiratory tract: nasal cavities and lung (Kasprzak *et al.*, 2003). Ni-compounds, however, have been shown to not be mutagenic in bacteria such as *E. coli* and *S. typhimurium*. It is believed that this is due to the presence of efficient uptake/export systems for nickel in them. Ni(II) usually competes with Fe(II) for DMT (divalent metal transporter) and other import/exporters. Hence Ni(II) is taken up by such mechanisms but also can be removed from cells.

The injection of a soluble Ni(II) sulfate into an experimental animal has been shown not to lead to a tumor, probably because it can be removed readily from the susceptible cells as described earlier. On the other hand, the injection of fine particles of solid Ni_3S_2 or NiS into animals resulted in tumors at the injection site (Kasprzak *et al.*, 2003). It turned out that the small particles of these sulfides are phagocytized into a vacuole localized near the nucleus. The nickel is then solubilized as Ni(II), which interacts with the nuclear components. Apparently this is the only route by which Ni(II) comes into contact with the nuclear DNA. Hence nickel has to be incorporated as an appropriate solid form such as NiS, which must be solubilized in the vacuole.

Nickel compounds are known to generate specific morphologic chromosomal damage. In addition, DNA-protein cross links and oxidative DNA base damage were also observed in Ni(II)-exposed cells (Kasprzak *et al.*, 2003). It also has been observed that Ni(II) enhances methylation of DNA, resulting in inactivation of gene expression. Ni(II) and some other heavy metals have been shown *in vitro* to impair the function of DNA polymerase and cause mistakes in base incorporation. Also it was found that Ni(II) and other metal cations inhibit base and nucleotide excision repair mechanism (Kasprzak *et al.*, 2003). A likely cause for these effects is that Ni(II) replaces the Zn(II) in DNA polymerase and repairs enzymes or interferes with the enzymatic reactions through binding to some crucial amino acid residues in these enzymes. The Ni-carcinogenicity is more fully reviewed by Kasprzak *et al.* (2003).

11.3.3.9. *Anions*

S^{2-} or S^--Cys, Se^{2-} or Se^--Cys, CN^- and similar entities bind to the catalytically active cations, and block their enzymatic activities. Arsenic, as AsO_4^{3-}, may replace the essential PO_4^{3-}. The adverse effects of this replacement would be wide-ranging. The acute and chronic toxicities of arsenic are reviewed by Ratnaike (2003).

11.3.4. **Alzheimer's Disease and Metals**

Pathology of Alzheimer's disease involves possibly a number of metal ions (Bush, 2003). A brief outline is given here to indicate the intricateness of metal toxicity. It has been agreed among researchers that the cause of Alzheimer's disease is related to the aggregation of a normal protein, β-amyloid in the neocortex.

β-amyloid in brain has been found in three fractions: membrane-associated, aggregated, and soluble. Most of β-amyloid is membrane-associated in healthy brains, but the other fractions increase significantly in the diseased brain (Bush, 2003). An *in vitro* study found that Zn(II) at low concentrations precipitated soluble β-amyloid to form aggregated β-amyloid, due to the formation of His-Zn-His bridge between β-amyloid molecules (Miura *et al.*, 2000). As discussed in Chapter 10, Zn(II) is released into the intersynaptic space along with a neurotransmitter. This Zn(II) normally will be absorbed back swiftly. However, if this retrieval system (ZnT) does not function properly, the remaining Zn(II) can aggregate β-amyloid. This effect seems to increase with age. Cu(II) and Fe(II) also aggregate β-amyloid. Mouse and rat β-amyloids lack the crucial bridging histidine residues and are not aggregated by Zn(II) or Cu(II) at physiological concentrations (Bush, 2003).

Cross-linking of β-amyloid seems to be induced also oxidatively by $Cu-H_2O_2$. The resulting oligomer is resistant to proteolysis. The amyloid itself seems to catalyze the generation of H_2O_2 through O_2/ Cu(II) or Fe(III) in conjunction with reducing agents such as ascorbic acid and catecholamines (Bush, 2003).

11.4. **BIOLOGICAL DEFENSES AGAINST TOXICITY**

Organisms have developed some mechanisms to cope with the toxic effects of elements to a certain extent; otherwise they would not have survived. The defense mechanisms employed are based on reasonable chemical principles, and are summarized in Table 11.2 (adapted from Ochiai, 1987). Several prominent examples of such biological

Table 11.2. Defense Mechanisms against Heavy Metals and Others (adapted from Ochiai, 1987)

Mechanism	Note; Examples
A. Extracellular but caused intracellularly	
1. Cell wall, adsorbing material	Nonspecific; mucus on the gill of fish, bacterial cell wall, hair
2. Secretion of chelating agents	Can be specific
B. Intracellular and internal	
1. Differential uptake of ions	No specific transport system for heavy metals
2. Removal by pumping out	None is known for heavy metals
3. Removal by deposition/sequestering in some locale where heavy metals are better tolerated	Nuclear inclusion body, cell wall, hair, vacuoles, etc.
4. Removal by converting a heavy metal into an innocuous form	Metallothionein, CuS, HgSe, etc.
5. Removal by converting it into a more readily excreted form	$(CH_3)_2Hg$, Hg(0), $(CH_3)_2As(OH)$, etc.
6. Increased production of an enzyme that is inhibited	
7. Alternative metabolic pathway bypassing the inhibited sites	

defense mechanisms will be discussed here. The molecular mechanisms of defense against ROSs (reactive oxygen species) were outlined earlier.

11.4.1. **Biological Defense against Mercury**

Mercury is one of the most toxic elements and has been studied most intensively among toxic elements. An outline of Hg-biotransformation and semiquantitative biogeochemical cycling processes is given in Figure A.4. The toxicity is based on the binding of Hg(II) to bioactive entities, especially thiol (sulfhydryl) group of enzymes, proteins, and small anti-oxidant molecules. Hence the way to reduce the Hg(II)-toxicity is to reduce the binding capacity of Hg(II). Three ways have been devised: (1) to convert Hg(II) to dimethyl mercury, $Hg(CH_3)_2$, which is coordinatively saturated and would not bind another entity; (2) to convert Hg(II) to Hg(0), the metallic state that has no binding capacity; and (3) to sequester Hg(II), that is, to bind Hg(II) tightly leaving no binding capacity. The first two types of mercury biotransformation are reviewed in detail by Barkay and Wagner-Döbler (2005).

Monomethyl mercury CH_3Hg^+ appears to be the initial product in the methylation process. It is further methylated to $(CH_3)_2Hg$, which can be degraded back to CH_3Hg^+. For an organism to render mercury to a harmless form, it has to produce the dimethyl species, because the monomethyl species is still quite toxic. Methylation of mercury in the environment is carried out mostly by sulfate reducing bacteria such as *Desulfovibrio desulfuricans*, though methane-bacteria seem to be able to do the same. It is believed that CH_3-tetrahydrofolate is the initial methylating agent that methylates Co-corrin to form CH_3-cobalamin. CH_3-cobalamin is known to methylate Hg(II) with or without an enzyme (methyltransferase). CH_3-transfer to Hg(II) from CH_3-cobalamin seems to be in competition with another crucial CH_3-transfer reaction, acetyl CoA synthesis (Barkay and Wagner-Döbler, 2005). $(CH_3)_2Hg$, as it is, is harmless and volatile, and hence it can escape through the bacterial membrane. $(CH_3)_2Hg$ can degrade readily in the environment to form CH_3Hg^+, which can be readily taken up by other organisms, including humans.

Some prokaryotes were found to have Hg-resistance factor (gene *mer*) on plasmid. It turned out to produce an enzyme mercuric reductase (MR), which coverts Hg(II) to Hg(0). Hg(0) will be removed from the bacterial cell because it is nontoxic and relatively volatile. Examples of microorganisms that have this capacity include bacteria such as *Pseudomonas aeruginosa, Escherichia coli, Staphylococcus aureus,* algae such as *Chlamydomonas,* and a yeast *Cryptococcus.* The enzyme is a FAD-dependent oxidoreductase and requires NADPH and some sulfhydryls. The operon *Mer* contains genes for several proteins to transport Hg(II) to MR, in addition to the enzyme itself. For example, MerP is the first protein to catch Hg(II) by two cysteine residues. MerP then interacts with MerT, which accepts the Hg(II) from MerP. MerT has two sets of two cysteines; Hg(II) seems to be relayed through these pairs of cysteine residues. MR then obtains Hg(II) from MerT. In other bacteria, other similar proteins such as products of *merC, merF,* and the like are involved (Barkay and Wagner-Döbler, 2005). The product of *merB* is an enzyme, organomercury lyase, that splits the C—Hg bond of organomercury compounds by addition of protons to the carbon, resulting in the formation of Hg(II). Hg(II) is then reduced by MR.

The expression pf *mer* operon is controlled by regulator proteins MerR and MerD. The operon *mer* is repressed in the absence of Hg(II) in the bacterial cell, but it is induced strongly in the presence of Hg(II). MerR responds to Hg(II) and induces the operon expression at the fairly low level of 10^{-9} mol/L of Hg(II) (Utschig *et al.,* 1995). It is also very selective for Hg(II) against similar cations such as Zn(II), Cd(II), and Ag(I). MerR binds tightly to DNA and suppresses the expression of *mer* operon. When Hg(II) binds to the MerR/DNA complex, it induces changes in the conformation of DNA allowing the transcription. A ^{199}Hg NMR study (Utschig *et al.,* 1995) indicated that the coordination structure about Hg(II) in MerR/DNA is a trigonal plane of three Ss (of cysteine). Its binding is stronger than the ordinary linear coordination of two ligands, and is advantageous in terms of entropy as well. The formation of this structure exerts a conformation-changing effect on the DNA.

The third defense mechanism against Hg(II) is metallothionein, which is widely used as defense against a number of heavy metals, and is discussed separately. The basis for this protective effect is the strong binding between Hg(II) and thiols. It turned out that Hg(II) has a higher affinity toward Se^{2-} (than S^{2-} or sulfhydryl). The K_{sp} for HgSe formation is extremely small, $10^{-58} \sim 10^{-65}$ (Raymond and Ralston, 2004). It has been known that selenide can antagonize the toxicity of Hg(II). Indeed many reports have been published that showed that Hg toxicity was not observed in many animals when a sufficient Se coexisted; and the mole ratio of Hg/Se was often close to 1.0, suggesting the formation of HgSe. As discussed in Chapter 9,

Se is now known to be essential to many biosystems. Se most often is found in the form of selenocysteine, which requires Se^{2-} as the starting entity. Hg(II), if present, will bind Se^{2-} and hence reduce the availability of Se^{2-} for the synthesis of necessary selenoproteins. This is another cause of Hg-toxicity, and will be prevented by a supplement of Se (Raymond and Ralston, 2004).

11.4.2. **Metallothioneins and Phytochelatins**

11.4.2.1. *Metallothioneins*

Metallothioneins and phytochelatins contain a number of thiols (cysteine) that strongly bind heavy metal cations and the like such as Hg(II), Cd(II), Zn(II), Cu(I), and Ag(I). They can thus sequester these metals, rendering them unavailable for toxic functions. Metallothioneins (MT) are found widely among many organisms: animals, plants, fungi, and cyanobacteria, and several classes are known: MT-1, MT-2, MT-3, and MT-4. In mammals, MT-1 and MT-2 are found mostly in liver but in all the organs as well, whereas MT-3 is expressed mainly in brain and MT-4 is found mostly in certain stratified tissues (Vaska and Haler, 2000). MT-1 and MT-2 are inducible by a variety of conditions and substances, including metal ions, ROSs, cytokines, and glucocorticoids. It has been demonstrated that the functions of MT-1 and MT-2 are (a) detoxification of metals (Hg(II), Cd(II), Ag(I), Zn(II), and Cu(I)); (b) regulation of homeostasis of Zn and Cu; and (c) the reduction of oxidative stress. This last function is very likely due to the abundant presence of thiols, which act as antioxidants. In function (b), MT-Zn or Cu plays storage and provider of metals for apo-proteins and enzymes. MTs strongly bind Zn or Cu, but they would release readily the metal ions when the cysteine thiols are oxidized by the oxidized form of glutathione.

MT-3 is present in the glutaminergic neuronal synapse (see Chapter 10). It has been demonstrated that mice deficient in the gene for MT-3 were highly susceptible to kainic acid-induced seizures and postsynaptic neuron injury (Tapiero and Tew, 2003). Hence it was suggested that MT-3 is involved in controlling the release of Zn(II) into the postsynaptic cleft.

Human MTs consist of 60 to 68 amino acids out of which 20 are cysteines and highly conserved. It consists of two domains; α-domain (N-terminal) binds four metal ions and β(C-terminal) domain binds three metal ions. The structure of rat MT-2 with Zn_2Cd_5 is shown in Figure 11.4 (PDB 4MT2; Braun *et al.*, 1992). Two Zns can

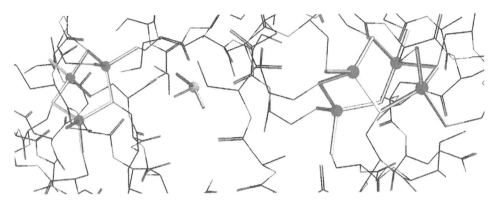

■ **Figure 11.4.** A metallothionein with 2Zn(II)s and 5Cd(II)s; α-domain contains the tetra-nuclear cluster (Cd$_4$), and β the tri-nuclear cluster (CdZn$_2$). (PDB 4MT2: Braun, W., Vasak, M., Robbins, A.H., Stout, C.D., Wagner, G., Kagi, J.H., Wuthrich, K. 1992. Comparison of the NMR solution structure and the x-ray crystal structure of rat metallothionein-2. *Proc. Natl. Acad. Sci. USA* **89**, 10124–10128.)

be replaced by two Cds, and the Cd$_7$-MT takes the same structure. As seen in Figure 11.4, metal ions are clustered in Cd$_4$ and Cd$_3$. In the Cd$_3$ cluster, each Cd is coordinated by two cysteine residues and each pair is bridged by a cysteine: Cd$_3$(Cys)$_9$. The Cd$_4$ cluster has the overall composition of Cd$_4$(Cys)$_{11}$ and contains two different types of Cd. Each of the first two Cds has two cysteine residues bound and is bridged to the Cd of the second type, and each of the Cds of this second type has one cysteine coordinated but is bridged to all three Cds. This type of binding is found with Hg(II), Zn(II), and Cd(II) in MT-1 and MT-2.

11.4.2.2. *Copper-Thionein (Cu-MT)*

Copper-binding thionein is slightly different from the other MTs, and hence designated as Cu-MT. It consists of 53 amino acids, of which 12 are cysteines, and binds six to eight Cu(I) (Calderone et al., 2005). A recent x-ray crystallographic study (Calderone et al., 2005) of yeast Cu-MT showed that a cluster consisting of eight Cu(I)s fits snuggly in the 53 amino acid polypeptide. All eight Cu(I) ions cluster through metal-metal bonds and bridges by cysteinyl thiols; the cluster has the composition of Cu$_8$(Cys)$_{10}$. As seen in Figure 11.5, the cluster looks like it is made of two tetrahedrons sharing one apex with one extra Cu being attached to one of the tetrahedrons. It suggests that this extra Cu may be released easily, forming a Cu$_7$ cluster, as has been observed in a number of previous studies. As Zn-MT,

■ **Figure 11.5.** Copper-thionein. (PDB 1RJU: Calderone, V., Dolderer, B., Hartmann, H.J., Echner, H., Luchinat, C., Del Bianco, C., Mangani, S., Weser, U. 2005. The crystal structure of yeast copper thionein: The solution of a long lasting enigma. *Proc. Natl. Acad. Sci. USA* **102**, 51–56.) Each Cu in the $[Cu_8(Cys)_{10}]$ cluster is bound to an adjacent Cu through a metal-metal bond and is bridged to adjacent Cu's by cysteine residues.

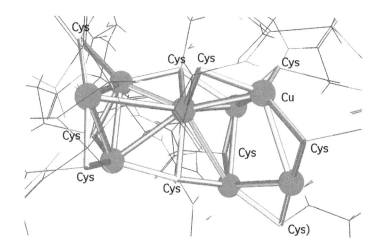

the Cu-MT plays the roles of a provider of Cu(I) for apo-proteins, and also likely that of a storage and of a defense mechanism.

11.4.2.3. *Phytochelatins*

Phytochelatins (PCs) are small peptides consisting of the repeating unit of γ-GluCys dipeptide and ending in Gly (see the structure below). The last portion γ-GluCysGly is the same as glutathione (GSH), and indeed GSH is the substrate of the enzymatic biosynthesis of PC. It was first discovered in yeast *Schizosaccharomyces pombe*, and is found in a wide variety of plants, and some microorganisms. The functional homologues also have been found in some animals (Cobbett and Goldsbrough, 2002). Some variants in which the last amino acid is β-Ala, Ser, or Glu instead of Gly have been found in some plant species.

phytochelatin

The enzyme PC synthase is activated by a range of metal ions; the best activator is Cd(II), followed by Ag(I), Bi(III), Pb(II), Zn(II), Cu(II), Hg(II), and Au(I) cations (Cobbett and Goldsbrough, 2002). The mechanism for activation turned out to be that a metal-thiolate of GSH complex rather than GSH itself acts as the substrate (Cobbett and Goldsbrough, 2002). Studies have indicated that PC synthase is expressed constitutively and is unaffected by exposure of cell cultures or intact plants to Cd(II) (and other metal cations).

In the case of Cd, the sequestered (into vacuole) entity is a CdS crystallite core coated with PCs; the presence of sulfide S^{2-} seems required for the formation of Cd-PC complex. It is not known whether sulfide is involved in the detoxification of other metal cations by PCs (Cobbette and Goldsbrough, 2002). PC-deficient mutants of *Arabidopsis* and *S. pombe* are highly sensitive to Cd(II) and arsenate. No or little sensitivity was observed with other metals including Cu, Hg, Ar, Zn, Ni, and selenite ions, despite the fact that Cu, for example, is a strong activator of PC biosynthesis and that PC forms complexes with Cu *in vivo*. PCs seem to provide a detoxification mechanism for Cd(II) and arsenate, but not for others.

11.4.2.4. *Use of Sulfide*

Sulfide itself rather than thiol may be used to trap metal cations. One of the defenses against Hg(II) is the formation of selenide HgSe (analogue of sulfide) as mentioned earlier. A copper-detoxifying mechanism found in many fungal species is the excessive production of H_2S, which combines with copper to form insoluble, hence inaccessible CuS or Cu_2S. Certain strains of yeast acquire a brown color when cultured in the presence of copper. The color is due to CuS/Cu_2S, and the solid is located mainly in and around the cell wall (Ashida, 1965). Some of the more recent studies in this regard are summarized by a recent review (Whiteley and Lee, 2006).

11.4.3. **Defense against Lead**

A characteristic of lead poisoning is the production of a protein-lead complex known as an inclusion body that appears in renal cells of poisoned animals, including humans. A similar inclusion body also has been observed to form in one type of plant, moss. Inclusion bodies first form in cytoplasm and then migrate to the nucleus (Waalkes *et al.*, 2004). The nuclear inclusion bodies isolated from the kidneys of lead-poisoned rats contained an average of 57 ppm (fresh basis) of Pb, whereas the lead content averaged 0.8 ppm in the whole kidneys of the poisoned rats, and only 0.009 ppm in the whole kidneys of the control rats (Goyer *et al.*, 1970). An inclusion body is approximately spherical and consists of an electron-dense core (Pb) with peripheral fibrous proteins. It has been suggested that the inclusion body is protective in that it may render Pb toxicologically inert and blocking interactions of Pb with sensitive cellar targets (Waalkes *et al.*, 2004).

Metallothionein (MT) provides a defense mechanism against a number of heavy metals as discussed earlier, but its role for Pb has not been well defined. As expected, Pb(II) can bind avidly MT *in vitro*, and some observations have been made that Pb(II) binds to MT in human erythrocytes (Waalkes *et al.*, 2004). Besides, it has also been demonstrated that MT mitigated Pb-toxicity in cultured cells, and that MT reduced Pb-inhibition of ALAD (see earlier). Therefore, MT should be able to bind Pb(II), but the MT's effect on Pb-toxicity *in vivo* has been equivocal. Mice with MT-1/2 knocked out showed dose-related kidney tumors and hyperplasias and modest reduction in renal function when they were exposed to lead, whereas such effects were not observed with the control animals under the same condition. It was discovered also that the mice without MT-1/2 accumulated less renal lead than the control, and that they did not produce Pb-induced inclusion body (Waalkes *et al.*, 2005). How MT affects the production of Pb-inclusion body has not been elucidated. MT has been shown to provide no protective effect on the acute lethal effects of injected Pb(II), but it reduces the subchronic and chronic toxic effects of Pb(II). MT seems to assist the formation of nuclear inclusion body, which sequesters Pb(II), rather than MT directly sequestering Pb(II).

11.4.4. **Biotransformation of Arsenic**

The most toxic form of arsenic seems to be As(III), though As(IV) as AsO_4^{3-} is the most common form found. In mammals and other organisms, AsO_4^{3-} is absorbed through the PO_4^{3-} absorption mechanisms, but is reduced to As(III), arsenite O=As(OH), or $As(OH)_3$ catalyzed by arsenate reductase. The reducing agent is believed to be glutathione. The reduction potential (at pH 0) for H_3AsO_4/H_3AsO_3 is $+0.56$ v compared to -0.3 v of H_3PO_4/H_3PO_3. The reduction potential of RS-SR to 2HSR is about -0.4 v, and hence RSSR (/2HSR) can reduce arsenate readily (in terms of thermodynamics) but not phosphate. This is the difference between As and P, and is a basic cause of arsenic's toxicity when it replaces P in biosystems.

Arsenite is more toxic than arsenate, as outlined earlier, and is to be methylated to monomethyl arsonate (MMA^V) by S-adenosylmethione (SAM; see Chapter 7) catalyzed by a methyltransferase.

$$As^{III}(OH)_3 + SAM\ (CH_3^+{-}R) \rightarrow CH_3As^VO_3H_2\ (MMA) + RH$$

This is an oxidative addition reaction, and nominally the As(III) is oxidized to As(V). MMA is then readily reduced again by glutathione to methyl arsonous acid (MMA^{III}), $CH_3As^{III}(OH)_2$. This is further methylated by S-adenosyl methionine to dimethyl arsenic acid (DMA^V) and reduced to arsenious acid (DMA^{III}) (Vahter, 2002; Aposhian and Aposhian, 2006). This methylation is also catalyzed by another methyltransferase:

$$CH_3As^{III}(OH)_2 \ (MMA^{III}) + SAM \rightarrow$$
$$(CH_3)_2As^VO_2H \rightarrow (CH_3)_2As^{III}(OH)$$

The methylation is effected mainly in the liver. MMA and DMA are more readily excreted in urine, and hence the formation of these methylated species constitutes a defense mechanism against As-toxicity (Aposhian and Aposhian, 2006), even though they are more toxic than nonmethylated As-species.

11.5. **BIOREMEDIATION OF METALS**

The use of bioremediation as a means to clean up contaminated environments has been increasingly researched and applied. Biological systems, plants, and microorganisms have abilities to render some organic compounds harmless or to absorb toxic metallic substances. These abilities are utilized to reduce the environmental burden in soil and water caused by pollutants. In some cases, cells or isolated enzymes may be used instead of the whole organisms. These are called bioremediation. When the remediating agent is plant, it often is called phytoremediation (e.g., see Suresh and Ravishanker, 2004). Bioremediation is generally the safest, least disruptive, and most cost-effective treatment, compared to the traditional physico-chemical methods.

The organic pollutants such as aromatic hydrocarbons and chlorinated organic compounds can be degraded into non- or less-toxic substances by biochemical reactions (Alcalde *et al.*, 2006). Many of such reactions are catalyzed by metalloenzymes and have been discussed throughout this discourse. The bioremediation of this kind, therefore, will not be discussed in this chapter. Our focus here is on the bioremediation of inorganic pollutants including heavy metals.

11.5.1. **Biosorption by Brown Algae and by Microbial Surfactants**

The simplest bioremediation would be absorption or adsorption of metal cations by nonliving biomass through a passive process, and would be inexpensive if the absorbent is cheap. It turned out that

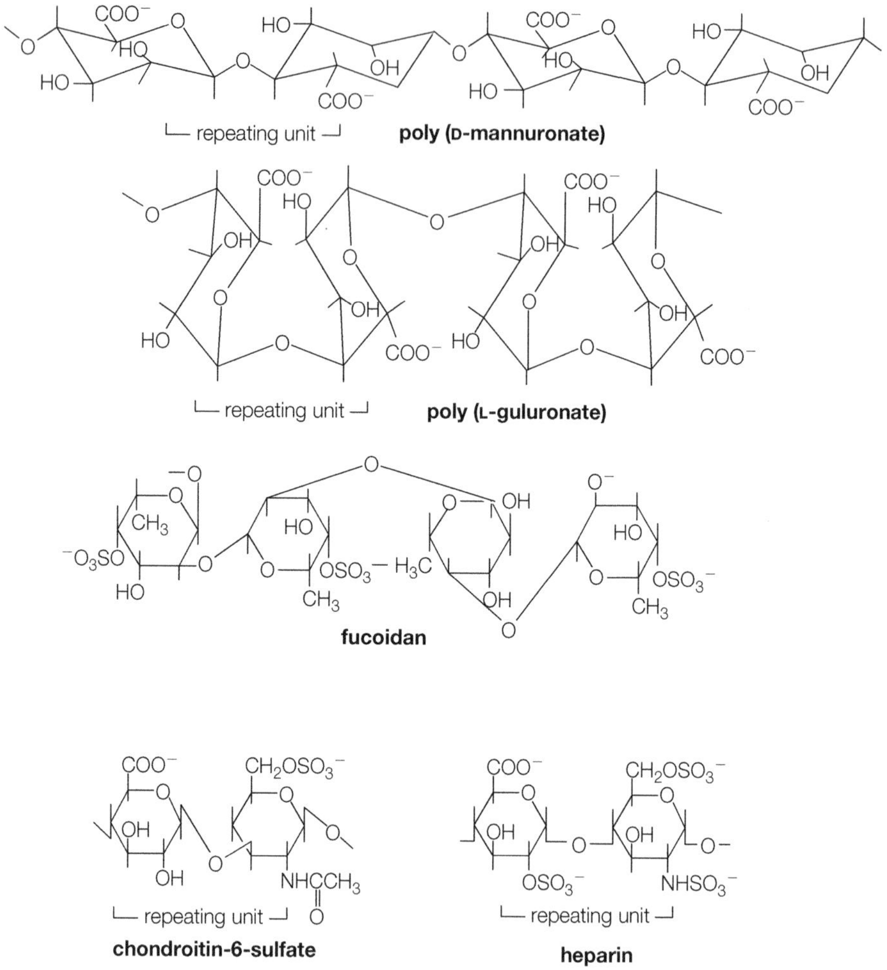

■ **Figure 11.6.** Alginates and fucoidan.

some brown algae (kelp) can bind a high level of metal cations such as Zn(II), Cu(II), Cd(II), and Pb(II) (Davis *et al.*, 2003). The cell wall of brown algae consists of cellulose, alginate, and sulfated mucopolysaccharides, mostly fucoidan. Alginate is either a mixture of poly(α-L-guluronate) and poly(β-D-mannuronate (see Figure 11.6) and/or a copolymer of the two repeating units. Fucoidan is a polymer of sulfate esters of fucose; the predominant linkage is $\alpha(1 \rightarrow 2)$ with an occasional $(1 \rightarrow 4)$, and the position and number of sulfate esters are somewhat irregular. It has been shown that the alginate binding of divalent cations such as Pb(II), Cu(II), Cd(II), Zn(II), and Ca(II) increased with the guluronate content (Davis *et al.*, 2003), suggesting that guluronate provides a better binding site for the divalent cations. The carboxylate and OH groups of two (or more) chains of guluronate are situated in such a way that they can provide appropriate coordinating ligands for cations. The negative charge on sulfonate in fucoidan and carboxylate of mannuronate can attract the cations, but the major binding ability of brown algae seems to be that of alginate, especially of guluronate. According to a set of data collected in the review by Davis *et al.* (2003), the extent (metal mmol/g of brown algae) of metal binding to several brown algae species is Cd 1.91 (*Ascophyllum nodosum*)~0.66 (*Sargassum filipendula*); Cu 1.59 (*Laminaria japonica*)~0.80 (*S. fluitans*); Ni 0.75 (*S. fluitans*)~0.39 (*Fucus vesiculosus*); Pb 1.31 (*A. nodosum*)~ 1.1 (*S. vulgare*); Zn (*L. japonica*)~0.8 (*F. vesiculosus*). No study of Hg was mentioned in this review.

Likewise, the bacterial cell wall/cell membrane material contains a number of metal-binding substances, including glycolipids, lipoproteins, phospholipids, polysaccharides, and fatty acids. These substances often are called biosurfactants. Rhamno-lipids produced by *Pseudomonas aeruginosa* strains have been shown to preferentially bind toxic metals such as Cd(II) and Pb(II) rather than normal soil metal cations such as Ca(II) and Mg(II) (Singh and Cameotra, 2004). Rhamnose and its derivatives are mentioned in Chapter 9. A 5 mM solution of rhamnolipids from *P. aeruginosa* ATCC9027 complexed 92% of Cd(II); Cd 22 μg/mg rhamnolipid. Surfactin from *Bacillus subtilis* showed a similar, though less effective capability (Singh and Cameotra, 2004). One technique used was a biosurfactant-based ultrafiltration of solution containing Cu(II), Zn(II), Cd(II), and Ni(II). Biosorption by other kinds of organisms also have been studied, which include fungi and yeast.

11.5.2. Phytoremediation (Phytoextraction of Metals from Soil)

If pollutants dispersed in the environment (e.g., soil) are absorbed and accumulated in the shoot portion of a live plant, then one simply harvests it and disposes of the accumulated pollutant appropriately. A significant portion of pollutants may be removed by several growth cycles of such plants. This is the idea of phytoextraction. Another strategy is phytostabilization; it provides a cover of vegetation for a contaminated site, thus preventing wind and water erosion. Plants suitable for phytostabilization are those that develop extensive root systems, provide good soil cover, tolerate the pollutants, and ideally, immobilize the contaminants in the rhizosphere (Krämer, 2005). This does not remove the contaminants and will not be discussed here. Only phytoextraction will be discussed next.

A number of plants (less than 0.2% of all plant species) have been found to accumulate one or more inorganic elements excessively; they are called hyperaccumulators (Brooks, 1998). Some examples of hyperaccumulators are listed in Table 11.3, and were collected from the book edited by Brooks (1998).

McGrath and Zhao (2003) did some calculation on how effective such phytoextraction might be. Two critical factors are metal bioconcentration (BC expressed by the ratio of plant shoot/metal concentration in the soil) and biomass production (biomass produced per crop per unit of the area: ton ha^{-1}). They estimated how many crops have to be harvested before the concentration of the metal in soil is halved. For example, if the BC factor is 40, and plant biomass is about 10 t ha^{-1} (which is a reasonable production rate), it takes about four crops to reduce the metal concentration by half. If the plant produces more biomass, for example, 20 t ha^{-1}, it requires only two crops. If the BC factor is 5 instead and biomass is 10 t ha^{-1}, it takes about 40 crops. Assuming again that the BC factor is 40 and

Table 11.3. Examples of Hyperaccumulators

Metal	Plant family/genus	Metal content observed (average)
Ni	*Berkeya coddii*	11,600 (μg/g dry weight (ppm))
	Senecio coronatus	24,000
	Peltaria emerginata	34,400
	Phyllanthus nummularioides	12,240–22,930
Zn	*Arenaria patula*	1.31%
	Haumaniastrum katangense	1.98
	Thlaspi calminare	3.96
	T. stenopterum	2.70
Cu	*Bulbostylis mucronata*	7783 (μg/g dry weight (ppm))
	Haumaniastrum katangense	8356
	Ipomoea alpina	12,300
	Lindernia perennis	9322
Se	*Astragalus bisulcatus*	2276 (μg/g dry weight (ppm))
	A. pattersoni	2696
	Neptunia amplexicaulis	2661

biomass production is $10\,t\,ha^{-1}$ and the removal process can be approximated as first order, it takes about 12 crops to remove 90% of the contaminants. If one crop takes a year, then it takes 12 years. If the growth period is much shorter and the plant can be grown a number of times in a year, the time required would be much shorter. Since this kind of calculation was done under a certain condition, it gives only a rough idea of how effective a phytoremediation would be. It very likely requires longer periods of time because in reality the supply of contaminants is not uniform and they can also be moved from the planted region.

The threshold value of hyperaccumulation has been defined as something like $10^4\,mg\,kg^{-1}$ (ppm) of Zn or Mn in the dry weight of shoot; 1000 ppm for Co, Cu, Ni, As, and Se; and 100 ppm for Cd (McGrath and Zhao, 2003). A fern, *Pteris vittata*, has been found to accumulate up to 2.2×10^4 ppm of As in the frond dry weight, though As toxicity becomes manifest at 1×10^4 ppm (quoted in McGrath and Zhao, 2003). When grown in pots on an As-contaminated soil (90 ppm), the bioconcentration factor became as high as 87, and the fern removed about 26% of As from the soil in 20 weeks (McGrath and Zhao, 2003). Several other fern species also have been found to be hyperaccumulators of As, but two species in *Pteris* genus, *P. straminea* and *P. tremula*, are not As-hyperaccumulators.

Many hyperaccumulators are low biomass producers; that is, they don't grow well. For example, an excellent hyperaccumulator of Zn/Cd, *Thlaspi caerulescens*, produces only $2{\sim}5\,t\,ha^{-1}$ of shoot dry matter. However, Ni hyperaccumulators, *Alyssum bertolonii* and *Berkheya coddii*, produce 9 and $22\,t\,ha^{-1}$, respectively. Few studies actually evaluated the feasibility of phytoextraction under practical conditions. It has been suggested that phytoremediation by *T. caerulescens* may be feasible for moderate levels of Zn-contamination, because the BC factor was shown to be as high as 30 under such conditions (McGrath and Zhao, 2003). Some variants of *T. caerulescens* have been found to be adequate for Cd-remediation.

The contaminants may be converted into volatile compounds by plants and then released into the atmosphere; this process is termed phytovolatilization (Krämer, 2005). One success story in this regard, according to Krämer, was to introduce *merA* and *merB* into some plants such as *Arabidopsis*, tobacco, and poplar. As mentioned earlier, these genes produce enzymes, organomercurial lyase and mercuric reductase. Thus, the first converts CH_3Hg^+, $(CH_3)_2Hg$, or other organomercury compounds to inorganic Hg(II) and the latter turns Hg(II) to volatile Hg(0). This may remove the pollutants from the soil but may disperse them into the environment; the overall effects may not be desirable. Another example of genetic engineering is *Arabidopsis* with overexpressed *E. coli*'s arsenate reductase. In this case, plants absorb arsenate and convert it to arsenite, which combines with thiol compounds in the plant cells and is fixed. A few other examples of genetically modified plants for phytoremediation purpose are mentioned in the review article by Krämer (2005).

The phenomenon of hyperaccumulation requires both excessive uptake mechanisms and tolerance mechanisms of excessive levels of toxic substances. Studies on Zn/Cd-hyperaccumulator *Arabidopsis thaliana* and others demonstrated that several metal homeostasis genes are constitutively expressed at much higher levels in accumulators than in nonaccumulators of the same genus (Krämer, 2005). Suggested tolerance mechanisms are: (a) sequestration in such locale as leaf vacuoles, (b) chelations of metals by nicotinamine, phytochelatin, phytosiderophores, and (c) converting metals to insoluble substances such CuS.

11.5.3. Phytoextraction by Microalgae (Remediation of Polluted Water)

Similar techniques can be used to clean polluted water by employing microalgae (Parales-Vela *et al.*, 2006). The most common setups are high rate algal ponds and algal turf scrubbers, which use suspended biomass of green algae or cyanobacteria or both. Tests were conducted with urban polluted water with trace concentrations of Zn(II), Cu(II), and Pb(II). It was found that the high rate algal pond setup was more efficient in removing the metals compared to the traditional method (waste stabilization ponds). It was reported that a *Chlorella* strain was able to sustain growth at 11 mg Cd(II)/L, and that it removed 65% of Cd(II) when exposed to 5.6 mg Cd(II)/L (quoted in Parales-Vela *et al.*, 2006). The metal cations are supposed to be dealt with by phytochelatins and/or metallothioneins in the algae.

11.5.4. Other Types of Bioremediation

A few other examples of attempts of bioremediation will be mentioned. Some strains of filamentous soil fungi isolated from the sediments of industrially polluted streams removed 60 to 70% of Cd during a 13-day growth period (quoted in Malik, 2004). It was found that growing cultures of marine algae such as *Tetraselmis suecica* showed a high resistance and Cd removal capacity. *Candida* species isolated from sewage samples was found to accumulate Ni(II) and Cu(II) and to increase resistance to these metals (Malik, 2004). Several yeast strains, *Kluveromyces marxianus*, *Candida*, and *Saccharomyces cerevisiae* were shown to be able to remove 73 to 90% of Cu during their growth. The first step of metal uptake by yeast is a fast adsorption of metal cations to the yeast cell wall, and it is followed by a slower, energy-dependent transport into the cell. The majority of intracellular metals are bound to polyphosphate granules in and near the vacuoles or bound to metallothioneins or phytochelatins. The review by Malik (2004) gives further discussion of the ability of other microbes to accumulate metals during their growth and the engineering issues (reactor design, etc.). A recent review (Whiteley and Lee, 2006) gives a more general discussion of bioremediation including those of organic substances with emphasis on enzymatic aspects.

REVIEW QUESTIONS

1. How is superoxide $\cdot O_2^-$ produced in the human body? List the processes and tissues in which it is produced.

2. Explore the same for hydrogen peroxide.

3. Iron is an essential element, but its excess accumulation (overload) is severely toxic. Describe the toxic effects of iron-overload.

4. Binding strength of anions to proteins is in the following order (known as *lyotropic series*): $ClO_4^- >$ $NCS^- \sim I^- > ClO_3^- \sim N_3^- > NO_3^- \sim NCO^- > Br^- > Cl^-$ $> CH_3COO^- \sim F^- \sim OH^- > SO_4^{2-} > HPO_4^{2-}$. What would be the major factor determining this order?

5. Toxic elements such as Hg(II) and Cd(II) exert their own specific toxic effects (such as inhibition of specific enzymes), but they are believed to be involved also in the general toxic effects of ROSs (reactive oxygen species). How?

6. List all the biological methods for detoxifying Hg(II).

7. Metallothionein plays a role in detoxification of heavy metals such as Cd and Hg, but it has been suggested that it also is involved in detoxifying ROSs. How?

8. List advantages and disadvantages of some of bioremediation procedures.

PROBLEMS TO EXPLORE

1. Cd has caused a serious pollution problem in Japan. The symptom is extremely fragile bones, which readily break with a lot of pain; hence it was called *Itai-Itai* (ouch-ouch) disease. Explore the literature for the cause of the disease.

2. Another serious pollution problem caused by heavy metal, mercury in this case, was called *Minamata* disease, taken from the name of the town where cats and people were afflicted severely by mercury poisoning. What was the cause? How does mercury cause the problem? Mercury pollution seems to be more widespread and serious now, even in the United States and Canada. Explore literature.

3. Many active large fish such as bonito and tuna usually contain high levels of mercury. This has been found to be true with old specimens preserved in museums and does not seem to have been caused by recent industrialization. Then why do they have high levels of mercury, and why don't they exhibit any symptoms of mercury poisoning?

4. Bioremediation is an intentional process of remedying environmentally noxious material, but Nature has been doing it throughout the history of the Earth—the natural cleansing effect. Discuss the natural cleansing effect using some concrete examples.

5. Hyperaccumulator plants (which can accumulate specific elements to a high degree) have been utilized as an indicator of the presence of metallic ores. Why are they there with a high level of a toxic element? What would you suggest the mechanism of accumulation be?

Medical Applications of Inorganic Compounds: Medicinal Inorganic Chemistry

12.1. **INTRODUCTION**

In the 1960s, B. Rosenberg was studying the effects of electric field on the growth of bacteria, using platinum wires as the electrodes. He noticed that bacteria stopped dividing and grew longer. He guessed that something inhibited the cell division. A long search identified a platinum complex as the cause. Apparently the platinum complex formed from the platinum electrode. He figured that the compound might inhibit the uncontrolled divisions and growth of cancer cells, and so applied $[Pt(NH_3)_2Cl_2]$ to tumors planted on rats, and observed a dramatic shrinkage of the tumors (Rosenberg *et al.*, 1965, 1969). The rest is history. This is one of the most interesting cases of serendipity. Today platinum compounds are one of the most widely used anticancer drugs.

The platinum compound is not the first inorganic compound that had found use for pharmacological purposes. Inorganic compounds have long been used for many therapeutic purposes. They were used or tried, perhaps because of the general notion that inorganic compounds are toxic and that a controlled use of such a compound may suppress some biological process. Arsenic compounds (for example, arsphenamine = salvarsan) used for the treatment of syphilis are a typical example of the result of such a guiding principle. Today, an increasing number of inorganic compounds has been and is being tried for pharmacological effects, in the hope of finding effective cures for a number of diseases.

The diversity of inorganic compounds and their applications in medicine encompass cancer chemotherapy, arthritis, antimicrobial agents, metalloenzyme inhibitors, antimanic agents, and many others.

Another area of use of inorganic substances for medical purposes is enhancement in diagnostic devices.

Inorganic elements may not be sufficiently provided for a body or may be ingested in excess. The former condition is deficiency (of essential elements) and the latter overdose, which may manifest toxic effects. Nutritional supplement can be administered in the case of the former condition, and therapeutic substances such as metal-chelating agents may be used to counter toxic effects in the overdose. These are medical uses of inorganic substances, but will not be dealt with here. These topics are implicitly discussed in Chapters 10 and 11.

In this chapter, the focus will be on the artificial, intentional intervention in diseases by use of inorganic substances, both for pharmacological and diagnostic purposes. Reviews on the topic are found in *Chem. Rev.* vol. 99 (issue 9, 1999), Roat-Malone (2002), and Shaw (2005).

12.2. **CANCER THERAPY**

In broad terms, cancer is defined as an abnormal cell behavior including irregular and rapid proliferation at the cellular level. In almost all cases, cancer is caused by mutation of cellular genes that control cell growth and division. In spite of years of intensive basic as well as clinical research, the treatment and cure of cancer still remains far from ideal.

All living cells (including cancer cells) contain genetic information necessary for replication and survival. When a cell is divided, its DNA is reproduced. Hence, one of the most intuitive approaches has been to inhibit the growth of cancer cells by disrupting the flow of their genetics information. Almost all of the currently used agents owe their useful but limited activities to an inhibition of DNA synthesis enzymes, direct chemical damage to the genomic DNA, or some alternative means of inhibiting the mechanics of cell division. Chemo- and radiation-therapies have been among the major strategies to treat cancers.

Chemotherapy uses drugs to treat cancer. The drug is referred as an anti-tumor agent, and it is targeted against a malignant tumor. A malignant tumor is the result of the abnormal cancer cell proliferation. Inorganic complexes have been employed in the chemotherapy of cancer in many different approaches, including DNA-adduct formation, and site-specific DNA strand cleavage.

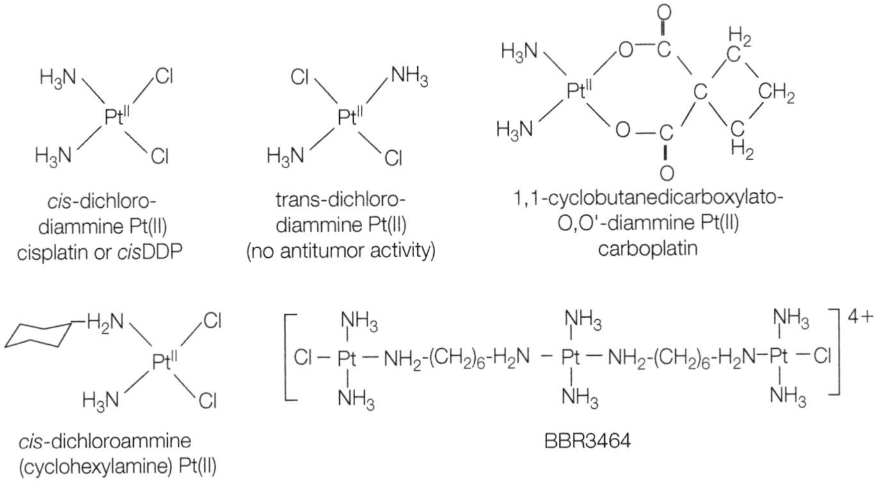
cis-dichloro-
diammine Pt(II)
cisplatin or cisDDP
trans-dichloro-
diammine Pt(II)
(no antitumor activity)
1,1-cyclobutanedicarboxylato-
O,O'-diammine Pt(II)
carboplatin
cis-dichloroammine
(cyclohexylamine) Pt(II)
BBR3464

Figure 12.1. Examples of Pt(II) antitumor agents.

Radiation therapy uses high-energy ionizing radiation to destroy cancer cells. About 60% of all people with cancer are treated with radiation therapy sometime during the course of their illness. Radiation therapy is considered a local therapy because the cancer cells are killed only in the anatomical area being treated. In all cases, frustrating to researchers and clinicians alike has been the severe side effects of the cancer treatment, because often it is not very tissue specific.

12.2.1. **Platinum Compounds**

Among the most common metal binding sites in DNA, the heteroatoms of the nucleoside bases form strong complexes with transition metal ions such as Pt(II) (see Fig. 12.1). *Cis*-diamminedichloroplatinum (II) was discovered in 1965 by Rosenberg (1965, 1969). This compound also is referred to as cisplatin and it is synthesized by treatment of $[PtCl_4]^{2-}$ ion with ammonia. Although cisplatin is not active against all types of cancer, it has been useful in treatment of testicular carcinomas, ovarian, head and neck, bladder, and lung cancers.

Cisplatin is first transported into the cell *via* diffusion subsequent to an intravenous administration. As an electrically neutral entity, it can diffuse into a cell relatively easily. The chloride ions then are replaced by reaction with water, yielding a positively charged complex (see Fig. 12.2). The rate constant (k) for the hydrolytic displacement of the first chloro ligand is $1.02 \times 10^{-4}\ s^{-1}$ at 37 °C with a half-life of about two hours. Thus, due to the positive charge(s)

■ **Figure 12.2.** Loss of 2Cl⁻ from cisplatin.

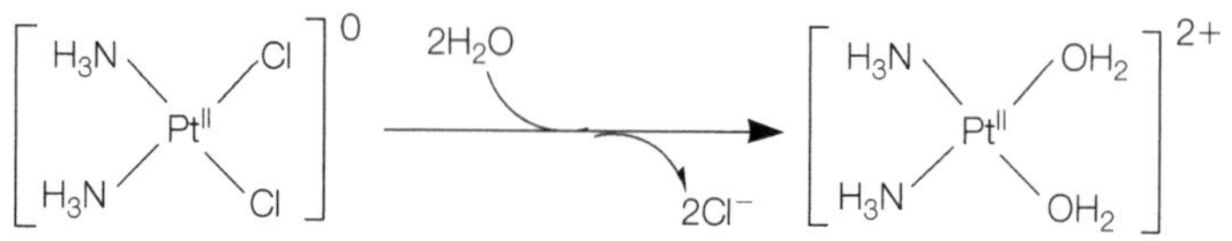

■ **Figure 12.3.** Pt(II) bridges two guanine nucleotides.

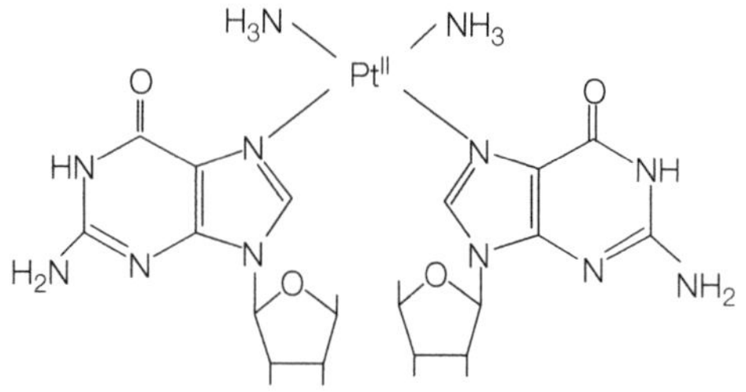

acquired during the activation reaction, Pt(II) complexes react readily with the purine bases of nucleotides. The majority of the products are formed under kinetic rather thermodynamic control. The major forms of adducts are intrastrand cross-links of the kind *cis*-[Pt(NH$_3$)^2{d(GpG)}] and *cis*-[Pt(NH$_3$)$_2${d(ApG)}], which together account for about 90% of the coordinatively bound platinum in the cell (Lippard and Berg, 1994). The N-7s of guanines are cross-linked by platinum (II) (see Fig. 12.3). The N-7 position of guanine is much more basic than that of adenine and therefore provides a stronger site for the attack by platinum.

The structure of a Pt(II) complex with a double-stranded oligonucleotide (d(CCTCTG*G*CTCC)/d(GGAGACCAGAGG)) was studied both by x-ray and NMR methods. Pt(II) cross-links the G*G* (in one of the strands); other GG segments were not cross-linked in this case. The double helix was linked by about 45° at the cross-linked position in the solid state, as revealed by x-ray crystallography (Takahara *et al.*, 1995), but the link was more pronounced in solution as revealed by NMR studies (Gelasco and Lippard, 1998).

Although the major adducts between Pt(II) and DNA are intrastrand G-Pt-G or A-Pt-G, other modes also have been found, including interstrand G-Pt-G (see, for example, Roat-Malone, 2002). One x-ray study showed that Pt(II) cross-links N7 of G5 of one strand of a double-stranded oligonucleotide and N7 of G15 of another (PDB 1A2E). Another study by NMR demonstrated that Pt(II) cross-link caused the double helix (of an oligonucleotide) to be locally reversed to left-handed, unwound and bent toward to the minor groove (Huang *et al.*, 1995). Deformations such as discussed here in DNA structure caused by Pt(II) cross-linking seem to be exaggerated further by a kind of protein called HMG (high mobility group) (Roat-Malone, 2002), and likely disrupt correct readings of information in DNA. And if the DNA is in a cancer cell, it may stop the proliferation of the cancer cells, because it may disrupt duplication of DNA.

A target of Pt(II) anticancer agent may be the telomeric regions of DNA, because they are usually guanine-rich; for example, 5'-TTAGGG-3' in human telomere. A telomere protects the end of a DNA from degradation. Telomerase, an enzyme, adds a telomere to a chromosome end and may enhance DNA replication, hence leading to proliferation of tumor cells (e.g., Roat-Malone, 2002).

Chemically speaking, Pt(II) is a very good soft Lewis acid, and hence would bind to various ligands containing N and/or S. Cancer cells tend to develop some resistance against Pt-agents. One reason for resistance seems to be that a special DNA polymerase η (Pol η) bypasses DNA lesions caused by agents such as cisplatin and UV radiation (Alt *et al.*, 2007). Besides, a chronic exposure to Pt-agents seems to lead to an elevation of cellular levels of glutathione and metallothionein. Both of these provide good S-ligands that may sequester Pt(II). Likewise the cancer cells that acquired resistance against Pt have been demonstrated to contain elevated levels of these sulfhydryl-containing compounds (Reedijk, 1999).

Various analogues of cisplatin such as *cis*-diammine(cyclobutane-1, 1-di-carboxylato)-platinum(II) or carboplatin, *cis*-dichloroammine (cyclohexylamine) Pt(II) (JM118), and their derivatives have been studied in tumors that have exhibited cisplatin resistance (see Fig. 12.1 for a few examples; other examples can be found in Shaw, 2005). These are called second-generation platinum compounds. For example, carboplatin that entered clinical trial in early 1980s was shown to be effective against ovarian cancer and undergoes a reaction similar to that of cisplatin. Its toxicology and kinetics are significantly different from other second-generation platinum complexes.

Dinuclear and trinuclear Pt(II) complexes, and some Pt(IV) complexes also have been studied for antitumor activity. The mechanism of anti-cancer activity of one such example designated as BBR3464 (see Fig. 12.1 for its structure) was studied (Zehnulova *et al.*, 2001). It carries a high electric charge (+4), and hence binds to DNA more quickly than cisplatin, and with long-range cross-links. A BBR3464-DNA

■ **Figure 12.4.** Bleomycin.

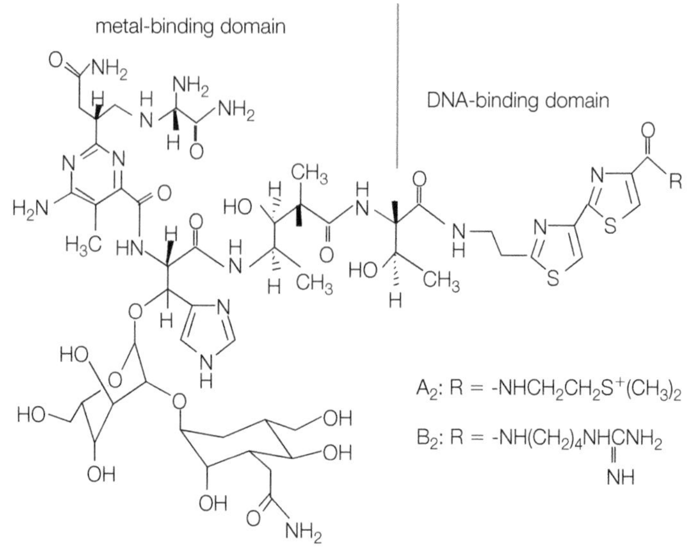

complex seems to contain more interstrand cross-link than that in the case of cisplatin. It turned out that these cross-links caused conformational distortions but did not create stable bends in the DNA (Zehnulova *et al.*, 2001), and that no HMG1 protein recognized these bends in the DNA. And yet, these adducts (BBR3464-DNA) were removed effectively through the nucleotide excision repair mechanism. Therefore, the mechanism by which DNA complexed with BBR3464 is removed seems to be different from that in the case of cisplatin and its analogues (Zehnulova *et al.*, 2001).

12.2.2. **Bleomycin**

A somewhat different strategy for inhibition of DNA replication is to cause fragmentation of DNA. Bleomycins (BLMs), a family of water soluble basic glycopeptides, have been an attractive antitumor agent for such a task. Bleomycins originally were discovered by Umezawa and coworkers (Umezawa *et al.*, 1966), as a fermentation product of *Streptomyces verticillus*. The drug currently used clinically is a mixture of BLM A$_2$ and BLM B$_2$ (see Fig. 12.4). The mixture has been demonstrated to be effective in treatment of testicular, neck, and head cancers as well as non-Hodgkin's lymphoma.

BLM has two domains: metal binding domain and DNA-binding domain as illustrated in Figure 12.4 (Kuroda *et al.*, 1982).

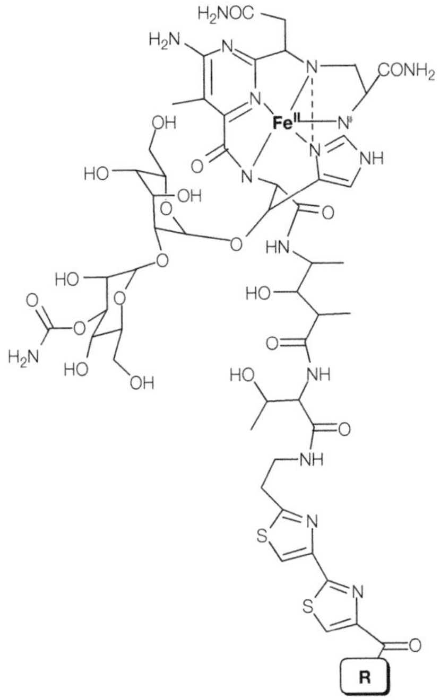

Figure 12.5. Bleomycin-Fe(II) complex.

The native form of BLM peptides are isolated as copper bound coordination complex, and the Cu(II) of Cu(II)-BLM is readily converted to Fe(II)-BLM by polythiol ligands such as metallothionein and free iron present in plasma and tissues. A structural model of Fe(II)-BLM is shown in Figure 12.5. Binding of BLM was suggested to occur through the intercalation of the two thiazole rings (of DA-binding portion) between bases in the DNA, and the positive charge on the $S(\text{-}S^+(CH_3)_2)$ anchors bleomycin through its interaction with the negative charge of phosphate on the backbone. A structure of a double-stranded oligonucleotide bound with a bleomycin-Co(III)-OOH complex (determined by NMR) indeed shows that the bi-thiazole ring region intercalates (see Fig. 12.6).

The therapeutic effect of BLM is believed to arise from its propensity to cause DNA scission based on a unique free radical mechanism. BLM requires molecular oxygen, a metal ion such as Fe(II) and a reducing agent in order to degrade DNA. The activated BLM is supposed to form a HOO-Fe(III) complex as the result of a reaction as:

$$\text{Fe(II)BLM} + O_2 + H^+ + e \longrightarrow \text{HOO-Fe(III)BLM (activated BLM)}$$

■ **Figure 12.6.** A model of bleomycin-DNA interaction (bleomycin-Co(III)) bound to a double-stranded deca-nucleotide (PDB 1GJ2: Hoehn, S.T., Junker, H.D., Bunt, R.C., Turner, C.J., Stubbe, J. 2001. Solution structure of Co(III)-bleomycin-OOH bound to a phosphoglycolate lesion containing oligonucleotide: Implications for bleomycin-induced double-strand DNA cleavage. *Biochemistry* **40**, 5894–5905.)

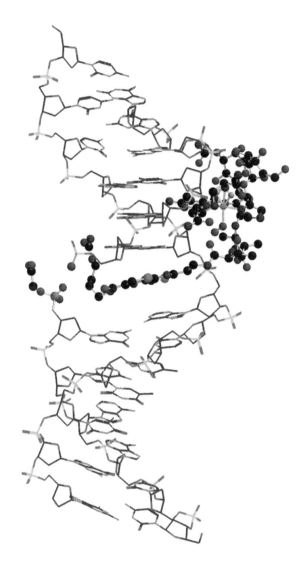

The activated BLM-Fe is structurally analogous to BLM-Co(III)-OOH, and hence would bind to DNA as illustrated in Figure 12.6 (Chen and Stubbe, 2004). The next step is strand cleavage of DNA. The mechanism seems to involve a homolytic cleavage of the O—O bond in the activated HOO-Fe(III)BLM (Chen and Stubbe, 2004). But studies have demonstrated that no OH free radical was formed, because there was no sign of random free radical reaction. It is believed that the homolytic splitting takes place in a concerted manner with abstraction of 4′-H of deoxyribose of DNA (Chen and

Stubbe, 2004). There are, however, a number of other modes of DNA degradation by the activated BLM. Other metals such as Cu(II), Co(II), and Zn(II) are capable of binding to BLM, but Fe(II) is the most potent cofactor in formation of the activated BLM and subsequent DNA strand scission.

12.2.3. **Radioactive Pharmaceuticals**

Radioactive nuclides, if deposited in the target area (tumor), can kill tumor cells by radiation, β or α. If the effect is sufficiently specific to the target, it can provide an effective therapy. β-Emitters are the most widely used radioactive nuclides in today's clinical applications (Volker and Hoffman, 1999). FDA-approved β-emitters include: ^{32}P (half-life = 14.3 days), ^{47}Sc (3.4), ^{67}Cu (2.6), ^{89}Sr (50.5), ^{131}I (8), and ^{153}Sm (1.9). Choice of a nuclide depends on many factors. Examples of practical use include ^{131}I (as NaI) for thyroid cancer; ^{32}P (as phosphate) for blood disorders; ^{32}P (Na_3PO_4), $^{89}SrCl_2$, ^{90}Y, and ^{135}Sm-EDTMP for pain control in bone cancers; and ^{131}I-mIBG for neuroendocrine tumors (Volker and Hoffman, 1999; EDTMP = ethylenediamine tetramethylenephosphonic acid, mIBG = *m*-iodobenzylguanidine).

An iodine compound is effective for thyroid because iodine is specifically absorbed by thyroid, because iodine constitutes the thyroid hormone, thyroxine (see Chapter 9). mBIG, being similar in structure to nor-epinephrine (nor-adrenaline), is selectively taken up by adrenergic neurons, adrenal medulla, and some neuroendocrine cancer cells. The other device is to bind a radionuclide to immunologically derived compound specific for target cells. Target specificity is, however, in general, difficult to provide for many radioactive nuclides for various targets. A result of weak selectivity is damage to tissues other than the target one. Readers are referred to Volker and Hoffman (1999) for further details.

12.3. **GOLD COMPOUNDS FOR RHEUMATOID ARTHRITIS**

The most commonly used gold complexes for chemotherapy of rheumatoid arthritis are of the type L-Au-X, where L is a neutral donor such as R_3P or R_2S and X is a halogen. Figure 12.7 shows the structures of some of the anti-arthritic gold-containing compounds. The gold compounds suppress the symptoms of arthritis but do

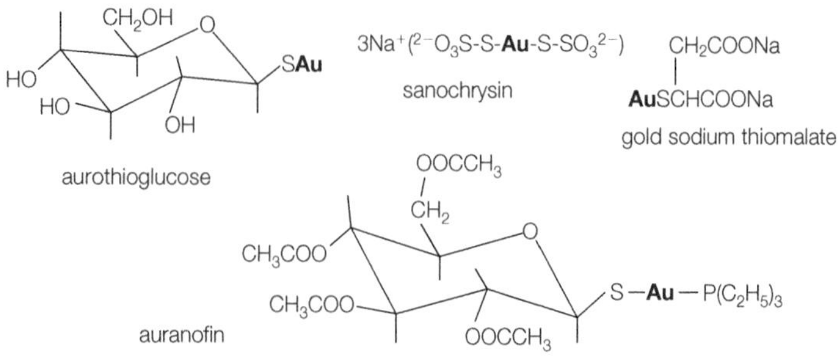

■ **Figure 12.7.** Examples of gold containing anti-arthritis agents.

aurothioglucose

sanochrysin $3Na^+ (^{2-}O_3S\text{-}S\text{-}\textbf{Au}\text{-}S\text{-}SO_3{}^{2-})$

gold sodium thiomalate CH_2COONa / $\textbf{Au}SCHCOONa$

auranofin

not cure it (see Shaw (2005) for a recent review). Gold compounds such as aurothioglucose, sanochrysin, and gold sodium thiomalate are injectable, whereas auranofin (2,3,4,6-tetra-O-acetyl-1-thio-β-D-glucopyranosato-S-(triethylphosphine)-gold(I) can be orally absorbed, and seems to be much less toxic than the injectable compounds.

Auranofin, being hydrophobic, passes readily through the intestinal wall, and is accumulated in the synovial membranes, liver, spleen, lymph nodes, and kidney. Once taken up in blood, auranofin loses the thioglucose portion in less than an hour, and the triethylphosphine portion in a day or so. Most of Au(I) or $[(C_2H_5)_3PAu]^+$ is bound to a cysteine S of plasma albumin during transport by the blood (Shaw, 2005).

The mechanism of anti-arthritic action of gold compounds has not been well understood. As manifested in its binding to albumin through cysteine S, Au(I) binds to S of cysteine residues or Se-selenocysteine of enzymes and proteins and inhibits their functions. It has been suggested that its anti-inflammatory effect (in arthritis) is due to its inhibition of enzymes involved in prostaglandin biosynthesis (Shaw, 2005).

In recent years, however, other organic-based, nonsteroidal anti-inflammatory drugs (NSAIDs) have entered the market, and these drugs exhibit lower toxicity and side effects than the gold compounds. Hence use of gold compounds has declined in recent years.

12.4. **VANADIUM COMPOUNDS FOR DIABETES**

Vanadium has been found in biospheres in, among others, the blood cells of sea squirts and some mushrooms (see Chapter 9). Vanadium-containing enzymes have been identified, including bromoperoxidase (see Chapter 9), and a certain type of nitrogenase

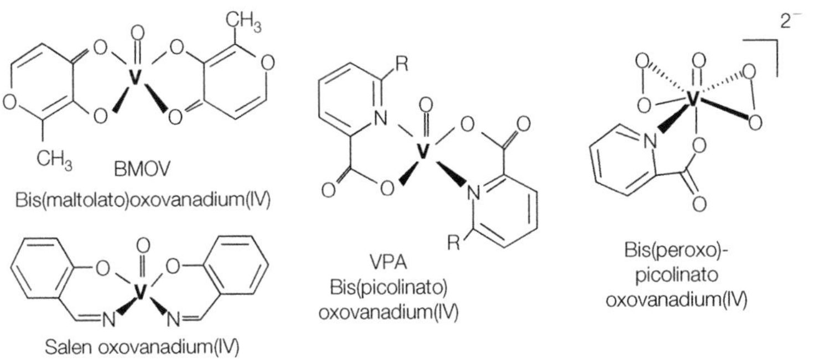

Figure 12.8. Examples of VO-containing insulin enhancers.

of nitrogen-fixing bacteria (see Chapter 8). Vanadium seems to be essential for the normal growth and survival of mammalian cells in culture, but it would be unnecessary to note that an excess of V (or any entity) is toxic.

Diabetes can be due to several factors including deficiency of insulin. Insulin stimulates muscle and fat cells to absorb glucose. In a diabetic individual, the insulin is either not secreted or it is not efficiently stimulating its target cells for glucose regulation. Thus, the glucose level in blood becomes so elevated that glucose shows up in urine; this provides a practical diagnostic test for the disease. The more common type of diabetes, the noninsulin-dependent diabetes, occurs in over 90% of the diagnosed cases. It is generally a result of genetic predisposition, obesity, and age (commonly after the age of 40).

A number of studies have demonstrated that some vanadium compounds mimic or enhance the insulin stimulation of glucose uptake and metabolism (see, e.g., Thompson and Orvig, 2001; Shaw, 2005; Roat-Malone, 2002). A careful evaluation of various results led Cam and coworkers to conclude that vanadium effect is not global but rather tissue-specific, and vanadium enhances, rather that mimics, the insulin function (Cam *et al.*, 2000).

A simplest compound, vanadyl sulfate $VO(SO_4)$, was found to have an anti-diabetic activity, but has gastrointestinal side effects. Hence, other compounds containing vanadyl (VO) have been devised. Figure 12.8 shows several vanadium complexes that have been reported for their insulin-like or insulin-enhancing properties. Inorganic chemistry of these compounds is detailed in a review by Thompson and Orvig (2001). V(III) compounds also have been shown to be somewhat effective, but they can readily be oxidized to V(IV) *in vivo*.

The mechanism of enhancement of insulin activity by V(IV) is obscure. One mechanism suggested is the inhibition of an enzyme protein phosphatase 1B (Shaw, 2005). It is the first enzyme in the insulin regulatory cascade, and its inhibition leads to an increase in insulin activity by preventing phosphorylation of the insulin receptor. Vanadyl complexes are competitive inhibitors for phosphatase, whereas peroxovanadate (see an example in Fig. 12.8) destroys the active cysteine residue of the enzyme. Another mechanism suggested is the interaction of vanadium with the cellular oxidative stress (see Chapter 11). Since the oxidative stress appears to play a role in development of diabetic conditions and vanadium can be subject to oxidation-reduction, vanadium may intervene somehow with the development of diabetes (Shaw, 2005).

12.5. LITHIUM COMPOUNDS FOR PSYCHIATRIC DISORDERS

Introduced in 1940s, lithium salts were employed in psychiatry disorders and in particular in the treatment of manic-depression. Convincing evidence for both the safety and the efficacy of lithium salts up to a specific dosage and its ability in treatment of manic depression have been provided (see, e.g., Shaw, 2005). In Britain, about one in every 1500 people take gram quantities of Li_2CO_3 (prepared as a tablet) daily for the treatment of manic depression. Li^+ is a labile metal ion with a water exchange rate of about 10^9 sec^{-1}. Due to its small size, lithium is distributed easily across the cellular membrane and completely absorbed from the gastrointestinal tract. About 95% of the administered lithium is excreted in the urine.

Although Li^+ is used in depression therapy, it differs from other alternative drugs since it is not a sedative. Several mechanisms have been proposed for the mode of action of lithium in the central nervous system. The effects of Li^+ on the cellular distribution of sodium, calcium, and magnesium as well as glucose metabolism have been attributed to its therapeutic symptoms. Na^+/K^+ channels and ATPases likely would be influenced competitively by Li^+. Li^+ has been shown to influence the regulation of norepinephrine and dopamine in the presynaptic neurons. Norepinephrine and dopamine serve as neurotransmitters. Another biogenic amine, serotonin (5-hydroxytryptamine), which is derived from the amino acid tryptophan, mediates muscle control and sensations. Li^+ also enhances the release of serotonin in particular in the hippocampus region of the brain that is crucial in the memory function. Moreover, monoamine transporters (or neurotransmitter carriers) that regulate the action of serotonin in the central nervous system could also be the targets of anti-depressant agents such as Li^+. Like many other treatments, lithium therapy is not free of side effects. There have been several reports documented on the dermatologic, hypotension, and lithium resistance during the treatment procedure.

12.6. **OTHER POTENTIAL DRUGS CONTAINING INORGANIC COMPOUNDS**

A review by Shaw (2005) and reviews in *Chem. Rev.* vol. 99 (issue 9, 1999) discuss other potential inorganic drugs. Readers are referred to these sources for further information.

12.7. **DIAGNOSTIC (IMAGING) AGENTS**

12.7.1. **Gd(III)-Contrasting Agents for MRI**

The behavior of a magnetic nucleus in NMR measurement is characterized by two relaxation times, T_1 (longitudinal) and T_2 (transverse). In a MRI (magnetic resonance imaging) experiment on a biological sample, the signal intensity is governed by the protons of the water molecules in the tissues and to a lesser degree, the —CH_2— moieties (ca. 1.2 ppm) from fat. Since water comprises about 70% of human body weight, the H_2O proton signal (ca. 4.7 ppm) dominates the MR images. In any imaging technique such as photography, obtaining a signal is not the only prerequisite to an image. Indeed, a clear image is obtained only when there is a contrast or a difference in signal and thus resolving one part of an object from another. In MRI, this task is accomplished by the density of magnetic nuclei in a given volume, and hence by T_1 and/or T_2 values. Variation of each of these parameters can optimize an image for a specific part of a human body. For example, water in tumor cells generally has a longer proton T_1 than a normal cell. The function of the so-called MRI contrast agents is to yield an enhancement in the MRI signal intensity; this is accomplished by a decrease in relaxation time due to the presence of the paramagnetic agent. That is, the fluctuating magnetic field of the paramagnetic entity enhances relaxation of the magnetically excited state of, for example, the proton (of water) located nearby, increasing $1/T_1$ and/or $1/T_2$ by varying degrees, and hence increasing the intensity of MRI signals. Pulse sequences (in NMR measurement) that emphasize changes in $1/T_1$ are referred to as T_1-weighted, and the alternative is T_2-weighted scans (Caravan *et al.*, 1999).

The contrast agents are chelated complexes of paramagnetic metal ions such as Gd(III), Fe(II), Mn(II), among others. Gd(III) increases $1/T_1$ and $1/T_2$ by similar degrees and hence a T_1-weighted method is appropriate. On the other hand, iron oxide particles generally have a much larger increasing effect on $1/T_2$ than on $1/T_1$, and hence its effect is best seen with T_2-weighted scans (Caravan *et al.*, 1999).

The general criteria of a contrast agent to enhance a detecting signal in MRI (magnetic resonance imaging) are high water solubility, paramagnetism, thermodynamic stability, efficient formulation, high tolerance (lack of toxicity), physiological inertness, specificity to the target tissue, and excretability subsequent to the examination.

A large number of Gd(III)-chelate compounds have been tried (Caravan *et al.*, 1999). Two such complexes are shown in Figure 12.9. One of the most widely used clinically useful agents is $[Gd(DTPA)(H_2O)]^{2-}$ (DTPA = diethylenetriaminepenta-acetate) (see Fig. 12.9). Gd(III) has an f^7 electronic configuration (with seven unpaired f-electrons) and is a potent promoter of both T_1 and T_2 relaxation. This complex is intravenously administered to the individual suffering from a cerebral tumor. Selective accumulation of the Gd(III) complex in the target tissue enables the detection of characteristic patterns inherent to a tumor. The details of structural data of Gd(III)-complexes are given in the review article by Caravan *et al.* (1999).

12.7.2. ^{99m}Tc-Radioactive Diagnostic Pharmaceuticals

A radionuclide that emits γ-rays, if concentrated at a specific diseased tissue, may allow detection of and provide diagnostic information

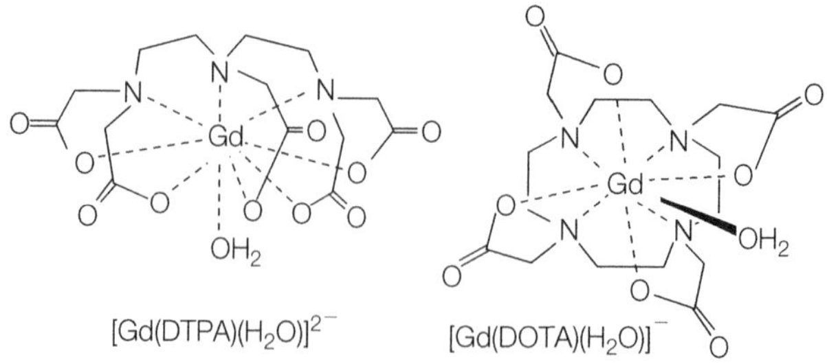

■ **Figure 12.9.** Examples of Gd(III)-MRI image enhancers (DTPA = diethylene triaminepenta-acetate: DOTA = 1,4,7,10-tetraazadodecane-N,N′ N″,N‴-tetraacetic acid).

$[Gd(DTPA)(H_2O)]^{2-}$ $[Gd(DOTA)(H_2O)]^{-}$

for the diseased state, because the penetrating γ can be monitored from the outside of the body; obviously α and β are not appropriate for this purpose. γ-emitting nuclides convenient for such a purpose include ^{64}Cu (half life = 12.8 hr), ^{67}Ga (78.3 hr), ^{99m}Tc (6.02 hr), ^{111}In (2.83 days), and ^{117m}Sn (14 days) (Liu and Edwards, 1999). ^{99m}Tc is most widely used today, because a small convenient generator has been made commercially available. The production of ^{99m}Tc is based on the following nuclear reaction:

$$^{99}\text{Mo}_{42} \longrightarrow {}^{99m}\text{Tc}_{43} + {}^{0}e_{-1}(\beta)$$

^{99m}Tc is in a meta-stable state and emits γ-radiation as it goes down to the ground state ^{99}Tc. But about 15% of ^{99}Mo decays also directly to the ground state ^{99}Tc, which is a β-emitter (^{99}Tc$_{43} \rightarrow$ $^{0}e_{-1}(\beta) + {}^{99}Ru_{44}$) with a long decay half life (2.13×10^{5} years).

A critical issue is how to bring a ^{99m}Tc-containing entity to a target tissue. Of course other pharmacokinetic factors, particularly its fast removal through blood, need to be attended to also. A way to accomplish the target-specificity is to make a composite molecule: (chelate containing ^{99m}Tc)–(conjugate group)–(receptor ligand for the target tissue) (Liu and Edwards, 1999). The chelate part has to have a high affinity for Tc and a kinetic inertness. A typical formula contains a low concentration of Tc at ca 5×10^{-7} mol/L (ca 100 mCi), and the whole complex concentration at 10^{-6} to 10^{-5} mol/L level. The receptor ligand is usually a peptide that binds specifically to the receptor molecule on the tissue; in other words, this is in essence a radiolabeled antibody.

Tc, a congener of Mn, can take various oxidation states, +VII to −I. The Tc used for this purpose has to withstand oxidation or reduction, as lability and stability of Tc complexes depend on the oxidation

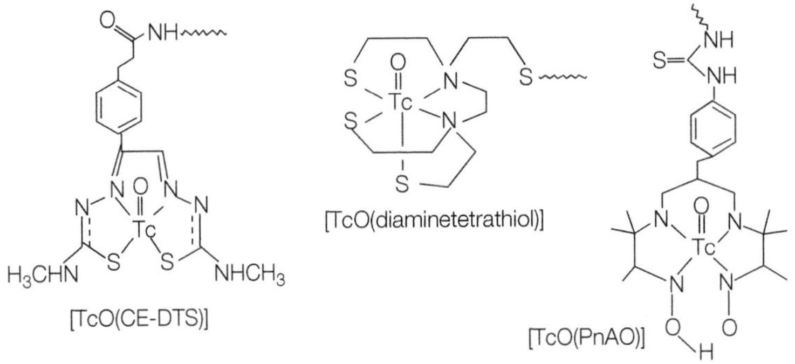

[TcO(CE-DTS)]

[TcO(diaminetetrathiol)]

[TcO(PnAO)]

■ **Figure 12.10.** Examples of Tc = O radio-labeling agents that are connected to peptide through the wavy conjugators (CE-DTS = *p*-Carboxyethylphenylglyoxal-di(N-methylthiosemicarbazone; PnAO = 3,3,9,9-tetramethyl-4,8-diazaundecane-2,10-dione dioxime).

state. Most Tc compounds for this purpose have Tc in $+III$ or $+IV$ oxidation states, but $[Tc = O]^{3+}$ (Tc(V)) is used most often. Examples of the Tc-chelate (with conjugate group to peptide) that have been tried are shown in Figure 12.10. Other examples including those of $(Tc = N)^{2+}$ and $(O = Tc = O)^+$ cores and others (instead of TcO) are found in the review by Liu and Edwards (1999).

REVIEW QUESTIONS

1. What advantage would a medicine based on an inorganic element have over one made of organic compounds?

2. Discussed in the text are pharmacological and diagnostic applications of substances based on inorganic elements. Are there any other possible medical applications of substances based on inorganic elements?

3. Gd(III) seems to be a best choice for enhancing MRI. Why? Explain fully.

4. What is the advantage of ^{99m}Tc for diagnostic purposes?

PROBLEMS TO EXPLORE

1. A critical problem of any medicine is its side effects. Would side effects be generally more problematic or less so in the case of an inorganic medicine as compared to an organic medicine? Discuss.

2. Explore how the gold compounds have been discovered that are effective against rheumatoid arthritis.

3. Vanadium compounds have been found to have an insulin-enhancing effect. Chromium is considered to be involved in GTF (glucose tolerance factor) as discussed in Chapter 9. That is an insulin mimic. Vanadium and chromium are located side-by-side in the periodic table. Would there be any connections between these functions?

4. Pick up a recent paper on medical applications of inorganic elements other than those discussed in the text, and discuss the possibility of their usefulness.

Appendix

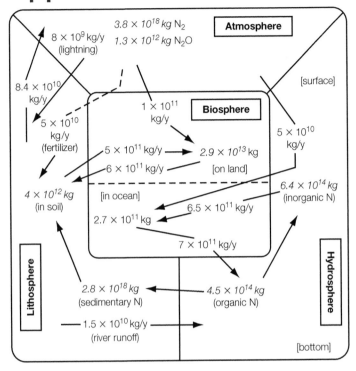

■ **Figure A.1.** The biogeochemical cycling of element nitrogen.

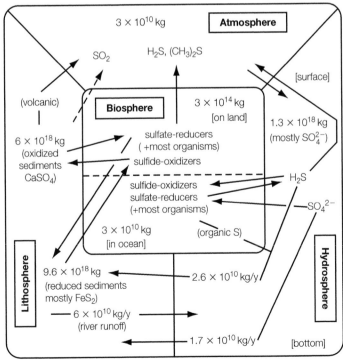

■ **Figure A.2.** The biogeochemical cycling of element nitrogen.

■ **Figure A.3.** The biogeochemical cycling of element nitrogen.

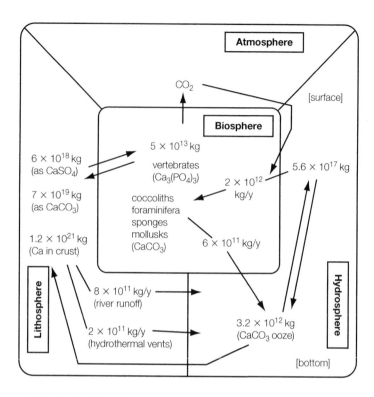

■ **Figure A.4.** The biogeochemical cycling of element nitrogen.

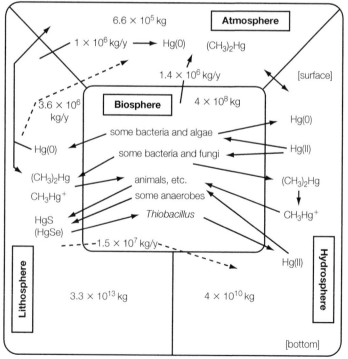

References

Aisen, P. 2004. Transferrin receptor 1. *Int. J. Biochem. Cell Biol.* **36**, 2137–2143.

Aisen, P., Enns, C., Wessling-Resnick, M. 2001. Chemistry and biology of eukaryotic iron metabolism. *Int. J. Biochem. Cell Biol.* **33**, 940–959.

Alcalde, M., Ferrer, M., Plou, F. J., Ballesteros, A. 2006. Environmental catalysis: From remediation with enzymes to novel green processes. *Trends Biotechnol.* **24**, 281–287.

Allemand, D., Ferrier-Pages, C., Furla, P., Houlbrèque, F., Puverel, S., Reynaud, S., Tambuttè, É., Tambuttè, S., Zoccola, D. 2004. Biomineralization in reef-building corals: Molecular mechanisms to environmental control. *C. R. Palevol.* **3**, 453–467.

Almeida, M., Filipe, S., Humanes, M., Maia, M. F., Melo, R., Severino, N., da Silva, J. A. L., Frusto de Silva, J. J. R., Wever, R. 2001. Vanadium haloperoxidases from brown algae of the Laminariaceae family. *Phytochemistry* **57**, 633–642.

Alonso, J. M., Stepanova, A. N. 2004. The ethylene signaling pathway. *Science* **306**, 1513–1515.

Alt, A., Lammens, K., Chiocchini, C., Lammens, A., Pieck, J. C., Kuch, D., Hopfner, K-P., Carrell, T. 2007. Bypass of DNA lesions generated during anticancer treatment with cisplatin by DNA polymerase η. *Science* **318**, 967–970.

Andrews, G. K. 2000. Regulation of metallothionein gene expression by oxidative stress and metal ions. *Biochem. Pharmacol.* **59**, 95–104.

Andrews, N. D. 2002. Metal transporters and disease. *Curr. Opin. Chem. Biol.* **6**, 181–186.

Aposhian, H. V., Aposhian, M. M. 2006. Arsenic toxicology: Five questions. *Chem. Res. Toxicol.* **19**, 1–15.

Arkowitz, R. A., Abeles, R. H. 1991. Mechanisms of action of clostridial glycine reductase: Isolation and characterization of a covalent acetyl enzyme intermediate. *Biochemistry* **30**, 4090–4097.

Armstrong, E. M., Collison, D., Ertock, N., Garner, C. D. 2000. NMR studies on natural and synthetic amavadin. *Talanta* **53**, 75–87.

Armstrong, F. A. 2004. Hydrogenases: Active site puzzles and progress. *Curr. Opin. Chem. Biol.* **8**, 133–140.

Arredondo, M., Núñez, M. T. 2005. Iron and copper metabolism. *Mol. Aspects Med.* **26**, 313–327.

Arteel, G. E., Sies, H. 2001. The biochemistry of selenium and the glutathione system. *Environ. Toxicol. Pharmacol.* **10**, 153–158.

Ashida, J. 1965. Adaptation of fungi to metal toxicants. *Ann. Rev. Phytopathol.* **3**, 153–174.

Asmus, K.-D., Henglein, A., Wigger, A., Beck, G. 1966. Pulsradiolytische Versuche zur elektrolytishen Dissoziation von aliphatische Alkoholradikalen. *Ber. Bunsenges. Phys. Chem.* **70**, 756–758.

Auling, G., Follman, H. 1994. Manganese-dependent ribonucleotide reduction and overproduction of nucleotides in coryneform bacteria. In: *Metal Ions in Biological Systems*, Vol. **30**, 131–161, Sigel, H., Sigel, A., eds. Marcel Dekker, New York.

Averill, B. A. 1996. Dissimilatory nitrite and nitric oxide reductases. *Chem. Rev.* **96**, 2951–2964.

Banci, L., Bertini, I., Cantini, F., Migliardi, M., Rosato, A., Wang, S. 2005. An atomic-level investigation of the disease-causing A629P mutant of the Menkes protein, ATP7A. *J. Mol. Biol.* **352**, 409–417.

Bansal, K. M., Patterson, L. K., Henglein, A., Janata, E. 1973. Polarographic and optical absorption studies of radicals produced in the pulse radiolysis of aqueous solutions of ethylene glycol. *J. Phys. Chem.* **77**, 16–19.

Barkay, T., Wagner-Döbler, I. 2005. Microbial transformations of mercury: Potentials, challenges, and achievements in controlling mercury toxicity in the environment. *Adv. Microbiol.* **57**, 1–51.

Barry, B. A., Babcock, G. T. 1987. Tyrosine radicals are involved in the photosynthetic oxygen-evolving system. *Proc. Natl. Acad. Sci.* **84**, 7099–7103.

Barry, B. A., El-Deeb, M. K., Sandusky, P. O., Babcock, G. T. 1990. Tyrosine radicals in photosystem II and related model compounds. Characterization by isotopic labeling and epr spectroscopy. *J. Biol. Chem.* **265**, 20139–20143.

Bazylinski, D. A., Richard B., Frankel, R. B. 2004. Magnetosome formation in prokaryotes. *Nature Rev. Microbiol.* **2**, 217–230.

Beinert, H., Holm, R. H., Münck, E. 1997. Iron-sulfur clusters: Nature's molecular, multipurpose structures. *Science* **277**, 653–659.

Beinert, H., Kennedy, M. C., Stout, D. D. 1996. Aconitase as iron-sulfur protein, enzyme, and iron-regulatory protein. *Chem. Rev.* **96**, 2335–2373.

Bekker, A., Holland, H. D., Wang, P.-L., Rumble, D. III, Stein, H. J., Hanna, J. L., Coetzee, L. L., Beukes, N. J. 2004. Dating the rise of atmospheric oxygen. *Nature* **427**, 117–120.

Benkovic, S. J., Hammes-Schiffer, S. 2003. A perspective on enzyme catalysis. *Science* **301**, 1196–1202.

Berg, J. M., Tymoczko, J. L., Stryer, L. 2002. *Biochemistry*, 5th ed. W. H. Freeman and Co., New York.

Berkovitch, F., Nicolet, Y., Wan, J. T., Jarrett, J. T., Drennan, C. L. 2004. Crystal structure of biotin synthase, an S-adenosylmethionine-dependent radical enzyme. *Science* **303**, 76–79.

Berner, R. A., VandenBrooks, J. M., Ward, P. D. 2007. Oxygen and evolution. *Science* **316**, 557–558.

Bernhard, M., Buhrke, T., Bleijlevens, B., De Lacey, A. L., Fernandez, V. M., Albracht, S. P. J., Friedrich, B. 2001. The H_2 sensor of *Ralstonia eutropha*. *J. Biol. Chem.* **276**, 15592–15597.

Berridge, M. J., Bootman, M. D., Roderick, H. L. 2003. Calcium signaling: Dynamics, homeostasis and remodeling. *Nature Rev. Mol. Cell Biol.* **4**, 517–529.

Berridge, M. J., Lipp, P., Bootman, M. D. 2000. The versatility and universality of calcium signaling. *Nature Rev. Mol. Cell Biol.* **1**, 11–21.

Biesiadka, J., Loll, B., Kern, J., Irrgang, K.-D., Zouni, A. 2004. Crystal structure of cyanobacterial photosystem II at 3.2 Å resolution: A closer look at the Mn cluster. *Phys. Chem., Chem. Phys.* **6**, 4733–4736.

Binda, C., Coda, A., Aliverti, A., Zanetti, G., Mattevi, A. 1998. Structure of the mutant E92 K of [2Fe-2S] ferredoxin I from *Spinacia oleracea* at 1.7 Å resolution. *Acta Crystallogr., Sect. D*, **54**, 1353–1358.

Birringer, M., Pilawa, S., Flohé, L. 2002. Trends in selenium biochemistry. *Nat. Prod. Rep.* **19**, 693–718.

Blevins, D. G., Lukaszewski, K. M. 1998. Boron in plant structure and function. *Ann. Rev. Plant Phys. Plant Mol. Biol.* **49**, 481–500.

Bock, C. W., Katz, A. K., Glusker, J. P. 1995. Hydration of zinc ions: A comparison with magnesium and beryllium ions. *J. Am. Chem. Soc.* **117**, 3754–3765.

Boehr, D. D., McElthey, D., Dyson, H. J., Wright, P. E. 2006. The dynamic energy landscape of dihydrofolate reductase catalysis. *Science* **313**, 1638–1642.

Bollinger, J. M., Jr., Edmonson, D. E., Huyn, B. H., Filley, J., Norton, J. R., Stubbe, J. 1991. Mechanism of assembly of the tyrosyl radical-dinuclear iron cluster cofactor of ribonucleotide reductase. *Science* **253**, 292–298.

Bortolini, O., Conte, V. 2005. Vanadium(V) peroxocomplexes: Structure, chemistry and biological implications. *J. Inorg. Biochem.* **99**, 1549–1557.

Bowen, R., Gunatilaka, A. 1977. *Copper: Its Geology and Economics*. Halstead Press.

Braun, V., Braun, M. 2002. Active transport of iron and siderophore antibiotics. *Curr. Opin. Microbiol.* **5**, 194–201.

Braun, W., Vasak, M., Robbins, A. H., Stout, C. D., Wagner, G., Kagi, J. H., Wüthrich, K. 1992. Comparison of the NMR solution structure and the x-ray crystal structure of rat metallothionein-2. *Proc. Natl. Acad. Sci.* **89**, 10124–10128.

Brini, M. 2003. Ca^{2+} signaling in mitochondria: Mechanism and role in physiology and pathology. *Cell Calcium* **34**, 399–405.

Brooks, R. R. 1998. Geobotany and hyperaccumulators. In: *Plants that Hyperaccumulate Metals: Their Role in Phytoremediation*, Brooks, R. R., ed. Oxford University Press, New York.

Bruckdorfer, R. 2005. The basics about nitric oxide. *Mol. Aspects Med.* **26**, 3–31.

Bugg, T. D. H. 2001. Oxygenases: Mechanisms and structural motifs for O_2-activation. *Curr. Opin. Chem. Biol.* **5**, 550–555.

Burgess, B. K., Lowe, D. J. 1996. Mechanism of molybdenum nitrogenase. *Chem. Rev.* **96**, 2983–3011.

Burk, R. F., Hill, K. E. 1993. Regulation of selenoproteins. *Ann. Rev. Nutr.* **13**, 65–81.

Bush, A. I. 2003. The metallobiology of Alzhemier's disease. *Trends Neurosci.* **26**, 207–214.

Calderone, V., Dolderer, B., Hartmann, H. J., Echner, H., Luchinat, C., Del Bianco, C., Mangani, S., Weser, U. 2005. The crystal structure of yeast copper thionein: The solution of a long lasting enigma. *Proc. Natl. Acad. Sci.* **102**, 51–56.

Cam, M. C., Brownsey, R. G., McNeill, J. H. 2000. Mechanisms of vanadium action: Insulin-mimetic or insulin-enhancing agent? *Can J. Physiol. Pharmacol.* **78**, 829–847.

Caravan, P., Ellison, J. J., McMurry, T. J., Randall, B., Lauffer, R. B. 1999. Gadolinium(III) chelates as MRI contrast agents: structure, dynamics, and applications. *Chem. Rev.* **99**, 2293–2352.

Carpenter, L. J., Malin, G., Liss, P. S., Küpper, F. C. 2000. Novel biogenic iodine-containing trihalomethanes and other short-lived halocarbons in the coastal East Atlantic. *Global Biogeochem. Cycles,* **14**, 1191–1204.

Carter, J. N., Beatty, K. E., Simpson, M. T., Butler, A. 2002. Reactivity of recombinant and mutant vanadium bromoperoxidase from the red alga *Corallina officinalis*. *J. Inorg. Biochem.* **91**, 59–69.

Cervantes, C., Campos-Garcia, J., Devars, S., Gutierrez-Corona, F., Loza-Tavera, H., Torres-Guzman, J. C., Moreno-Sanchez, R. 2001. Interactions of chromium with microorganisms and plants. *FEMS Microbiol. Rev.* **25**, 335–347.

Cha, J. N., Shimizu, K., Zhou, Y., Christiansen, S. C., Chemlka, B. F., Stucky, G. D., Morse, D. E. 1999. Silicatein filaments and subunits from a marine sponge direct the polymerization of silica and silicones in vitro. *Proc. Natl. Acad. Sci. USA* **96**, 361–365.

Chameides, W. L., Perdue, E. M. 1997. *Biogeochemical Cycles—A Computer-Interactive Study of Earth System Science and Global Change.* Oxford University Press.

Changela, A., Chen, K., Xue, Y., Holschen, J., Outten, C. E., O'Halloran, T. V., Mondragon, A. 2003. Molecular basis of selectivity and sensitivity in metal-ion recognition by CueR. *Science* **301**, 1383–1387.

Chapman, E. R. 2002. Synaptotagmin: A Ca^{2+} sensor that triggers exocytosis?. *Nature Rev. Mol. Cell Biol.* **3**, 498–508.

Chen, C-C., Liao, S-L. 2003. Neurotrophic and neurotoxic effects of zinc on neonatal cortical neurons. *Neurochem. Int.* **42**, 471–479.

Chen, J., Stubbe, J-A. 2004. Bleomycins: New methods will allow reinvestigation of old issues. *Curr. Opin. Chem. Biol.* **8**, 175–181.

Cheng, Y., Zak, O., Aisen, P., Harrison, S. C., Walz, T. 2004. Structure of the human transferrin receptor-transferrin complex. *Cell* **116**, 565–576.

Choi, G., Choi, S-C., Galan, A., Wilk, B., Dowd, P. 1990. Vitamin B_{12s}-promoted model rearrangement of methylmalonate to succinate is not a free radical reaction. *Proc. Natl. Acad. Sci.* **87**, 3174–3176.

Cobbett, C., Goldsbrough, P. 2002. Phytochelatins and metallothioneins: Roles in heavy metal detoxification and homeostasis. *Ann. Rev. Plant Biol.* **53**, 159–182.

Colin, C., Leblanc, C., Wagner, E., Delago, L., Leize-Wagner, E., Van Dorsselaer, A., Kloare, B., Potin, P. 2003. The brown algal kelp *Laminaria digitata* features distinct bromoperoxidase and iodoperoxidase. *J. Biol. Chem.* **278**, 23525–23545.

Daly, M. J., Gaidamakova, E. K., Matrosova, V. Y., Vasilenko, A., Zhai, M., Venkateswaran, A., Hess, M., Omelchenko, M. V., Kostandarithes, H. M., Makarova, K. S., Wackett, L. P., Frederickson, J. K., Ghosal, D. 2004. Accumulation of Mn(II) in *Deinococcus radiodurans* facilitates gamma-radiation resistance. *Science* **306**, 1025–1028.

Davis, T. A., Volesky, B., Mucci, A. 2003. A review of the biochemistry of heavy metal biosorption by brown algae. *Water Res.* **37**, 4311–4330.

Decker, A., Solomon, E. I. 2005. Dioxygen activation by copper, heme and non-heme iron enzymes: Comparison of electronic structures and reactivities. *Curr. Opin. Chem. Biol.* **9**, 152–163.

Dellis, O., Dedos, S. G., Tovey, S. C., Taufiq-Ur-Rahman, Dubel, S. J., Taylor, C. W. 2006. Ca^{2+} entry through plasma membrane IP_3 receptors. *Science* **313**, 229–233.

Dey, A., Glaser, T., Couture, M. M.-J., Eltis, L. D., Holm, R. H., Hedman, B., Hodgson, K. O., Solomon, E. I. 2004. Ligand K-edge x-ray absorption spectroscopy of $[Fe_4S_4]^{1+,2+,3+}$ clusters: Changes in bonding and electronic relaxation upon redox. *J. Am. Chem. Soc.* **126**, 8320–8324.

Dietz, R., Nastainczyk, W., Ruf, H. H. 1988. Higher oxidation states of prostaglanding H synthase. Rapid electronic spectroscopy detected two spectral intermediates during the peroxidase reaction with prostaglandin G2. *Eur. J. Biochem.* **171**, 321–328.

Dismukes, G. C. 1996. Manganese enzymes with binuclear active sites. *Chem. Rev.* **96**, 2909–2926.

Dolphin, D., Felton, R. H. 1974. The biochemical significance of porphyrin pi cation radical. *Acc. Chem. Res.* **7**, 26–32.

Dooley, D. M., McGuirl, M. A., Brown, D. E., Turowski, P. N., McIntire, W. S., Knowles, P. F. 1991. A Cu(I)-semiquinone state in substrate-reduced amine oxidase. *Nature* **349**, 262–264.

Doukov, T. I., Iverson, T. M., Seravalli, J., Ragsdale, S. W., Drennan, C. L. 2002. A Ni-Fe-Cu center in a bifunctional carbon monoxide dehydrogenase/acetyl Co-A synthase. *Science* **298**, 567–572.

Drennan, C. L., Heo, J., Sintchak, M. D., Schreiter, E., Ludden, P. W. 2001. Life on carbon monoxide: X-ray structure of *Rhodospirillum rubrum* Ni-Fe-S carbon monoxide dehydrogenase. *Proc. Natl. Acad. Sci.* **98**, 11973–11978.

Durrant, M. C. 2002. An atomic-level mechanism for molybdenum nitrogenase. Part I. Reduction of dinitrogen. *Biochemistry* **41**, 13934–13945.

Eady, R. R. 1996. Structure-function relationships of alternative nitrogenases. *Chem. Rev.* **96**, 3013–3030.

Eaton, J. W., Qian, M-W. 2002. Molecular bases of cellular iron toxicity. *Free Rad. Biol. Med.* **32**, 833–840.

Eichler, J. 1976. In: *Handbook of Stratabound and Stratiform Ore Deposits*, Vol. 7, 157–201, Wolfe, K. H., ed. Elsevier.

Einsle, O., Tezcan, A., Andrade, S. L. A., Schmidt, B., Yoshida, M., Howard, J. B., Rees, D. C. 2002. Nitrogenase MoFe-protein at 1.16 Å resolution: A central ligand in the FeMo cofactor. *Science* **297**, 1969–1700.

Eisenmesser, E. Z., Bosco, D. A., Akke, M., Kern, D. 2002. Enzyme dynamics during catalysis. *Science* **295**, 1520–1523.

Ekdahl, A., Pedersen, M., Abrahamsson, K. 1998. A study of the diurnal variation of biogenic volatile halocarbons. *Mar. Chem.* **63**, 1–8.

Ermler, U. 2005. On the mechanism of methyl-coenzyme M reductase. *Dalton Trans.* **2005**, 3451–3458.

Ermler, U., Grabarse, W., Shima, S., Goubeaud, Thauer, R. K. 1998. Active sites of transition-metal enzymes with a focus on nickel. *Curr. Opin. Struct. Biol.* **8**, 749–758.

Erskine, P. T., Duke, E. M., Tickle, I. J., Senior, N. M., Warren, M. J., Cooper, J. B. 2000. MAD analyses of yeast 5-aminolaevulinate dehydratase: Their use in structure determination and in defining the metal-binding sites. *Acta Crystallogr. Sect. D* **56**, 421–430.

Erskine, P. T., Newbold, R., Brindley, A. A., Wood, S. P., Shoolingin-Jordan, P. M., Warren, M. J., Cooper, J. B. 2001. The x-ray structure of yeast 5-aminolaevulinic acid dehydratase complexed with substrate and three inhibitors. *J. Mol. Biol.* **312**, 133–141.

Eshaghi, S., Niegowski, D., Kohl, A., Molina, D. M., Lesley, S. A., Nordlund, P. 2006. Crystal structure of a divalent metal ion transporter Cora at 2.9 angstrom resolution. *Science* **313**, 354–357.

Evenäs, J., Malmendal, A., Forsen, S. 1998. Calcium. *Curr. Opin. Chem. Biol.* **2**, 293–302.

Exley, C. 1998. Silicon in life: A bioinorganic solution to bioorganic essentiality. *J. Inorg. Biochem.* **69**, 139–144.

Ferguson, S. J. 1998. Nitrogen cycle enzymology. *Curr. Opin. Chem. Biol.* **2**, 182–193.

Ferreira, K. N., Iverson, T. M., Maghlaoui, K., Barber, J., Iwata, S. 2004. Architecture of the photosynthetic oxygen-evolving center. *Science* **303**, 1831–1838.

Fiedler, T. J., Davey, C. A., Fenna, R. E. 2000. X-ray crystal structure and characterization of halide-binding sites of human myeloperoxidase at 1.8 Å resolution. *J. Biol. Chem.* **275**, 11964–11971.

Finke, R. G., Schiraldi, D. A., Mayer, B. J. 1984. Toward the unification of coenzyme B_{12}-dependent dio dehydratase stereochemical and model studies: The bound radical mechanism. *Coordin. Chem. Rev.* **54**, 1–22.

Finney, L. A., O'Halloran, T. V. 2003. Transition metal speciation in the cell: Insights from the chemistry of metal ion receptors. *Science* **300**, 931–936.

Flint, D. H., Allen, R. M. 1996. Iron-sulfur proteins with nonredox functions. *Chem. Rev.* **96**, 2315–2334.

Franceschi, V. R., Nakata, P. A. 2005. Calcium oxalate in plants: Formation and function. *Ann. Rev. Plant Biol.* **56**, 41–71.

Frausto da Silva, J. J. R., Williams, R. J. P. 2001. *The Biological Chemistry of the Elements*, 2nd ed. Oxford University Press, Oxford.

Frazzon, J., Dean, D. R. 2003. Formation of iron-sulfur clusters in bacteria: An emerging field in bioinorganic chemistry. *Curr. Opin. Chem. Biol.* **7**, 163–177.

Frey, P. A. 1990. Importance of organic radicals in enzymic cleavage of unactivated carbon-hydrogen bonds. *Chem. Rev.* **90**, 1343–1357.

Frey, P. A. 2001. Radical mechanisms of enzymatic catalysis. *Ann. Rev. Biochem.* **70**, 121–148.

Fu, L-H., Wang, X-F., Eyal, Y., She, Y-M., Donald, L. J., Standing, K. G., Ben-Hayyim, G. 2002. A selenoprotein in the plant kingdom. *J. Biol. Chem.* **277**, 25983–25991.

Furtmüller, P. G., Zederbauer, M., Jantschko, W., Helm, J., Bogner, M., Jakopitsch, C., Obinger, C. 2006. Active site structure and catalytic mechanisms of human peroxidase. *Arch. Biochem. Biophys.* **445**, 199–213.

Gaetke, L. M., Chow, C-K. 2003. Copper toxicity, oxidative stress, and antioxidant nutrients. *Toxicology* **189**, 147–163.

Gamer, C. D., Armstrong, E. M., Berry, R. E., Beddoes, R. L., Collison, D., Cooney, J. J. A., Ertok, S. N., Helliwell, M. 2000. Investigations of amavadin. *J. Inorg. Biochem.* **80**, 17–20.

Ganther, H. E., Oh, S. H., Schaich, E., Hoekstra, W. G. 1974. Studies on selenium in glutathione peroxidase. *Fed. Proc.* **33**, 694.

Ganz, T. 2003. Hepcidin, a key regulator of iron metabolism and mediator of anemia inflammation. *Blood* **102**, 783–788.

Garcia-Viloca, M., Gao, J., Karplus, M., Truhlar, D. G. 2004. How enzymes work: Analysis by modern rate theory and computer simulations. *Science* **303**, 186–195.

Garcin, E., Vernede, X., Hatchikian, E. C., Volbeda, A., Frey, M., Fontecilla-Camps, J. C. 1999. The crystal structure of a reduced [NiFeSe] hydrogenase provides an image of the activated catalytic center. *Structure Fold. Des.* **7**, 557–566.

Gaudemer, A., Zybler, J., Zybler, N., BaranMarszac, M., Hull, W. E., Fountoulakis, M., Koenig, A., Woelf, K., Retry, J. 1981. Reversible cleavage of the cobalt-carbon bond to coenzyme B12 catalyzed by methylmalonyl-CoA mutase from *Propionibacterium shermanii*. The use of coenzyme B12 stereospecifically deuterated on position 5′. *Eur. J. Biochem.* **119**, 279–285.

Gelasco, A., Lippard, S. J. 1998. NMR solution structure of a DNA dodecamer duplex containing a cis-diammineplatinum (II) d(GpG) intrastrand cross-link, the major adduct of the anticancer drug cisplatin. *Biochemistry* **37**, 9230–9239.

Gill, D. L., Spassova, M. A., Soboloff, J. 2006. Calcium entry signals—Trickles and torrents. *Science* **313**, 183–184.

Godwin, H. A. 2001. The biological chemistry of lead. *Curr. Opin. Chem. Biol.* **5**, 223–227.

Godwin, H. A. 2005. Reexamination of lead(II) coordination preferences in sulfur-rich sites: Implications for a critical mechanism of lead poisoning. *J. Am. Chem. Soc.* **127**, 9495–9505.

Gong, W., Hao, B., Mansy, S. S., Gonzalez, G., Gilles-Gonzalez, M. A., Chan, M. K. 1998. Structure of a biological oxygen sensor: A new mechanism for heme-driven signal transduction. *Proc. Natl. Acad. Sci. USA* **95**, 15177–15182.

Gonzalez, J. M., Kiene, R. P., Moran, M. A. 1999. Transformation of sulfur compounds by an abundant lineage of marine bacteria in the α-subclass of the class Proteobacteria. *Appl. Envrion. Microbiol.* **65**, 3810–3819.

Gouaux, E., MacKinnon, R. 2005. Principles of selective ion transport in channels and pumps. *Science* **310**, 1461–1465.

Goyer, R. A., Leonard, D. L., Moore, J. F., Rhyne, B., Krigman, M. R. 1970. Lead dosage and the role of the intranuclear includion body. *Arch. Environ. Health*, **20**, 705–711.

Grabarse, W., Mahlert, F., Duin, E. C., Goubeaud, M., Shima, S., Thauer, R. K., Lamzin, V., Ermler, U. 2001. On the mechanism of biological methane formation: Structural evidence for conformational changes in methylcoenzyme M reductase upon substrate binding. *J. Mol. Biol.* **309**, 315–330.

Greenwood, N. N., Earnshaw, A. 1997. *Chemistry of the Elements*, 2nd ed. Elsevier, 983–994

Goves, J. T. 2005. Models and mechanisms of cytochrome P450 action. In: *Cytochrome P450: Structure, Mechanism, and Biochemistry*, 3rd ed. 1–44, Ortiz de Montellano, P. R., ed. Springer Verlag.

Guallar, V., Gherman, B. F., Lippard, S. J., Friesner, R. A. 2002. Quantum chemical studies of methane monooxygenase: Comparison with P450. *Curr. Opin. Chem. Biol.* **6**, 236–242.

Haddad, J. J. 2004. Oxygen sensing and oxidant/redox-related pathways. *Biochim. Biophys. Res. Commun.* **316**, 969–977.

Hagino, K., Okada, H. 2006. Intra- and infra-specific morphological variation in selected coccolithophore species in the equatorial and subequatorial Pacific ocean. *Mar. Micropaleont.* **58**, 184–206.

Hakansson, K. O. 2003. The crystallographic structure of Na, K-ATPase N-domain at 2.6 Å resolution. *J. Mol. Biol.* **332**, 1175–1182.

Hamberg, M., Samuelsson, B. 1967. On the mechanims of the biosynthesis of prostaglandins E_1 and $F_1\alpha$. *J. Biol. Chem.* **242**, 5336–5343.

Hartmanis, M. G. N., Stadtman, T. C. 1987. Solubilization of a membrane-bound diol dehydratase with retention of epr g = 2.02 signal by using 2-(N-cyclohexyl-amino)ethane sulfonic acid buffer. *Proc. Natl. Acad. Sci.* **74**, 76–79.

Haumann, M., Liebisch, P., Muller, C., Barra, M., Grabolle, M., Dau, H. 2005. Photosynthetic O_2 formation tracked by time-resolved X-ray experiments. *Science* **310**, 1019–1021.

He, C., Mishina, Y. 2004. Modeling non-heme iron proteins. *Curr. Opin. Chem. Biol.* **8**, 201–208.

Hecky, R. D., Mopper, K., Kilham, P., Degens, E. T. 1973. The amino acid and sugar composition of diatom cell walls. *Mar. Biol.* **19**, 323–331.

Hille, R. 1996. The mononuclear molybenum enzymes. *Chem. Rev.* **96**, 2757–2816.

Hirano, S., Kobayashi, Y., Cui, X., Kanno, S., Hayakawa, T., Shraim, A. 2004. The accumulation and toxicity of methylated arsenicals in endothelial cells: Important roles of thiol compounds. *Toxicol. Appl. Pharm.* **198**, 458–467.

Hoch, M. 2001. Organotin compounds in the environment—An overview. *Appl. Geochem.* **16**, 719–743.

Hoehn, S. T., Junker, H. D., Bunt, R. C., Turner, C. J., Stubbe, J. 2001. Solution structure of Co(III)-bleomycin-OOH bound to a phosphoglycolate lesion containing oligonucleotide: Implications for bleomycin-induced double-strand DNA cleavage. *Biochemistry* **40**, 5894–5905.

Holland, H. D. 1984. *The Chemical Evolution of the Atmosphere and Oceans.* Princeton University Press.

Holloway, M., White, H. A., Joblin, K. N., Johnson, A. W., Lappert, W. F., Wallis, O. C. 1978. A spectrophotometric rapid kinetic study of reactions catalysed by coenzyme-B_{12}-dependent ethanolamine ammonia-lyase. *Eur. J. Biochem.* **82**, 143–154.

Holmgren, A. 2000. Antioxidant function of thioredoxin and glutaredoxin systems. *Antioxid. Redox Signaling* **2**, 811–882.

Howard, E. C., Henriksen, J. R., Alison, B., Reisch, C. R., Bürgmann, H., Welsh, R., Ye, W., Gonzalez, J. M., Mace, K., Joyce, S. B., Kiene, R. P., Whitman, W. B., Moran, M. A. 2006. Bacterial taxa that limit sulfur flux from the ocean. *Science* **314**, 649–652.

Howard, J. B., Rees, D. C. 1996. Structural basis of biological nitrogen fixation. *Chem. Rev.* **96**, 2965–2982.

Huang, H., Zhu, L., Reid, R. B., Drobny, G. P., Hopkins, P. B. 1995. Solution structure of a cisplatin-induced DNA interstrand cross-link. *Science* **270**, 1842–1845.

Hubregtse, T., Neeleman, E., Maschmeyer, T., Sheldon, R. A., Anefeld, U., Arends, I. W. C. E. 2005. The first enantioselective synthesis of the amavadin ligand and its complexation to vanadium. *J. Inorg. Biochem.* **99**, 1264–1267.

Igarashi, R. Y., Seefeldt, L. C. 2003. Nitrogen fixation: The mechanism of the Mo-dependent nitrogenase. *Crit. Rev. Biochem. Mol. Biol.* **38**, 351–384.

Isaacs, Neil, S. 1987. *Physical Organic Chemistry.* Longman Scientific and Technical. Chapter 15.

Ito, N., Phillips, S. E. V., Stevens, C., Orgel, Z. B., McPherson, M. J., Keen, J. N., Yadav, K. O. S., Knowles, P. F. 1991. Novel thioether bond revealed at a 1.7 Å crystal structure of galactose oxidase. *Nature* **350**, 87–90.

Iwata, S., Barber, J. 2004. Structure of photosystem II and molecular architecture of the oxygen-evolving centre. *Curr. Opin. Struct. Biol.* **14**, 453–457.

Jacquamet, J., Sun, Y., Hatfield, J., Gu, W., Cramer, S. P., Crowder, M. W., Lorigan, G. A., Vincent, J. B., Latour, J-M. 2003. Characterization of chromodulin by x-ray absorption and electron paramagnetic resonance spectroscopies and magnetic susceptibility measurements. *J. Am. Chem. Soc.* **125**, 774–780.

Jaffe, E. K., Martins, J., Li, J., Kervinen, J., Dunbrack, R. L., Jr. 2001. The molecular mechanism of lead inhibition of human porphobilinogen synthase. *J. Biol. Chem.* **276**, 1531–1537.

Janes, S. M., Mu, D., Wemmer, D., Smith, A. K., Kaut, S., Maltby, D., Burlingame, A. L., Klinman, J. P. 1990. A new redox cofactor in eukaryotic enzymes: 6-hydroxydopa at the active site of bovine serum amine oxidase. *Science* **248**, 981–987.

Jencks, W. P. 1975. Binding energy, specificity and enzymatic catalysis: The Circe effect. *Adv. Enzymol.* **43**, 219–410.

Jeoung, J-H. and Dobbek, H., 2007. Carbon dioxide activation at the Ni-Fe-cluster of anaerobic carbon monoxide dehydrogenase. *Science* **318**, 1461–1464.

Jian, W., Yun, D., Saleh, L., Barr, E. W., Xing, G., Hoffart, L. M., Maklak, M-A., Krebs, C., Bollinger, J. M., Jr. 2007. A manganese(IV)/iron(III) cofactor in Chlamydia trachomatis ribonucleotide reductase. *Science* **316**, 1188–1191.

Johnson, M. K., Rees, D. C., Adams, M. W. W. 1996. Tungstoenzymes. *Chem. Rev.* **96**, 2817–2839.

Karthein, R., Dietx, R., Nastainczyk, W., Ruff, H. H. 1988. Higher oxidation states of prostaglandin H synthase. EPR study of a transient tyrosyl radical in the enzyme during the peroxidase reaction. *Eur. J. Biochem.* **171**, 313–320.

Kasprzak, K. S., Sunderman, F. W., Jr., Salnikow, K. 2003. Nickel carcinogenesis. *Mutation Res.* **533**, 67–97.

Kaufman, A. J., Johnston, D. T., Farquar, J., Masterson, A. L., Lyons, T. W., Bates, S., Anbar, A. D., Arnold, G. L., Garvin, J., Buick, R. 2007. Late archaean biospheric oxygenation and atmospheric evolution. *Science* **317**, 1900–1903.

Kim, H. J., Graham, D. W., DiSpirito, A. A., Alterman, M. A., Galeva, N., Larive, C. K., Asunskis, D., Sherwood, P. M. A. 2004. Methanobactin, a copper-acquisition compound from methane-oxidizing bacteria. *Science* **305**, 1612–1615.

Kisker, C., Schindelin, H., Rees, D. C. 1997. Molybenum-cofactor-containing enzymes: Structure and mechanism. *Ann. Rev. Biochem.* **66**, 233–267.

Kitajima, N., Fujisawa, K., Moro-oka, Y., Toriumi, K. 1989. μ-Peroxobinuclear copper complex, $[Cu(HB(3,5-(Me_2CH)_2pz)_3)_2](O_2)$. *J. Am. Chem. Soc.* **111**, 8975–8976.

Klein Gebbink, R. J. M., Martens, C. F., Kenis, P. J. A., Jansen, R. J., Nolting, H-F., Solé, V. A., Feiters, M. C., Karlin, K. D., Nolte, R. J. M. 1999. Synthesis, structure and reactivity of copper dioxygen complexes derived from molecular receptor ligands. *Inorg. Chem.* **38**, 5755–5768.

Knowles, P. F., Dooley, D. M. 1994. Amine oxidase. In: *Metal Ions in Biological Systems*, Vol. **30**, 361–403, Sigel, H., Sigel, A., eds. Marcel Dekker, New York.

Knutson, M., Wessling-Resnick, M. 2003. Iron metabolism in the reticuloendothelial system. *Curr. Rev. Biochem. Mol. Biol.* **38**, 61–88.

Komeili, A., Li, Z., Newman, D. K., Jensen, G. J. 2006. Magnetosomes are cell membrane invaginations organized by actin-like protein MamK. *Science* **311**, 242–245.

Kosman, D. J. 2003. Molecular mechanisms of iron uptake in fungi. *Mol. Microbiol.* **47**, 1185–1197.

Kovaleva, E. G., Lipscomb, J. D. 2007. Crystal structures of Fe^{2+} dioxygenase superoxo, alkylperoxo, and bound product intermediates. *Science* **316**, 453–457.

Krämer, U. 2005. Phytoremediation: Novel approaches to cleaning up polluted soils. *Curr. Opin. Biotechnol.* **16**, 133–141.

Kröger, N., Deutzmann, R., Sumper, M. 2001. Silica-precipitating peptides from diatoms, the chemical structure of silaffin-1A from *Cylindrotheca fusifomis*. *J. Biol. Chem.* **276**, 26066–26070.

Kühnel, K., Blankenfeldt, W., Terner, J., Schlichting, I. 2006. Crystal structures of chloroperoxidase with its bound substrates and complexed with formate, acetate, and nitrate. *J. Biol. Chem.* **281**, 239–290.

Kulmacz, R. J. 2005. Regulation of cyclooxygenase catalysis by hydroperoxides. *Biochem. Biophys. Res. Comm.* **338**, 25–33.

Kuroda, R., Neidel, S., Riordan, J. M., Sakai, T. T. 1982. X-ray crystallographic analysis of 3-(2'-phdnyel-2,4'-bithiazole-4-carboxamide) propyldimethyl-sulphonium iodide, an analogue of the DNA-binding portion of bleomycinn A_2. *Nucl. Acid Res.* **15**, 4753–4763.

Lahiri, S., Roy, A., Baby, S. M., Hoshi, T., Semenza, G. L., Prabhakar, N. R. 2006. Oxygen sensing in the body. *Prog. Biophys. Mol. Biol.* **91**, 249–286.

Lakshiminarayanan, R., Valiyaveettil, S., Raol, V., Kini, R. M. 2003. Purification, characterization, and *in vitro* mineralization studies of a novel goose eggshell protein, ansocalcin. *J. Biol. Chem.* **278**, 2928–2946.

Lane, T. W. W., Morel, F. M. M. 2000. A biological function for cadmium in marine diatoms. *Proc. Natl. Acad. Sci. USA* **97**, 4627–4631.

Lane, T. W., Saito, M. A., George, G. N., Pickering, I. J., Prince, R. C., Morell, F. M. M. 2005. Biochemistry: A cadmium enzyme from a marine diatom. *Nature* **435**, 42.

Lange, S. J., Que, L., Jr. 1998. Oxygen activating nonheme iron enzymes. *Curr. Opin. Chem. Biol.* **2**, 159–172.

Lanzilotta, W. N., Schuller, D. J., Thorsteinsson, M. V., Kerby, R. L., Roberts, G. P., Poulos, T. L. 2000. Structure of the CO sensing transcription activator CooA. *Nat. Struct. Biol.* **7**, 876–880.

Larsen, T. M., Benning, M. M., Raymen, I., Reed, G. H. 1998. Bis Mg-ATP-K-oxalate complex of pyruvate kinase. *Biochemistry* **37**, 6247–6255.

Lassmann, G., Odenwaller, R., Curtis, J. F., DeGray, J. A., Mason, R. P., Marnett, L. J., Eling, T. E. 1991. Electron spin resonance investigation of tyrosyl radicals of prostaglandin H synthase. Relation to enzyme catalysis. *J. Biol. Chem.* **266**, 20045–20055.

Li, Y., Dutta, S., Doublie, S., Bdour, H. M., Taylor, J. S., Ellenberger, T. 2004. Nucleotide insertion opposite a cis–syn thymine dimer by a replicative DNA polymerase from bacteriophage T7. *Nat. Struct. Mol. Biol.* **11**, 784–790.

Liang, Z-X., Klinman, J. P. 2004. Structural bases of hydrogen tunneling in enzymes: Progress and puzzles. *Curr. Opin. Struct. Biol.* **14**, 648–655.

Lieutaud, C., Nitschke, W., Vermeglio, A., Parot, P., Schoepp-Cothenet, B. 2003. HiPIP in *Rubrivivax gelatinosus* is firmly associated to the membrane in a conformation efficient for electron transfer towards the photosynthetic reaction centre. *Biochim. Biophys. Acta*, **1557**, 83–90.

Lipscomb, W. M., Sträter, N. 1996. Recent advance in zinc enzymology. *Chem. Rev.* **96**, 2375–2433.

Liu, S., Edwards, D. S. 1999. ^{99m}Tc-labeled small peptides as diagnostic radiopharmaceuticals. *Chem. Rev.* **99**, 2235–2268.

Liuzzi, J. P., Cousins, R. J. 2004. Mammalian zinc transporters. *Ann. Rev. Nutr.* **24**, 151–172.

Lowry, T. H., Richardson, K. S. 1981. *Mechanism and Theory in Organic Chemistry*, 2nd ed. Chapter 9. Harper & Row.

Lu, M., Fu, D. 2007. Structure of the zinc transporter YiiP. *Science* **317**, 1746–1748.

Magyar, J. S., Wenf, E.-C., Stern, C. M., Dye, D. F., Rous, B. W., Payne, J. C., Bridgewater, B. M., Mijovilovich, A., Parkin, G., Zaleski, J. M., Penner-Hahn, J. E., Malik, A. 2004. Metal bioremediation through growing cells. *Environ. Int.* **30**, 261–278.

Malin, G. 2006. New pieces for the marine sulfur cycle zigsaw. *Science* **314**, 607–608.

Malkowski, M. G., Ginell, S. L., Smith, W. L., Garavito, R. M. 2000. The productive conformation of arachidonic acid bound to prostaglandin synthase. *Science* **289**, 1933–1937.

Mancia, F., Keep, N. H., Nakagawa, A., Leadlay, P. F., McSweeney, S., Rasmussen, B., Bösecke, P., Diat, O., Evans, P. R. 1996. How coenzyme B_{12} radicals are generated: The crystal structure of methylmalonyl-coenzyme A mutase at 2 Å resolution. *Structure* **4**, 339–350.

Mann, G. J., Gräslund, A., Ochiai, E-I., Ingemarson, R., Thelander, L. 1991. Purification and characterization of recombinant mouse and herpes simplex virus ribonucleotide reductase R2 subunit. *Biochemistry* **30**, 1939–1947.

March, J. 1985. *Advanced Organic Chemistry*, 3rd ed., Chapters 5 and 14. John Wiley & Sons.

Marcus, R. A. 1964. Chemical and electrochemical electron transfer theory. *Ann. Rev. Phys. Chem.* **15**, 155–196.

Marin, F., Luquet, G. 2005. Molluscan biomineralization: The proteinaceous shell constituents of *Pinna nobilis*. *Mater. Sci. Eng.* **C25**, 105–111.

Marnett, L. J. 2000. Cyclooxygenase mechanisms. *Curr. Opin. Chem. Biol.* **4**, 545–552.

Maroney, M. J. 1999. Structure/function relationships in nickel metallobiochemistry. *Curr. Opin. Chem. Biol.* **3**, 188–199.

Marsh, E. N., Drennan, C. L. 2001. Adenosylcobalamin-dependent isomerases: New insights into structure and mechanism. *Curr. Opin. Chem. Biol.* **5**, 499–501.

Marsh, M. E. 2003. Regulation of $CaCO_3$ formaton in coccollithophores. *Comp. Biochem. Physiol.* **Part B 136**, 734–754.

Martens, S., Kozlov, M. M., McMahon, H. T. 2007. How synaptotagmin promotes membrane fusion. *Science* **316**, 1205–1208.

Mayer, S. M., Lawson, D. M., Gormal, C. A., Roe, S. M., Smith, B. E. 1999. New insights into structure-function relationships in nitrogenase: A 1.6Å resolution X-ray crystallographic study of *Klebsiella pneumoniae* MoFe-protein. *J. Mol. Biol.* **292**, 871–891.

McDonough, M. A., Li, V., Flashman, E., Chowdhury, R., Mohr, C., Lienard, B. M., Zondlo, J., Oldham, N. J., Clifton, I. J., Lewis, J., McNeill, L. A., Kurzeja, R. J., Hewitson, K. S., Yang, E., Jordan, S., Syed, R. S., Schofield, C. J. 2006. Cellular oxygen sensing: Crystal structure of hypoxia-inducible factor prolyl hydroxylase (PHD2). *Proc. Natl. Acad. Sci. USA* **103**, 9814–9819.

McGrath, S. P., Zhao, F-J. 2003. Phytoextraction of metals and metalloids from contaminated soils. *Curr. Opin. Biotechnol.* **14**, 277–282.

McIntire, W. S., Wemmer, D. E., Chistoserdov, A., Lidstrom, M. E. 1991. A new cofactor in a prokaryotic enzyme: Tryptophylquinone as the redox prosthetic group in methylamine dehydrogenase. *Science* **252**, 817–823.

McKie, A. T., Latunde-Dada, G. O., Miret, S., McGregor, J. A., Anderson, G. J., Vulpe, C. D., Wrigglesworth, J. M., Simpson, R. J. 2002. Molecular evidence for the role of a ferric reductase in iron transport. *Biochem. Soc. Trans.* **30**, 722–724.

McMaster, J., Enemark, J. H. 1998. The active sites of molybdenum- and tungsten-containing enzymes. *Curr. Opin. Chem. Biol.* **2**, 201–207.

McMurry, T. J., Groves, J. T. 1986. Models and mechanisms of cytochrome P450 action. In: *Cytochrome P-450*; Chapter 1, Ortiz de Montellano, P. R., ed. Plenum Press, New York.

Mennier, B., de Visser, S. P., Shaik, S. 2004. Mechanism of oxidation reactions catalyzed by cytochrome P450 enzymes. *Chem. Rev.* **104**, 3947–3980.

Mesecar, D., Stoddart, B. L., Koshland, D. E., Jr. 1997. Orbital steering in the catalytic power of enzymes: Small structural changes with large catalytic consequences. *Science* **277**, 202–226.

Messerschmidt, A., Prade, L., Wever, R. 1997. Implications for the catalytic mechanism of the vanadium-containing enzyme chloroperoxidase from the fungus *Curvularia inaequalis* by X-ray structures of the native and peroxide form. *J. Biol. Chem.* **378**, 309–315.

Miller, A-F. 2004. Superoxide dismutases: Active sites that save, but a protein that kills. *Curr. Opin. Chem. Biol.* **8**, 162–168.

Mirica, L. M., Vance, M., Rudd, D. J., Hedman, B., Hodgson, K. O., Solomon, E. I., Stack, T. D. P. 2005. Tyrosinase reactivity in a model complex: An alternative hydroxylation mechanism. *Science* **308**, 1890–1892.

Miura, T. 2000. Metal binding modes of Alzheimer's amyloid β-peptide in insoluble aggregates and soluble complexes. *Biochemistry* **39**, 7024–7031.

Miyamoto, T., Ogino, N., Yamamoto, S., Hayaishi, O. 1976. Purification of prostaglandin endoperoxide synthetase from bovin vesicular gland microsomes. *J. Biol. Chem.* **251**, 2629–2636.

Miyamoto, T., Yamamoto, S., Hayaishi, O. 1974. Prostaglandin synthetase system-resolution into oxygenase and isomerase components. *Proc. Natl. Acad. Sci.* **71**, 3645–3648.

Møller, J. V., Nissen, P., Sørensen, T. L-M., le Maire, M. 2005. Transport mechanism of the sarcoplasmic reticulum Ca^{2+}-ATPase pump. *Curr. Opin. Struct. Biol.* **15**, 387–393.

Mora, C. V., Davidson, M., Wild, J. M., Walker, M. M. 2004. Magnetoreception and its trigeminal mediation in the homing pigeon. *Nature* **432**, 508–511.

Moura, I., Moura, J. J. G. 2001. Structural aspects of denitrifying enzymes. *Curr. Opin. Chem Biol.* **5**, 168–175.

Müller, W. E. G., Krasko, A., Le Pennec, G., Schröder, H. C. 2003. Biochemistry and cell biology of silica formation in sponges. In: *Biology of Silica Deposition in Sponges*, Uriz, Maria-J., ed. Wiley-Liss.

Munro, A. W., Taylor, P., Walkinshaw, M. D. 2000. Structures of redox enzymes. *Curr. Opin. Biotechnol.* **11**, 369–376.

Neta, P. 1976. Application of radiation techniques to the study of organic radicals. *Adv. Phys. Org. Chem.* **12**, 223–297.

Newcomb, M., Le Tadic-Bladatti, M. H., Chestney, D. L., Roberts, E. S., Hollenberg, P. F. 1995. A non-synchronous concerted mechanism for cytochrome P450 catalyzed hydroxylation. *J. Am. Chem. Soc.* **117**, 12085–12091.

Nicolet, Y., Piras, C., Legrand, P., Hatchikian, C. E., Fontecilla-Camps, J. C. 1999. *Desulfovibrio desulfuricans* iron hydrogenase: The structure shows unusual coordination to an active site Fe binuclear center. *Structure Fold. Des.* **7**, 13–23.

Nissen, P., Hansen, J., Ban, N., Moore, P. B., Steitz, T. A. 2000. The structural basis of ribosome activity in peptide bond synthesis. *Science* **289**, 920–930.

Noodleman, L., Peng, C. Y., Case, D. A., Mousesca, J-M. 1995. Orbital interactions, electron delocalization and spin coupling in iron-sulfur clusters. *Coord. Chem. Rev.* **144**, 199–244.

Nordlund, P., Sjöberg, B-M., Eklund, H. 1990. Three-dimensional structure of the free radical protein of ribonucleotide reductase. *Nature* **345**, 593–598.

Nouailler, M., Bruscella, P., Lojou, E., Lebrun, R., Bonnefoy, V., Guerlesquin, F. 2006. Structural analysis of the HiPIP from the acidophilic bacteria: *Acidithiobacillus ferrooxidans*. *J. Extremophiles* **10**, 191–198.

Ochiai, E-I. 2004a. Biogeochemical cycling of macronutrients. In: *Encyclopedia of Life Support System* (UNESCO).

Ochiai, E-I. 2004b. Biogeochemical cycling of micronutrients and other elements. In: *Encyclopedia of Life Support System* (UNESCO).

Ochiai, E-I. 1997. Global metabolism of elements. *J. Chem. Ed.* **74**, 926–930.

Ochiai, E-I. 1995a. Toxicity of heavy metals and biological defense. *J. Chem. Ed.* **72**, 479–484.

Ochiai, E-I. 1995b. Prebiotic metal ion-ligand interactions and the origin of life. In: *Handbook of Metal-Ligand Interactions in Biological Fluids, Bioinorganic Medicine*, Vol. **1**, 1–9, Berthon, G., ed. Marcel Dekker, Inc., New York.

Ochiai, E-I. 1994a. Free radical and metalloenzymes: General considerations. In: *Metal Ions in Biological Systems*, Vol. **30**, 1–24, Sigel, H., Sigel, A., eds. Marcel Dekker, New York.

Ochiai, E-I. 1994b. Adenosylcobalamin (vitamin B_{12} coenzyme)-dependent enzymes. In: *Metal Ions in Biological Systems*, Vol. **30**, 255–278, Sigel, H., Sigel, A., eds. Marcel Dekker, New York.

Ochiai, E-I. 1992. Chemical logic of life and biosphere on the earth. *J. Chem. Ed.* **69**, 356–357.

Ochiai, E-I. 1991. Why calcium? Principles and applications in bioinorganic chemistry-IV. *J. Chem. Ed.* **68**, 10–12.

Ochiai, E-I. 1987. *General Principles of Biochemistry of the Elements*. Plenum Press, New York, pp. 227–234.

Ochiai, E-I. 1983. Inorganic chemistry of earliest sediments: Bioinorganic chemical aspects of the origin and evolution of life. In: *Cosmochemistry and the Origin of Life*, 235–276. Ponnanperuma, C., ed. D. Reidel Publishing Co.

Ochiai, E-I. 1978a. The evolution of the environment and its influence on the evolution of life. *Origins of Life* **9**, 81–91.

Ochiai, E-I. 1978b. Principles in the selection of inorganic elements by organisms. *Biosystems* **10**, 329–337.

Ochiai, E-I. 1977. *Bioinorganic Chemistry—An Introduction*, Chapter 11. Allyn and Bacon, Boston.

Ochiai, E-I. 1974. Environmental bioinorganic chemistry. *J. Chem. Ed.* **51**, 235–238.

Ochiai, E-I. 1973. Oxygenation of cobalt(II) complexes. *J. Inorg. Nucl. Chem.* **35**, 3375–3389.

Ochiai, E-I., Loope, C. E. 1989. 4th Int. Conf. Bioinorg. Chem. (Boston). *J. Inorg. Chem.* **36**, 194.

Ochiai, E-I., Mann, G. J., Gräslund, A., Thelander, L. 1990. Tyrosyl free radical formation in the small subunit of mouse ribonucleotide reductase. *J. Biol. Chem.* **265**, 15758–15761.

Ohmoto, H., Felder, R. P. 1987. Bacterial activity in the water, sulphate-bearing, Archaean oceans. *Nature* **328**, 244–247.

Ohmoto, H., Watanabe, Y., Ikemi, H., Poulson, S. R. Taylor, B. E. 2006. Sulfur isotope evidence for an oxic Archaean atmosphere. *Nature* **442**, 908–911.

Ohshiro, T., Littlechild, J., Garcia-Rodriguez, E. N., Isupov, M. N., Iida, Y., Kobayashi, K., Izumi, Y. 2004. Modification of halogen specificity of a vanadium-dependent bromoperoxidase. *Protein Sci.* **13**, 1566–1571.

Olesen, C., Sorensen, T. L. S., Nielsen, R. C., Moller, J. V., Nissen, P. 2004. Dephosphorylation of the calcium pump coupled to counterion occlusion. *Science* **306**, 2251–2255.

O'Neill, M., Eberhard, S., Darvill, A. 2006. Plant cell walls, http://cell.ccrc.uga.edu/~mao/cellwall/main.htm.

Ortiz de Montellano, P. R., Stearns, R. A. 1987. Timing of the radical recombination step in cytochrome P-450 catalysis with ring-strained probes. *J. Am. Chem. Soc.* **109**, 3415–3420.

Page, C. C., Moser, C. C., Dutton, P. L. 2003. Mechanisms for electron transfer within and between proteins. *Curr. Opin. Chem. Biol.* **7**, 551–556.

Palmiter, R. D. 2004. Protection against zinc toxicity by metallothrionein and zinc transporter 1. *Proc. Natl. Acad. Sci. USA* **101**, 4918–4923.

Papagerakis, P., Berdal, A., Mesbah, M., Peuchmaur, M., Malaval, L., Nydegger, J., Simmer, J., MacDougall, M. 2002. Investigation of osteocalcin, osteonectin, and dentin dailophosphoprotein in developing human teeth. *Bone* **30**, 377–385.

Papanikolaou, G., Pantopoulos, K. 2005. Iron metabolism and toxicity. *Toxicol. Appl. Pharm.* **202**, 199–211.

Parales-Vela, H. V., Pena-Castro, J. M., Canizares-Villanueva, R. O. 2006. Heavy metal detoxification in eukaryotic microalgae. *Chemosphere* **64**, 1–10.

Patrick, L. 2004. Selenium biochemistry and cancer: A review of literature. *Alternat. Med. Rev.* **9**, 239–258.

Perry, C. C., Keeling-Tucker, T. 1998. Aspects of the bioinorganic chemistry of silicon in conjunction with the biometals calcium, iron and aluminium. *J. Inorg. Biochem.* **69**, 181–191.

Peter, A. L. J., Viraraghavan, T. 2005. Thallium: A review of public health and environmental concerns. *Environ. Int.* **31**, 493–501.

Peters, M., Lanzillotta, W. N., Lemon, B. J., Seefeld, L. C. 1998. X-ray crystal structure of the Fe-only hydrogenase (CpI) from *Clostridium pasteurianum* to 1.8 Å resolution. *Science* **282**, 1853–1858.

Petrak, J., Vyoral, D. 2005. Hephaestin-a ferroxidase of cellular iron export. *Int. J. Biochem. Cell Biol.* **37**, 1173–1178.

Philippot, P., Van Zuilen, M., Lepot, K., Thomazo, C., Farquhar, J., Van Kranendonk, M. J. 2007. Early archaen microorganisms preferred elemental sulfur, not sulfate. *Science* **317**, 1534–1537.

Pickering, I. J., Wright, C., Bubner, B., Ellis, D., Persans, M. W., Yu, E. Y., George, G. N., Prince, R. C., Salt, D. E. 2003. Chemical form and distribution of selenium and sulfur in the selenium hyperaccumulator *Astragalus bisulcatus*. *Plant Physiol.* **131**, 1460–1467.

Picot, D., Loll, P. J., Garavito, R. M. 1994. The X-ray crystal structure of the membrane protein prostaglandin H2 synthase-1. *Nature* **367**, 243–249.

Pignatti, E., Mascheroni, L., Sabelli, M., Barelli, S., Biffo, S., Pietrangelo, A. 2006. Ferroportin is a monomer *in vivo* in mice. *Blood Cells Mol. Diseases*, **36**, 26–32.

Pomposiello, P., Demple, B. 2001. Redox-operated genetic switches: The SoxR and OxyR transcription factors. *Trends Biotechnol.* **19**, 109–114.

Prigge, S. T., Eipper, B. A., Mains, R. E., Amzel, L. M. 2004. Dioxygen binds end-on to mononuclear copper in a precatalytic enzyme complex. *Science* **304**, 864–867.

Provot, S., Schipani, E. 2005. Molecular mechanisms of endochondral bone development. *Biochem. Biophys. Res. Comm.* **328**, 658–665.

Puig, S., Thiele, D. J. 2002. Molecular mechanisms of copper uptake and distribution. *Curr. Opin. Chem. Biol.* **6**, 171–180.

Pyle, A. M. 2002. Metal ions in the structure and function of RNA. *J. Biol. Inorg. Chem.* **7**, 679–690.

Que, L., Jr., Ho, R. Y. N. 1996. Dioxygen activation by enzymes with mononuclear non-heme iron active site. *Chem. Rev.* **96**, 2607–2624.

Ragsdale, S. W., Kumar, M. 1996. Nickel-containing carbon monoxide dehydrogenase/acetyl-CoA synthetase. *Chem. Rev.* **96**, 2515–2538.

Ratnaike, R. N. 2003. Acute and chronic arsenic toxicity. *Postgrad. Med. J.* **79**, 391–396.

Raymond, L., Ralston, N. V. C. 2004. Mercury: Selenium interactions and health implications. *SMDJ Seychelles Med. Dental J., Special Issue* **7**, 72–77.

Reed, G. H. 2004. Radical mechanisms in adenosylcobalamin-dependent enzymes. *Curr. Opin. Chem. Biol.* **8**, 448–477.

Reedijik, J. 1999. Why does cisplatin reach guanine-N7 with competing S-donor ligands available in the cell? *Chem. Rev.* **99**, 2499–2510.

Rees, D. C., Howard, J. B. 2000. Nitrogenase: Standing at the crossroads. *Curr. Opin. Chem. Biol.* **4**, 559–566.

Rehder, D. 2000. Vanadium nitrogenase. *J. Inorg. Biochem.* **80**, 133–136.

Rehder, D., Santoni, G., Licine, G. M., Schulzke, C., Meier, B. 2003. The medicinal and catalytic potential of model complexes of vanadate-dependent haloperoxidases. *Coord. Chem. Rev.* **237**, 53–63.

Rice, W. J., Young, H. S., Martin, D. W., Sachs, J. R., Stokes, D. L. 2001. Structure of Na^+, K^+-ATPase at $11\,Å$ resolution: Comparison with Ca^{2+}-ATPase in E_1 and E_2 States. *Biophys. J.* **80**, 2187–2197.

Ridarco, A., Carrigan, M. A., Olcott, A. N., Benner, S. A. 2004. Borate minerals stabilize ribose. *Science* **303**, 196.

Roat-Malone, R. M. 2002. Metals in medicine. *Bioinorganic Chemistry, a Short Course*, 265–335. Wiley-Interscience.

Roberts, G. P., Youn, H., Kerby, R. L. 2004. CO-sensing mechanisms. *Microbiol. Mol. Biol. Rev.* **68**, 453–473.

Rodriguez, F. I., Esch, J. J., Hall, A. E., Binder, B. M., Schaller, G. E., Bleecker, A. B. 1999. A copper cofactor for the ethylene receptor ETR1 from *Arabidopsis. Science* **283**, 996–998.

Ronconi, L., Sadler, P. J. 2007. Using coordination chemistry to design new medicines. *Coord. Chem. Rev.* **251**, 1633–1648.

Rosenberg, B., Van Camp, L., Krigas, T. 1965. Inhibition of cell division in Escherichia coli by electrolysis products from a platinum electrode. *Nature* **205**, 698–699.

Rosenberg, B., Van Camp, L., Trosko, J. E., Mansour, V. H. 1969. Platinum compounds: A new class of potent antitumour agents. *Nature* **222**, 385–387.

Ruf, J., Carayon, P. 2006. Structural and functional aspects of thyroid peroxidase. *Arch. Biochem. Biophys.* **445**, 269–277.

Rutherford, A. W., Boussac, A. 2004. Water photolysis in biology. *Science* **303**, 1782–1784.

Ryle, M. J., Hausinger, R. P. 2002. Non-heme iron oxygenases. *Curr. Opin. Chem. Biol.* **6**, 192–201.

Sauer, K., Yachandra, V. K. 2002. A possible evolutionary origin for the Mn_4 cluster of the photosynthetic water oxidation complex from natural MnO_2 precipitates in the early ocean. *Proc. Natl. Acad. Sci.* **99**, 8631–8636.

Schindelin, H., Kisker, C., Schlessman, J. L., Howard, J. B., Rees, D. C. 1997. Structure of ADP x AlF_4($^-$)-stabilized nitrogenase complex and its implications for signal transduction. *Nature* **387**, 370–376.

Schlesinger, W. H. 1997. *Biogeochemistry—An Analysis of Global Change*, 2nd ed., Academic Press.

Schofield, C. J., Zhang, Z. 1999. Structural and mechanistic studies on 2-oxo-glutarate-dependent oxygenases and related enzymes. *Curr. Opin. Struct. Biol.* **9**, 722–731.

Scientific American, September, 1983. "The Dynamic Earth."

Selvendran, R. R., O'Neill, M. 1985. Isolation and analysis of cell walls from plant material. In: *Methods of Biochemical Analysis*, Vol. **32**, 25–133, Glick, D., ed. John Wiley & Sons.

Senthilmohan, R., Kettle, A. J. 2006. Bromination and chlorination reactions of myeloperoxidase at physiological concentrations of bromide and chloride. *Arch. Biochem. Biophys.* **445**, 235–244.

Shanker, A. K., Cervantes, C., Loza-Tavera, H., Avudainayagam, S. 2005. Chromium toxicity in plants. *Environ. Int.* **31**, 739–753.

Shaw, F. C., 3rd. 2005. Metal-based drugs. In: *Encyclopedia of Inorganic Chemistry*, 2nd ed, Vol. **5**, 2992–3020, King, B., ed. John Wiley and Sons.

Shimizu, H., Obayashi, E., Gomi, Y., Arakawa, H., Park, S. Y., Nakamura, H., Adachi, S., Shoun, H., Shiro, Y. 2000. Proton delivery in NO reduction by fungal nitric-oxide reductase. Cryogenic crystallography, spectroscopy, and kinetics of ferric-NO complexes of wild-type and mutant enzymes. *J. Biol. Chem.* **275**, 4816–4826.

Shimizu, K., Cha, J., Stucky, G. D., Morse, D. E. 1998. Silicatein α: Cathepsin L-like protein in sponge biosilica. *Proc. Natl. Acad. Sci. USA* **95**, 6234–6238.

Shimokawa, T., Kulmacz, R. J., Dewitt, D. L., Smith, W. L. 1990. Tyrosine 385 of prostaglandin endoperoxide synthase is required for cyclooxygenase catalysis. *J. Biol. Chem.* **265**, 20073–20076.

Siegbahn, P. E. M. 2003. Mechanisms of metalloenzymes studied by quantum chemical methods. *Q. Rev. Biophys.* **36**, 91–145.

Singh, P., Cameotra, W. S. 2004. Enhancement of metal bioremediation by use of microbial surfactants. *Biochem. Biophys. Res. Comm.* **319**, 291–297.

Sivaraja, M., Goodwin, D. B., Smith, M., Hoffman, B. M. 1989. Identification by endor of trp191 as the free-radical site in cytochrome c peroxidase compound ES. *Science* **245**, 738–740.

Solomon, E. I., Sundaram, U. M., Machonkin, T. E. 1996. Multicopper oxidases and oxygenases. *Chem. Rev.* **96**, 2563–2605.

Sommerfeldt, D. W., Rubin, C. T. 2001. Biology of bone and how it orchestrates the form and function of the skeleton. *Eur. Spine J.* **10**, S86–S95.

Spalteholz, H., Panasenko, O. M., Arnhold, J. 2006. Formation of reactive halide species by myeloperoxidase and eosinophil peroxidase. *Arch. Biochem. Biophys.* **445**, 225–234.

Stahley, M. R., Stroble, S. A. 2005. Structural evidence for a two-metal-ion mechanism of group I intron splicing. *Science* **309**, 1587–1590.

Stein, M., Lubitz, W. 2002. Quantum chemical calculations of [NiFe] hydrogenase. *Curr. Opin. Chem. Biol.* **6**, 243–249.

Steinke, M., Wolfe, G. V., Kirst, G. O. 1998. Partial characterization of dimethyl sulfoniopropionate (DMSP) lyase isozymes in 6 strains of *Emiliania huxleyi. Mar. Ecol. Progr. Series,* **175**, 215–225.

Stubbe, J. 2003. Di-iron-tyrosyl radical ribonucleotide reductases. *Curr. Opin. Chem. Biol.* **7**, 183–188.

Stubbe, J. 1990. Ribonucleotide reductases. *Adv. Enzymol.* **63**, 349–419.

Stubbe, J. 1989. Protein radical involvement in biological catalysis? *Ann. Rev. Biochem.* **58**, 257–285.

Stubbe, J. 1988. Radicals in biological catalysis. *Biochemistry* **27**, 3893–3900.

Suresh, B., Ravishanker, G. A. 2004. Phytoremediation—A novel and promising approach for environmental cleanup. *Crit. Rev. Biotechnol.* **24**, 97–124.

Takahara, P. M., Rosenzweig, A. C., Frederick, C. A., Lippard, S. J. 1995. Crystal structure of double-stranded DNA containing the major adduct of the anticancer drug cisplatin. *Nature* **377**, 649–652.

Tapiero, H., Tew, D. D. 2003. Trace elements in human physiology and pathology: Zinc and metallothioneins. *Biomed. Pharmacotherapy* **57**, 399–411.

Teplova, M., Wallace, S. T., Tereshko, V., Minasov, G., Symons, A. M., Cook, P. D., Manoharan, M., Egli, M. 1999. Structural origins of the exonuclease resistance of a zwitterionic RNA. *Proc. Natl. Acad. Sci. USA* **96**, 14240–14244.

Tezcan, F. A., Kaiser, J. T., Mustafi, D., Walton, M. V., Howard, J. B., Rees, D. C. 2005. Nitrogen complexes: multiple docking sites for a nucleotide switch protein. *Science* **309**, 1377–1380.

Thakurta, P. G., Choudhury, D., Dasgupta, R., Dattagupta, J. K. 2004. Tertiary structural changes associated with iron binding and release in hen serum transferrin: A crystallographic and spectroscopic study. *Biochem. Biophys. Res. Comm.* **316**, 1124–1131.

Thakurta, G. P., Choudhury, D., Dasgupta, R., Dattagupta, J. K. 2003. Structure of diferric hen serum transferrin at 2.8Å resolution. *Acta Crystallogr. Section D* **59**, 1773–1781.

Thamdup, B. 2007. New players in an ancient cycle. *Science* **317**, 1508–1509.

Thelander, L., Reichard, P. 1979. Reduction of ribonucleotides. *Ann. Rev. Biochem.* **48**, 133–158.

Thompson, K. H., Orvig, C. 2001. Coordination chemistry of vanadium in metallopharmaceutical candidate compounds. *Coordin. Chem. Rev.* **219–221**, 1033–1053.

Tocheva, E. I., Rosell, F. I., Mauk, G., Murphy, M. E. P. 2004. Side-on copper nitrosyl coordination by nitrite reductase. *Science* **304**, 867–870.

Toyoshima, C., Nakasako, M., Nomura, H., Ogawa, H. 2000. Crystal structure of the calcium ion pump of sarcoplasmic reticulum at 2.6 Å resolution. *Nature* **405**, 647–655.

Trudinger, P. A., Swaine, D. J., eds. 1979. *Biogeochemical Cycling of Mineral-forming Elements*. Elsevier, Amsterdam.

Umezawa, H., Maeda, K., Takeuchi, T., Okami, Y. 1966. New antibiotics, bleomycin A and B. *J. Antibiot.* **19**, 200–209.

Urich, T., Gomes, C. M., Kletzin, A., Frazao, C. 2006. X-ray structure of a self-compartmentalizing sulfur cycle metalloenzymes. *Science* **311**, 996–1000.

Utschig, L. M., Bryson, J. W., O'Halloran, T. V. 1995. Mercury-199 nmr of the metal receptor site in MerR and its protein-DNA complex. *Science* **268**, 380–385.

Vahter, M. 2002. Mechanisms of arsenic biotransformation. *Toxicology* **181–182**, 211–217.

Valentine, J. S., Wertz, D. L., Lyons, T. J., Liou, L-L., Goto, J. J., Gralla, E. B. 1998. The dark side of dioxygen biochemistry. *Curr. Opin. Chem. Biol.* **2**, 253–262.

Valko, M. I., Morris, H., Cronin, M. T. D. 2005. Metals, toxicity and oxidative stress. *Curr. Med. Chem.* **12**, 1161–1208.

Vallee, B. L., Willams, R. J. P. 1968. Metalloenzymes: The entatic nature of their active sites. *Proc. Natl. Acad. Sci., USA* **59**, 498–505.

Vaska, M., Hasler, D. W. 2000. Metallothioneins: New functional and structural insights. *Curr. Opin. Chem. Biol.* **4**, 177–183.

Vendruscolo, M., Dobson, C. M. 2006. Dynamic visions of enzymatic reactions. *Science* **313**, 1586–1587.

Verga Falzacappa, M. V., Muckenthaler, M. U. 2005. Hepcidin: Iron-hormone and anti-microbial peptide. *Gene* **364**, 37–44.

Villa, J., Warshel, A. 2001. Energetics and dynamics of enzymatic reactions. *J. Phys. Chem. B.* **105**, 7887–7907.

Vincent, J. B. 2004. Recent advances in the nutritional biochemistry of trivalent chromium. *Proc. Nutr. Soc.* **63**, 41–47.

Vincent, J. B. 2001. The bioinorganic chemistry of chromium (III). *Polyhedron* **20**, 1–26.

Volbeda, A., Martin, L., Cavazza, C., Matho, M., Faber, B. W., Roseboom, W., Albracht, S. P. J., Garcin, E., Rousset, M., Fontecilla-Camps, J. C. 2005. Structural difference between the ready and unready oxidized states of [NiFe] hydrogenase. *J. Biol. Inorg. Chem.* **10**, 239–249.

Volkert, W. A., Hoffman, T. J. 1999. Therapeutic radiopharmaceuticals. *Chem. Rev.* **99**, 2269–2292.

Waalkes, M. P., Liu, J., Goyer, R. A., Diwan, B. A. 2004. Metallothionein-I/II double knockout mice are hypersensitive to lead-induced kidney carcinogenesis: Role of inclusion body formation. *Cancer Res.* **64**, 7766–7772.

Wächterhäuser, G. 2000. Origin of life: Life as we don't know it. *Science* **289**, 1308–1309.

Wagner, A. F. V., Frey, M., Neugebauser, F. A., Schaefer, W., Knappe, J. 1992. The free radical in pyruvate formate-lyase is located on glycine-734. *Proc. Natl. Acad. Sci.* **89**, 996–1000.

Wallar, B. J., Lipscomb, J. D. 1996. Dioxygen activation by enzymes containing bionuclear non-heme iron clusters. *Chem. Rev.* **96**, 2625–2657.

Wang, A-Y., Stoltenberg, M., Huang, L., Danscher, G., Dahlström, A., Shi, Y., Li, J-Y. 2005. Abundant expression of zinc transporters in Bergman glia of mouse cerebellum. *Brain Res. Bull.* **64**, 441–448.

Wang, J., Slungaard, A. 2006. Role of eosinophil peroxidase in host defense and disease pathology. *Arch. Biochem. Biophys.* **445**, 256–260.

Westbroek, P. 1991. *Life as Geological Force, Dynamics of the Earth.* W. W. Norton, New York.

Westhof, E., Dumas, P., Moras, D. 1988. Restrained refinement of two crystalline forms of yeast aspartic acid and phenylalanine transfer RNA crystals. *Acta Crystallogr. Sect. A* **44**, 112–123.

Weyand, M., Hecht, H., Kiess, M., Liaud, M., Vilter, H., Schomburg, D. 1999. X-ray structure determination of a vanadium-dependent haloperoxidase from *Ascophyllum nodosum* at 2.0Å resolution. *J. Mol. Biol.* **293**, 595–611.

Whiteley, C. G., Lee, D-J. 2006. Enzyme technology and biological remediation. *Enz. Microb. Technol.* **38**, 291–316.

Whittaker, J. W. 1994. The free radical-coupled copper active site of galactose oxidase. In: *Metal Ions in Biological Systems*, Vol. **30**, 315–360, Sigel, H., Sigel, A., eds. Marcel Dekker, New York.

Whittaker, M. M., Whittaker, J. M. 1990. Tyrosine-derived free radical in apogalactose oxidase. *J. Biol. Chem.* **265**, 9610–9613.

Wilce, M. C., Dooley, D. M., Freeman, H. C., Guss, J. M., Matsunami, H., McIntire, W. S., Ruggiero, C. E., Tanizawa, K., Yamaguchi, H. 1997. Crystal structures of the copper-containing amine oxidase from *Arthrobacter globiformis* in the holo and apo forms: Implications for the biogenesis of topaquinone. *Biochemistry* **36**, 16116–16133.

Williams, L. E., Pittman, J. K., Hall, J. L. 2000. Emerging mechanisms for heavy metal transport in plants. *Biochem. Biophys. Acta* **1465**, 104–126.

Wilmot, C. M., Hajdu, J., McPherson, M. J., Knowles, P. F., Phillips, S. E. V. 1999. Visualization of dioxygen bound to copper during enzyme catalysis. *Science* **286**, 1724.

Wilson, E. K. 2000. Enzyme dynamics. *C&E News* July/17, 42–45.

Wilt, F. H. 2005. Developmental biology meets material science: Morphogenesis of biomineralized structures. *Developmental Biol.* **280**, 15–25.

Winkelmann, G. 2002. Microbial siderophore-mediated transport. *Biochem. Soc. Trans.* **30**, 691–696.

Winslow, M. M., Crabtree, G. R. 2005. Decoding calcium signaling. *Science* **307**, 56–57.

Wu, L. J-C., Leeneders, A. G. M., Cooperman, S., Meyron-Holtz, E., Smith, S., Land, W., Tasi, R. Y. L., Berger, U. V., Sheng, Z-H., Rouault, L. A. 2004. Expression of the iron transporter ferroportin in synaptic vesicles and the blood-brain barrier. *Brain Res.* **1001**, 108–117.

Yamamoto, A., Wada, O., Ono, T. 1984. Distribution and chromium-binding capacity of a low-molecular-weight chromium-binding substance in mice. *J. Inorg. Biochem.* **22**, 91–102.

Yanchandra, V. K., Sauer, K., Klein, M. P. 1996. Manganese cluster in photosynthesis: Where plants oxidize water to dioxygen. *Chem. Rev.* **96**, 2927–2950.

Yandulov, D. M., Schrock, R. S. 2003. Catalytic reduction of dinitrogen to ammonia at a single molybdenum center. *Science* **301**, 76–78.

Yano, J., Kern, J., Sauer, K., Latimer, M. J., Pushkar, Y., Biesiadka, J., Loll, B., Saenger, W., Messinger, J., Zouni, A., Yachandra, V. K. 2006. Where water is oxidized to dioxygen: Structure of the photosynthetic Mn_4Ca cluster. *Science* **314**, 821–825.

Yoshihra, M., Ueki, T., Watanabe, T., Yamaguchi, N., Kamino, K., Michibata, H. 2005. Vanabin P, a novel vanadium-binding protein in the blood plasma of an ascidian, *Ascidia sydneiensis samea. Biochem. Biophys. Acta,* **1730**, 206–214.

Zalewski, P. D., Truong-Tran, A. Q., Grosser, D., Jayaram, L., Murgia, C., Ruffin, R. E. 2005. Zinc metabolism in airway epithelium and airway inflammation: Basic mechanisms and clinical targets. A review, *Pharm. Therapeut.* **105**, 127–149.

Zawia, N. H., Crumpton, T., Brydie, K., Reddy, G. R., Razmiafshari, M. 2000. Disruption of the zinc finger domain: A common target that underlies many of the effects of lead. *Neurotoxicology* **21**, 1069–1080.

Zayzafoon, M. 2005. Calcium/calmodulin signaling controls osteoblast growth and differentiation. *J. Cell. Biochem.* **97**, 56–70.

Zehnulova, J., Kasparkova, J., Farrell, N., Brabec, V. 2001. Conformation, recognition by high mobility group domain proteins, and nucleotide excision repair of DNA intrastrand cross-links of novel antitumor trinuclear platinum complex BBR3464. *J. Biol. Chem.* **276**, 22191–22199.

Zeth, K., Offermann, S., Essen, L. O., Oesterhelt, D. 2004. Iron-oxo clusters biomineralizing on protein surfaces: Structural analysis of *Halobacterium salinarum* DpsA in its low- and high-iron states. *Proc. Natl. Acad. Sci. USA* **101**, 13780–13785.

Zumft, W. G., Mortenson, L. E., Palmer, G. 1974. Electron paramagnetic studies on nitrogenase. *Eur. J. Biochem.* **46**, 525–535.

Index

Printed and bound by CPI Group (UK) Ltd, Croydon, CR0 4YY

03/10/2024

01040318-0003